Dwelling in Resistance

Nature, Society, and Culture

Scott Frickel, Series Editor

A sophisticated and wide-ranging sociological literature analyzing nature-society-culture interactions has blossomed in recent decades. This book series provides a platform for showcasing the best of that scholarship: carefully crafted empirical studies of socio-environmental change and the effects such change has on ecosystems, social institutions, historical processes, and cultural practices.

The series aims for topical and theoretical breadth. Anchored in sociological analyses of the environment, Nature, Society, and Culture is home to studies employing a range of disciplinary and interdisciplinary perspectives and investigating the pressing socio-environmental questions of our time—from environmental inequality and risk, to the science and politics of climate change and serial disaster, to the environmental causes and consequences of urbanization and war making, and beyond.

Available titles in the Nature, Society, and Culture series:

Diane C. Bates, *Superstorm Sandy: The Inevitable Destruction and Reconstruction of the Jersey Shore*

Cody Ferguson, *This Is Our Land: Grassroots Environmentalism in the Late Twentieth Century*

Stefano B. Longo, Rebecca Clausen, and Brett Clark, *The Tragedy of the Commodity: Oceans, Fisheries, and Aquaculture*

Stephanie A. Malin, *The Price of Nuclear Power: Uranium Communities and Environmental Justice*

Chelsea Schelly, *Dwelling in Resistance: Living with Alternative Technologies in America*

Diane Sicotte, *From Workshop to Waste Magnet: Environmental Inequality in the Philadelphia Region*

Sainath Suryanarayanan and Daniel Lee Kleinman, *Vanishing Bees: Science, Politics, and Honeybee Health*

Dwelling in Resistance

Living with Alternative
Technologies in America

CHELSEA SCHELLY

Rutgers University Press

New Brunswick, Camden, and Newark, New Jersey, and London

978-0-8135-8650-2
978-0-8135-8651-9
978-0-8135-8652-6
978-0-8135-8653-3

Cataloging-in-Publication data is available from the Library of Congress.

A British Cataloging-in-Publication record for this book is available from the British Library.

∞ The paper used in this publication meets the requirements of the American National
Standard for Information Sciences—Permanence of Paper for Printed Library Materials, ANSI
Z39.48–1992.

www.rutgersuniversitypress.org

Manufactured in the United States of America

For Avery, always

Contents

Dwelling in Resistance

1

What Does It Mean to Dwell in Resistance?

● ●

Residential houses come in all shapes and sizes and are located in communities both large and small. Yet modern residential dwellings largely share a set of common technological systems—electricity, water, waste removal—and these technologies involve common sets of practice, unthinking habits that we engage in throughout the day, every day. These shared practices are the little active rituals of residential life. Most Americans largely take for granted much of what is materially involved in the daily rituals of dwelling.[1] There is often very little reason to question whether the lights will turn on when we flip the appropriate switch, hot water will be delivered momentarily after turning the appropriate valve, and the garbage man will come on Tuesday. In many parts of America, life is almost impossible without a personal vehicle, and we rarely question why a family may own more than one.

Residential dwelling in America involves almost constant interaction with technological systems that we hardly ever even contemplate. We learn how to interact with these systems through our bodily practices, flipping light switches and opening garage doors and flushing toilets and cooking for families of four with ingredients from the grocery store. Dwelling involves bodily rituals that take place in concurrence with the material systems around us.

Technologies that support daily life in the American home enroll us in both techniques of the body[2] and particular patterns of thought regarding the material and social world. The systems of electricity, domestic heat, water provision, waste removal and treatment, transportation, and food production are just some of the massive material infrastructures that provide for the needs and comforts of modern human dwellers. Yet technologies are not merely material; they are "at the same time a mode of organizing and perpetuating (or changing) social relationships, a manifestation of prevalent thought and behavior patterns, an instrument for control and domination."[3] Technologies are not static material systems; they are "a social process in which technics proper (that is, the technical apparatus of industry, transportation, communication) is but a partial factor."[4] Material systems shape how we act, and think, in our dwelling lives.

This study involves conscious attention to the ways the dominant technological systems that support residential dwelling in America shape social thought and practice. I draw on classical thinkers like Pierre Bourdieu, Michel Foucault, Herbert Marcuse, and Lewis Mumford to suggest that material systems operate to shape both action and thought. These insights inform a more recent body of scholarship on theories of practice,[5] and this book explicitly pays attention to how the sociotechnical networks that support residential dwelling encourage particular forms of practice, rendering most options for alternative forms of technology and practice invisible.[6]

The technologies that meet the needs and comforts of modern residential dwellers are based on a scaffold of political structures (like building codes and zoning regulations), economic institutions (like utility companies that demand monthly payments for their services), and scientific ideals (about the safety and suitability of various technologies). These technologies are also based on a particular conception of what normal residential life is and should be, involving monthly utility bills and an unending source of electric power, weekly trash service and immediate wastewater removal, and the largely unthinking reliance on infrastructures, bureaucracies, and individualized economies to meet our needs and comforts. Residential dwelling, then, involves the interplay of political, economic, material, and cultural structures that shape conceptualizations of normal and acceptable ways of living, and the rituals of social practice, the intersection of norms and action where individuals habitually enact shared patterns of daily life.

The question then becomes—and this question shapes the empirical investigations of the case studies presented in this book—how do some Americans come to adopt alternative forms of residential dwelling technologies? I do not ask "why" people choose to adopt alternative technologies because, like other sociologists, I recognize that the conceptualization of and language used to describe motivation may be more cognitively, socially, and temporally complicated than the simple formulaic that yesterday's motivation leads to today's action.[7] While I, like other sociologists, am often tempted to ask why, here I attempt to focus on how—how the people I've met who dwell differently than the average American understand their own dwelling choices, how their practices differ from other dwellers, how they understand the choices (and compromises) they've made to dwell in resistance.

The chapters in this book present four diverse cases of alternative technology adoption at the residential scale through a theoretical lens that focuses on how alternative material systems involve alternative forms of social organization and social practice. The cases are: (1) an intentional community largely centered on alternative education and alternative birthing options in Tennessee (The Farm, chapter 3); (2) an intentional community based on communal labor and income sharing in Virginia (Twin Oaks, chapter 4); (3) a third intentional community, located in rural Missouri, that blends shared resource use with private economic systems (Dancing Rabbit, chapter 5); and (4) the case of Earthships, radically efficient off-grid homes made out of tires (in New Mexico, chapter 6). Three of these cases involve communal dwelling, where people are living in formally identified intentional communities.[8] The Earthship community is not formally an intentional community, but is similar in that the people who live in the neighborhoods of Earthships around Taos, New Mexico, and the people who come to work for and volunteer with the company that builds these homes, are all there with a shared purpose—to pursue radically alternative and sustainable dwelling through Earthships.

This introductory chapter, which reviews the intellectual history, scholarly questions, and academic insights that frame my approach to studying these four diverse case studies, is intended to examine how technological systems act to maintain and secure existing understandings and enactments of so-called normal life. In this, I hope to make myself clear straight away: it is my view that power in the modern world is not held but exercised—it acts, and in doing so, it actively shapes the actions of individuals and groups and societies writ large.[9] Technological systems—like electricity; transportation; water and wastewater collection, distribution, and treatment; modern home heating; modern food systems; modern reproductive technologies; educational systems; and so on—shape and are shaped by relationships of power, through the political, economic, and cultural structures under which we dwell. As elements of active power, technologies shape the very ways we humans think and act within these structures.

Historical examination helps to demonstrate this, and chapter 2 examines the history of technological development to define and clarify the technologies that support "normal" contemporary residential life. Yet American history also provides examples of alternatives, attempts to utilize technological systems that involve different understandings of normal life as well as different forms of residential practice. This book traces this history forward to the present, examining some of the many possible ways that individuals are choosing to pursue technological alternatives in their residential lives and how these alternative technological systems are related to alternative ways of living, knowing, and being.

Technologies as Strategies: Power, Action, and Thought

Lewis Mumford, an infamous critic of modern technology, once wrote, "The brute fact of the matter is that our civilization is now weighted in favor of the use of mechanical instruments, because the opportunities for commercial production and for the exercise of power lie there: while all the direct human reactions or the personal arts which require a minimum of mechanical paraphernalia are treated as negligible."[10] Technological systems act as modes of political and social organization and as discursive constitutions regarding normal, acceptable, or appropriate ways of life and living. In other words, modern technological systems work to reinforce existing relations of power.

The work of Michel Foucault and Pierre Bourdieu helps us make sense of the role of technology in perpetuating particular power dynamics, forms of social organization, and mental conceptions. Foucault recognized that power is not held; instead, power is a verb, it is active, and it is exercised incessantly. In modern societies, power works through "procedures which allowed the effects of power to circulate in a manner at once continuous, uninterrupted, adapted, and 'individualized' throughout the entire social body."[11] Thus, power can be seen as diffuse and constantly exercised. Power is no longer held in the hands of a centralized state administration (this is what Foucault meant when he quipped that it was time to 'cut off the king's head').[12] Power does not sit in Washington, DC, or on Wall Street, but flows throughout and constantly within society.[13]

Furthermore, this power is exercised through individual bodies. Foucault terms this "biopower." Biopower is "that which is exercised on people as both a general category of population (nation, social groups of one sort or another) and individual bodies (people in their roles as individual citizens, wives, farmers, etc.)."[14] Biopolitics serve to categorize, classify, and contain; they construct and articulate groups and individuals through modes of power internalized through daily practice.

The technological systems that support residential dwelling and dwelling practice are strategies for deploying this kind of power. Technologies contribute to the "ceaseless temporal and spatial interweaving of . . . physical components . . . individual inhabitants, and the *concrete practices* of . . . families."[15] Technological systems, as strategies of power, help define what is normal and abnormal in our daily practices, from where we get our water to how we use it in our homes to where our water goes when we're done with it.[16] Technologies govern individual practice and social organization. The technologies that support residential dwelling are sustained by the constant maintenance of the boundaries between technologies considered acceptable for use and legitimated through dominant practices, and technologies considered alternative, unusual, or unacceptable.

For example, electricity is a strategy of power because its source (the fuel source used and the forms of generation) and use (including who uses it, how and how much) help define and shape a population of members, either as community members or citizens. It is common, shared knowledge that modern American citizens largely rely on fossil fuels for their electricity, that each home demands a practically endless supply of electricity, and that this power is provided by a large utility company. This example of electricity demonstrates how relations of power are embedded in and maintained by the technologies used in a given society. With this insight, residential dwellers are arguably enacting embedded relations of power with every turn of the faucet and every flip of the light switch.

What are the consequences of this view of power? When thinking about power, it is difficult to escape the categories of thought produced by power. Pierre Bourdieu, in characteristically convoluted language, put it this way: "To endeavor to think the state is to take the risk of taking over (or being taken over by) a thought of the state, i.e., of applying to the state categories of thought produced and guaranteed by the state and hence to misrecognize its most profound truth."[17] Foucault argued that these categories of thought are produced through our disciplined daily practices, including our use of dominant technologies. Randall Collins similarly claims that these cognitive categories are shaped through ritual practices.[18] Karl Marx suggested that ways of thinking are shaped by technologies themselves.[19]

Thinking this way, there is no one actor or group of actors we can point to in a quest to overturn relations of power; instead, we have to grapple with the way active, capillary, and materially embedded flows of power shape even our own thoughts about normal or acceptable ways of acting and being and knowing. Turning to Bourdieu again can add yet another layer of depth to our understanding of how technologies shape cognition and thus contribute to thinking like the state (in other words, using categories of thought created by relations of power). Bourdieu writes, "By realizing itself in social structures and in the mental structures adapted to them, the instituted institution makes us forget that it issues out a long series of acts of *institution* (in the active sense) and hence has all the appearances of the *natural*."[20] Thus, power is exercised through the shaping of our categories of perception and thought. The "instituted institution" (the incessantly active and productive forms of power that flow throughout society to produce the categories of thought used by members of society) does this through the naturalization of what are actually instituted (constructed) institutions.

Technology is one such institution. The technologies we use, which are by no means socially or technologically determined, come to be seen as natural, inevitable, and normal through our very use of these technologies and the categories of thought we come to associate with this use. Through our every-day use of the material systems that support residential life, power is actively involved in daily practices and the categories of thought that correspond with daily practices. Technology operates as a strategy of power through its active involvement in shaping both practice and constructions of normal prac-tice. Power is thus exercised through us as individuals, rendering a particular constellation of dwelling practices "normal," while alternatives are rendered invisible.

Studying Alternative Technology Adoption

Viewing technological systems as strategies of power that support particular kinds of practice and ideas about that practice while rendering alternatives invisible is congruent with the approach I take to understanding motivations to change the relationships individuals and communities have to the domi-nant forms of technologies that support residential life in modern America. Through this conceptual lens, I can ask, if the use of technological systems is a form of power such that we are blinded to the social construction of these systems and see them instead as totally natural and normal, how can we under-stand the choice to live with alternative kinds of residential dwelling tech-nologies? Understanding power as actively and incessantly enrolling us as individuals in particular concepts of normal living through our very rituals of daily life brings attention to rituals of social practice and the ideas offered by theories of social practice, and understandings of both power and of practice are central to understanding motivations to adopt alternatives through the framework of "orientations" as developed throughout this book.

Theories of practice focus on how patterns of consumption, including the kinds of consumption involved in using the technologies that support residential life, involve largely unthinking, habitual action, standardized through internalized ideas about socially acceptable patterns of activity.[21] Instead of seeing the social world as organized by variables and attributes, this emergent line of scholarship sees constellations of practice as the essen-tial unit of analysis.[22] This work largely focuses on how practices become nor-malized and stable, such that people don't see their consumption choices as choices at all. It also draws attention to how stability in practices is supported by infrastructures and policies that encourage certain kinds of consumption habits while making alternatives difficult, impossible, or invisible.[23] Further, this body of work is based on eliminating the dualism between action and thought, recognizing that what we do as individuals and social groups is an enactment of what we think about the kinds of things we do—like flushing toilets, taking daily showers, or paying monthly bills in exchange for socio-technical systems to support the import and removal of the things that sustain residential life.[24]

The scholarship associated with the diverse body of work in practice theory is consistent with the conceptualization of power outlined above. Power works by shaping what we do with our bodies, actively enforcing use of particular kinds of material systems that are perceived as normal and inevitable, and in using these dominant technologies we engage in particular patterns of practice. The question becomes, if modern residential dwellers are unthinkingly playing out patterns of practice that simultaneously support and are supported by particular constellations of power involving technological, economic, political, spatial, and social relationships, how can we understand the individuals and groups who are stepping outside these dominant sociotechnical systems and the associated practices to engage in alternative forms of residential dwelling?

Understanding Motivations

I began this research wanting to understand what motivates adoption of alternative, renewable energy technologies for residential use. Recognizing that many of these technologies—like solar electric technology, solar thermal technology, rainwater collection and filtration for use, home-scale treatment of waste, home-scale gardening and animal husbandry, and electric vehicles, to name a few—are still relatively unfamiliar to most, I first thought that previous research on the adoption and diffusion of innovations could offer some insight.[25] This research suggests that networks, neighbors and other methods of getting the information out there help to educate people, and that "early adopters" of new technology are often motivated by perceptions of themselves as innovative or cutting edge. Based on the widely held perception that renewable energy technology adoption is motivated by environmental values, I also thought that people who live with alternative technologies in their residential lives would likely be motivated by environmental concerns. However, a lot of researchers have spent a lot of time studying how environmental values influence behavior, and the results are pretty ambiguous.[26] Often, abstract concerns don't change specific behaviors.[27] So I wondered, are there specific environmental values shared by alterative technology adopters that help to motivate their decision to adopt?

Yet my sociological training has taught me to be skeptical of our ability to understand motivations from the outside in, with a direct connection between the past and the present.[28] Motivation for behavior is often explored based on the dualistic understanding of means and ends, the incessant separation of thought from action, that underlies much conventional economic thinking. As economist Juliet Schor writes:

> One of the hallmarks of the standard economic model, which hails from the nineteenth century, is that people are considered relatively unchanging. Basic preferences, likes and dislikes, are assumed to be stable, and don't adjust as a result of the choices people make or the circumstances in which they find themselves. People alter their behavior in response to changes in prices and incomes, to be sure, and sometimes rapidly. But there are no feedback loops from today's choices to tomorrow's desires. This accords with an old formulation of human nature as fixed, and this view still dominates the policy conversation.[29]

Human action is simply more contextually, temporally, cognitively, and socially complicated than this understanding suggests. As C. Wright Mills, one of sociology's founding fathers, wrote long ago, "A man may begin an act for one motive. In the course of it, he may adopt an ancillary motive. . . . The vocalized expectation of an act, its 'reason,' is not only a mediating condition of the act but it is a proximate and controlling condition for which the term 'cause' is not appropriate."[30] He went on to claim, "Motives are of no value apart from the delimited societal situations for which they are the appropriate vocabularies. They must be situated."[31] In other words, "We can change, too. This has profound implications for our ability to shift from one way of living to another, and to be better off in the process."[32] How, then, can we move beyond the dominating means-ends schema that economists and policy makers use to understand motivation[33] to best understand the processes of human decision making that lead individuals and groups to pursue change in residential life through changed technologies and changed practices?

Pierre Bourdieu is helpful for understanding material technologies as cultural forms of power, but his perspective on agency, "(perhaps the dominant one in contemporary American sociology) sees human agency as habitual, repetitive, and taken for granted."[34] Michel Foucault's perspective of power as productive, as a creative force in the construction of the individual, leaves little room for understanding choice. Similarly, theories of practice often focus on the stabilization of practices, and there is currently little theorizing on how practices change.[35] For intellectual assistance here, I turn to understandings of action based on the theory of pragmatism and its sociological heritage in the Chicago School.[36]

Pragmatism as a theory of action recognizes the inherent connections among thought, action, and social context. Simply put, the human ability to think and act develops within a social environment.[37] Further, means are not stagnant precursors to ends. Sometimes, we only discover potential goals in the context of action after recognizing available means that we did not see before. "Actors develop their deliberative capacities as they confront emer-gent situations."[38] Thus, there is a constant interaction between goals and their realization; "ends and means develop coterminously within contexts that are themselves ever changing and thus always subject to reevaluation and reconstruction."[39]

Spatial context, temporal context, bodily habits, and social relationships are all part of understanding the social conditioning of thought and action, which are really two aspects of the same process. The foundational premises of pragmatism provide one way "to sidestep many of the conundrums that dom-inate sociological thought and to lay the foundations for a theory of action that analyzes the 'conditions of possibility' for the evaluative, experimental, and constructive dimensions of perception and action, within the contexts of social experience."[40]

Pragmatism, viewing thought and action as irrevocably connected and intertwined, fundamentally challenges the Cartesian dualism between thought and action and the prevalent means-ends model for understand-ing action, "replacing it with an account of the situational and corporeal embeddedness of action."[41] Thus, pragmatism and its sociological heritage in the Chicago School of sociology suggests that instead of asking why people adopt alternative dwelling technologies (assuming that fixed means oper-ate to accomplish fixed ends), we can instead ask how people come to adopt such technologies, exploring the cultural context and socially derived under-standings of behavior that shape action. Andrew Abbott claims that Chicago School sociologists utilized research methods that "converge on the direct analysis of patterns of social activity in temporal and social context. . . . They tell us what are the actual patterns, not what are the crucial variables."[42] In this way, the Chicago School sociologists were themselves focused on the analysis of practice.

The pragmatist approach to understanding action is also consistent with understanding technologies as instruments of power. While pragmatists and Chicago School sociologists rarely show explicit interest in relations of power, the understandings of action provided by pragmatism are consistent with a cultural, contextual, and processual understanding of power. As the pragma-tist and Chicago School sociologist W. I. Thomas once wrote, "the positive limitations of evolution which society imposes upon the individual by putting him into a determined frame of organized activities . . . establishes a regularity of periodical alternations of work and play, food and sleep, etc., and with the help of economic, legal, and moral sanctions prescribes and excludes certain forms of behavior."[43]

In other words, our shared social practices are shaped by a multitude of factors, including the technological systems that dominant our daily lives. A "determined frame of organized activities" shapes what we do and how we see the world, including our residential dwelling patterns and our perceptions of these patterns. Based on this intellectual lineage, the four case studies presented in this book suggest that the motivations and patterns of practice that shape alternative forms of residential dwelling are shaped by what I call orientations. The concept of orientations helps connect thought and action, the normalization of behaviors and the actual patterns of behaviors themselves, in order to understand what motivates individuals and communities to actively engage in dwelling in resistance.

Dwelling in Resistance

The concept of orientation guides us to focus on both the perspectives and practices of the people examined in this book who are living with alternative technology. Orientations can be understood as organized around a particular set of ethics. I see these orientations or orienting ethics as conceptually similar to Aldo Leopold's land ethic[44] (in that they are too broad and abstract to offer specific, consistent guides for every action), Pierre Bourdieu's notion of habitus[45] (although in our highly differentiated society, these metaphorical spaces are no longer organized simply based on class or status placement), and Max Weber's discussion of practical ethics. For Weber, and for me, the most important aspect of an orienting ethic is how it guides action.[46] These orientations help to highlight how individuals understand their relationships, including relationships among individuals, between humans and the natural world, and between the social and the material world. Based on this conceptualization of orientations, people think and act based understandings about relationships among people, nature, and the technological systems that support residential life.

I argue that individuals organize their life preferences and actions based on these orienting ethics, although they are not necessarily conscious positions. Orientations shape action, the patterns of practice involved in daily living. In each of the cases described in this book, people are engaging in alternative kinds of technological practice—they are, for example, sharing residential living spaces with people beyond their nuclear families; organizing life so they can live where they work; using renewable energy systems for electricity; harvesting and filtering their own water; treating their own waste; and engaging in alternative economies, kinds of diets, food production, and systems of childbirth and education. Yet upon close examination of the relationship between their practices and the language they use to describe their motivations for these practices, it becomes clear that the orientations that motivate these alternative kinds of residential technology use are not themselves all that alternative.

While the orientations that operate in each case are unique, they demonstrate a common theme. Focusing on residential dwelling as a domain of practice provides a means of moving past purely individualist, value-oriented attempts at explaining behavior.[47] Adopting solar energy technologies, constructing a home with natural building techniques, and living a voluntarily simplified lifestyle in terms of resource consumption are all often described as environmentally motivated behaviors, suggesting that individuals who espouse environmental values or internalize environmental concerns are the ones who are likely to adopt alternative technologies that lessen their impacts on the natural world. Yet labeling something an environmental choice actually limits adoption of that choice among many American consumers.[48] Research suggests that people adopt technologies and practices that are considered environmentally responsible for a multitude of reasons,[49] and decades of environmental psychology research has failed to demonstrate that environmental values consistently predict environmentally responsible behavior.[50]

Moving beyond an analysis focused on the individual values that predict individual behaviors allows us to recognize that residential dwelling involves a constellation of practices, bodily habits, techniques, and behaviors that individual dwellers engage in throughout their daily lives, and that these behaviors are shared among groups of people within particular social groups and cultures and in particular times.[51] Adopting alternative technologies involves a change in bodily practices and routines, suggesting that household practices are socially embedded and socially structured consequences of visible technical and political structures as well as invisible cultural expectations and understandings of normality.[52]

The case studies presented in this book all involve adoption of radically different technologies and forms of technological, spatial, and social organization that reduce the negative impact of human activities on the natural world. Yet the motivations to use less fossil fuel–generated electricity, go off-grid, or share resources communally were not universally or exclusively discussed in terms of environmental responsibility. I argue that orientations, operating as ethics to frame the way individuals and groups see the world and the priorities they emphasize in terms of their relationship to the natural world and relationships among people, shape the use of material systems and the corresponding bodily behaviors. Arguably, the orientations operating in each of these case studies reflect cultural values and tendencies rooted in longstanding American values.

Early American culture inherited a form of antifederalism that was "anti-state and anti-urban, idealizing the yeoman farmer in all his independence."[53] Arguably, the American tradition is deeply rooted in an individualism that "values independence and self-reliance above all else."[54] Yet for many modern Americans, this independence involves freedom, and "freedom turns out to mean being left alone by others, not having other people's values, ideas, or styles of life forced upon one, being free of arbitrary authority in work, family, and political life. What it is that one might do with that freedom is much more difficult for Americans to define."[55]

For Americans living with alternative technologies and in alternative communities, independence is an important motivational factor. Discussions with individuals who choose to live in communities where they produce at least some of their own the electricity and grow some of their own food, or who live in entirely off-grid homes, suggest that some Americans are actively engaging with the question of what to do with their freedom and are pursuing freedom in a positive sense. This means that instead of thinking of the American value of freedom as a "freedom from," these Americans are pursuing a "freedom to"—a freedom to use financial resources for something other than necessary monthly bills, to produce what they can for themselves with the resources available where they are, and to live in accordance with both environmental and economic values.

Furthermore, these case studies demonstrate that individual freedom does not have to be pursued at the expense of community mindedness; arguably, recognizing the connection between the two is part of the original conception of freedom rooted in America's value system. The individuals and communities represented in the case studies presented here embrace the congruence between individual freedom and community well-being. These communities, in various ways, reorganize the scale and patterns of residential practice so that individuals can be freed from the limitations of economic burdens like debt, and free to pursue a good life and to participate more freely in communal life. Furthermore, these communities call into question the recently solidified division between public and private life in America:

> The most distinctive aspect of twentieth-century American society is the division of life into a number of separate functional sectors: home and workplace, work and leisure, white collar and blue collar, public and private. This division suited the needs of the bureaucratic industrial corporations that provided the model for our preferred means of organizing society by the balancing and linking of sectors as 'departments' in a functional whole, as in a great business enterprise. Particularly powerful in molding our contemporary sense of things has been the division between the various 'tracks' to achievement laid out in schools, corporation, government, and the professions, on the one hand, and the balancing life-sectors of home, personal ties, and 'leisure,' on the other. All this is in strong contrast to the widespread nineteenth-century pattern in which, as on the often-sentimentalized family farm, these functions had only indistinct boundaries.[56]

Although most modern Americans arguably would not recognize it, alternative communities reorganize life in a way that actually corresponds to traditional American values, understood historically. The individuals choosing to live with alternative technologies in alternative communities recognize the value of reintegrating the alienated spheres of life, and the necessity of doing so with technological systems that allow individuals and social groups to meet their needs and comforts without the drastically negative impacts on the natural world, the drastic inequities in terms of access and impact, and the drastic reliance on money and experts that accompany the dominant forms of technology that currently meet these needs.

American society is understood as valuing individualism more than all else. Yet in terms of a national value system, old traditions continue to influence modern life. The cases of alternative technology adoption and alternative living presented here provide one glimpse of how traditional and longstanding American values remain alive in perhaps unexpected forms. Instead of a blind commitment to freedom from, these cases suggest how American traditions can motivate behaviors based on a broad conception of both environmental and social values that shape understandings of the good life and what to do with a freedom to. The case studies presented here all address big, cultural values about freedom, self-sufficiency, and community resiliency. These values are deeply rooted in American history.

This is not to suggest that alternative technologies and alternative communities are themselves uniquely American. There are intentional communities and eco-villages, where people are choosing to live together with different technological and material arrangements, around the world.[57] Other countries far surpass the United States in terms of what can be called alternative technology adoption, from solar electric panels to residential grey water treatment systems. The point is not that these practices and the orientations that motivate them are uniquely American; the point is that these alternative choices in material and social arrangement are not based on alternative values, but values that are congruent with longstanding cultural traditions. Participants in this study, however, suggest that some of these historical American values that recognize the interconnection between individual freedom and community well-being are currently lost behind technological and economic dependence, the emphases on isolationism, and consumerism as self-expression.

Thus, the people living in each of the communities described here dwell (live their residential lives) in resistance, although they are not necessarily consciously resisting or resisting by adopting incomprehensibly alternative values. Their actions contradict the ideals embedded within and perpetuated by mainstream dwelling technologies, but these actions are organized around abstract ethical principles, orientations, that in may ways correspond to values at the core of America's foundation. The alternative practices supported by the alternative technologies and forms of material organization represented in each case correspond to traditional American values related to the freedom to support oneself in ways that do not mitigate the ability of your neighbor to do the same. These values provide one organizing principle by which we can rethink residential and community life, and these case studies suggest that alternative technologies can promote alternative practices that correspond to a set of political forms, environmental impacts, and community values worthy of pursuit.

Dwelling in resistance is not just about where your electricity comes from. It is about dislodging oneself from the capillaries of power that pulse throughout the modern world. It is about reshaping mental attitudes and social arrangements through the very active practices of living, with alternative technological systems including electricity but also water provision, sanitation, food production and distribution, education, reproduction and childbirth, and even death. Americans are shifting their relationships with technology, the environment, and economics in myriad ways by using and experimenting with alternative technological systems that reshape residential dwelling. They are dwelling differently, dwelling in resistance. As they dwell, they demonstrate the possibilities for alternative technological systems to foster alternative ways of thinking, acting, and knowing.

In the cases presented here as in your own life, there are an infinite number of ways to go about dwelling differently. Many involve unplugging from the massive material systems that we depend on for the very things of sustenance—electricity, heat, water, food, sanitation—and instead adopting distributed (i.e., noncentralized; such as energy that is produced where it is used) residential technologies like solar or wind electricity, rainwater collection, wood-burning stoves, or "humanure" (composting toilet) systems. These rely on a synergistic relationship with the local natural environment and provide an economic alternative to monthly utility fees.

For example, the case studies of alternative dwelling presented here all involve some consideration of alternative relationships and practices when it comes to food; many Americans are already working to be more self-reliant when it comes to food production at an individual, community, or local level. Others are pursuing alternative methods of transportation, from biking to hybrid and electric cars to community car sharing. Many of these alternative practices are environmentally beneficial, opening up possibilities for a new relationship between humans and the natural world. Many reshape the modern economy. These alternative technological systems lessen the isolation and dependence that result from our current sociotechnical structures and forms of organization, as well as lowering monthly bills and decreasing the use of finite natural resources. There are so many forms of dwelling in resistance, a resistance that opens up possibilities for new futures from within the existing networks of power. It is in dwelling differently, through those who are living life with alternative technological systems of various kinds, that we can see the ways in which technological systems, as strategies, shape—and importantly, have the potential to reshape—thought, practice, and the limits of what is thought possible.

This work suggests that consciousness, conscious awareness of the forms of power operating in society or the ways in which personal actions resist this power, is not a prerequisite for action. Consciousness is much like an onion rather than a veil: we are not either aware or unaware but spend much of our time conscious of one layer without realizing there is another underneath. This work also suggests that moving beyond a mind-body dualism allows us to more accurately understand the inherent connection between these two realms of life that are too often considered separated poles. When people live with different technologies, their actions and their minds change in ways that they did not foresee or intend; their ethics and their habits emerge as meaningful in conjunction with their interaction with alternative material systems. The concept of orientation helps to connect body and mind, value and practice, in nondualistic terms. This work also provides relevant insights into the structural conditions that contribute to the adoption of alternative technologies, technologies that are environmentally beneficial but that also, arguably, reshape social organization in positive and beneficial ways.

Where We're Going

As I visited the places and collected the stories presented in this book, I also experienced firsthand some of the myriad possibilities for unplugging from technological dependency and dwelling in resistance as well as the social, emotional, and personal shifts that occur along the way. While I continue to explore sociological issues related to technological systems, relationships between society and nature, connections to space, bodily practice, and how all these are imbued in and connected to relations of power, I am also engaging in my own self-exploration. My work, like for many academics, is deeply personal; my interest in environmental sociology began with the idea that until we learn to treat the natural world with respect, we will be unable to treat one another with humanity. I am now convinced that the material systems that support residential life are implicated in both, and that examining the relationship between technology, the environment, social organization, and categories of thought offers unending opportunities to consider and hopefully contribute to reshaping how we use, understand, and interact with material systems, the natural world, and one another.

In order to set the stage for the case studies that follow, chapter 2 examines the historical development of modern dwelling technologies in America, focusing on how the technological systems that support residential dwelling shape action and practice in ways that make modern dwellers both isolated and dependent. In this chapter, we also see that history offers examples of others who sought to reshape personal practice and social organization through alternative technology adoption during the appropriate technology movement. This movement's influence is recognizable in many of the case studies that follow.

Chapter 3 presents the case study of The Farm. Located in rural Tennessee, The Farm is a product of the 1960s counterculture and is perhaps most well known for natural birth and midwifery. Today, approximately 200 adults and their children live in this community, where there are several community businesses, an active midwifery clinic, a center dedicated to appropriate and alternative technology, and an alternative school. The Farm has gone through significant transition both structurally and demographically in the generations since its founding. Today, the community emphasizes the value of their shared land and their role as stewards for life's big transitions, including birth, schooling, and death. This chapter examines the orientation of a custodial ethic that motivates the alternative material and social practices organizing life at The Farm.

Chapter 4 takes us to Twin Oaks. Here, approximately 100 adults and their children live, work, and play together in rural Virginia. An egalitarian income and labor sharing community, members of Twin Oaks are all required to work the same number of hours for the same small monthly stipend. Basic needs like housing, food, and clothing are provided to all community members, who also share access to the community's fleet of vehicles, bicycles, musical instruments, and other resources. Here, the community is organized by an orientation that I call the plenitude ethic, recognizing the real sense of abundance that comes from a radical sharing economy with shared access to material resources. While it may seem like the level of sharing that takes place at Twin Oaks involves personal sacrifice, community members are not oriented by a sense of sacrifice but by a sense of abundance.

Chapter 5 explores Dancing Rabbit Ecovillage. Founded in 1997, this community is located in rural Missouri. Here, members do not own land individually and can access multiple shared systems like a community fleet of vehicles and shared kitchens, laundry and shower facilities. However, individuals are responsible for their own housing, livelihoods, and subsistence needs, although the community requires that individuals live within a set of shared ecological commitments and it is much easier to live on less economic income here because of shared access to material systems. Members of this community are motivated by an orientation that I call symbiotic individualism, recognizing the inherent connections among humans and between human and natural systems while also valuing individuality and individual freedom. It is the dance between these two elements that characterizes the orientation of this community and its members.

In chapter 6, we turn to the case of Earthships and Earthship Biotecture, examining the social justice orientation that motivates their emphasis on self-sufficiency. Earthships are radically self-sufficient and sustainable off-grid homes. Made with discarded tires and rammed earth, these passive solar homes are intended to provide for almost all the needs of residential dwellers with the resources available on site through off-grid energy production, rainwater collection and filtration and sophisticatedly efficient water-use systems, and waste treatment. Earthship Biotecture operates a for-profit architectural firm, a non-profit development organization, an educational facility, and there are three subdivisions of Earthship homes around Taos, New Mexico. The individuals who live in Earthships as well as those who work for and volunteer with Earthship Biotecture share an orientation that I call self-sufficiency as social justice, recognizing that technological systems that allow individuals to meet their own needs and comforts through their own physical engagement with the systems that they require is the only way to achieve global social justice.

Each chapter presents the case study in detail, outlining the kinds of technologies used in each example of alternative residential dwelling and discussing the orientations, shared experiences, and commonalities among the individuals I met and interviewed. The conclusion of the book (chapter 9) presents some specific comparisons, considers general findings and their theoretical implications, and offers some policy-relevant insights.

Throughout the book, the language of "alternative technologies" and "alternative lifestyles" is used with intention to describe these case studies and the forms of technology adoption and material organization they entail. Using the word *alternative* is meant to highlight the normative nature of technology use in everyday residential life. Residential dwellers in America are surrounded by relatively invisible technological systems that they rely on for their subsistence, which also "solidify culturally homogenous notions of what normal residential life should look like, and individuals who adopt technologies outside of the dominant mode are indeed choosing an alternative. It is important to take seriously the cultural consequences of technology in creating standard practices that are more or less ubiquitous and unquestionably accepted."[58] Finally, the language of alternative is "intended to highlight the importance of looking at technology as involving active practice," as individuals who live alternative lifestyles in alternative communities with alternative technologies "have reconfigured the practices of everyday life in ways that diverge from mainstream residential dwelling."[59] In doing so, they are actively dwelling in resistance, demonstrating the possibilities for alternatives to reconfigure practice in ways that lessen the negative environmental impacts of residential life while also recognizing the connection between technology, social practice, social organization, and power, and allowing us to pursue technologies and practices that correspond to the value orientations we share and wish to honor through everyday life.

2

What "Normal" Dwelling Looks Like

• •

The History of Home Technologies

Two general features characterize the relationship between human beings and the technologies that support residential dwelling in the modern world: dependence and isolation. We are dependent on the technological systems that provide electric power, heating and cooling, water, sanitation, and comfort in our lives, as well as the political arrangements that support them and the experts that construct, legitimate, and maintain them. The infrastructures that support modern residential dwelling have "effectively led to a dependence culture, which means that our homes are, on their own, weak and vulnerable without their various connections."[1] Human beings dwelling in modernity are also isolated from one another and from the natural world.[2] The technological systems that sustain residential dwelling in its current form support and recreate these general characteristics.

There are myriad technological systems that support residential dwelling, including the infrastructures that provide electricity, heating and cooling, water filtration and distribution, waste and wastewater removal, transportation, and food production. Building technologies, economic technologies (the material and nonmaterial relationships that shape income earning), and technologies of labor (the material conditions of organizing and accomplishing work), education, and care arguably largely share similar sociopolitical features and consequences.[3] While these technologies differ in multiple ways, their shared attributes represent a common logic. They are monolithic technological systems, enormous and entrenched infrastructures, and the sociopolitical systems maintaining these "hard"[4] systems rely on similarly hard organizational principles, where users are a group distinct (and alienated)[5] from the group of political, economic, and technical experts constructing and maintaining the technological systems that make residential life comfortable, even livable. These principles contribute to conditions of dependence and isolation, consequences of how the technological and material systems that meet residential needs and comforts in the modern world are organized. These principles and their consequences also shape how we think about and interact with the natural environment and one another.[6]

As one example of how technological systems influence social thought and practice, consider Lewis Mumford's discussion of the steam engine. According to Mumford, there is "an important difference" in the consequences of the steam engine compared to the technologies that came before it:

> [T]he steam engine tended toward monopoly and concentration. Wind and water power were free; but coal was expensive and the steam engine itself was a costly investment; so, too, were the machines that it turned. . . . [S]team power fostered the tendency toward large industrial plants already present in the subdivision of the manufacturing process. Great size, forced by the nature of the steam engine, became in turn a symbol of efficiency. . . . With the big steam engine, the big factory, the big bonanza farm, the big blast furnace, efficiency was supposed to exist in direct ratio to size. Bigger was another way of saying better. . . . [T]he steam engine accentuated and deepened that quantification of life which had been taking place slowly and in every department during the three centuries that had preceded its introduction.[7]

Mumford's discussion of the steam engine is meant to demonstrate that dominant technological systems both shape and reflect not only important social arrangements and institutions but also important ways of thinking and being.[8]

Seeing technologies as merely material renders them politically and ideologically neutral, but, as Lewis Mumford understood, technological systems are imbued with social values, power relations, and politics. Further, technology and society are co-constituted; technologies not only shape societies, but also the ways of being, thinking, and knowing embraced and practiced by individuals within a society. As other environmental sociologists have noted, "What we take to be 'physical facts' are likely to be strongly shaped by social construction processes, and at the same time, what we take to be 'strictly social' will often have been shaped in part by taken-for-granted realities of the physical world. Technology offers important opportunities for tracing these interconnections, being an embodiment of both the physical and the social."⁹

The technological systems that support residential dwelling in American society are socially consequential; they establish social conditions of isolation and dependence, conditions that shape the lived experiences of modern residential dwellers. Examining the histories of the technological systems that support residential dwelling sheds light on how these technologies have come to shape our society in terms of both thought and action, mind-sets and practice—and demonstrates that changes in the technological systems used to support residential life contribute to reshaping social practice, as well as social organization and social thought.

Looking at the history of the appropriate technology movement, a set of ideas and activities initiated in the 1960s that sought to challenge the ever-expanding size and scale of the technological infrastructures that support everyday life, brings history forward to the present. Ideas from this movement continue to influence, in various ways and to various degrees, those who seek to move away from the dependence and isolation of modern residential dwelling by adopting alternative technologies, those whose stories are considered in the four case studies presented in the chapters that follow. These case studies allow us to see the potential for dwelling in resistance by changing both thought and practice in conjunction with changes in the use of technological systems and forms of material organization to support residential life.

Technology and Infrastructure in Residential Life

In common nomenclature, the word *technology* likely elicits imageries of "high tech" or of particular technologies like computers and iPads—how many of us think of automobiles, electricity, or toilets? When it comes to thinking about technology, "the fact is that mature technological systems—cars, roads, municipal water supplies, sewers, telephones, railroads, weather forecasting, buildings, even computers in the majority of their uses—reside in a naturalized background, as ordinary and unremarkable to us as trees, daylight, and dirt."¹⁰ Yet in the most general sense of the word, technology refers to the materials and material systems that contribute to practical human activity. Thus, technology is an incredibly broad term. Technologies are the diverse array of materials, material systems, and forms of material organization that human beings employ to get things done, to meet their practical needs and comforts and their impractical desires, to employ and organize the world around us.

The meaning of the word technology is a social product, with its own history shaping the associated ideas and imageries.¹¹ Many modern dwellers most likely associate the word technology with progress—technology as something that contributes to the good life—and with experts—technology as something known, constructed, and maintained by a specialized group of trained individuals. Technology is both "a means to an end" as well as "a human activity," and it is important to recognize both of these aspects of technology; the "two definitions of technology belong together."[12] Technology refers to a material or material system that serves as a means to an end for practical human activity. Yet technology is also the result of human activity and the social conditions in which that activity is embedded.

The technologies that support residential dwelling are actually systems, or infrastructures, in that they involve vast technological arrays and organizational arrangements that link time, space, and social organization. The President's Commission on Critical Infrastructural Protection defined infrastructure as "a network of independent . . . man-made systems and processes that function collaboratively and synergistically to produce and distribute a continuous flow of essential goods and services."[13] Infrastructures, and even the infrastructures within infrastructures (such as, for example, the transportation, food, energy, and water infrastructures, among others, which support the infrastructure of an airport), characterize modern society.[14] "To be modern is to live within and by means of infrastructures" and "infrastructures simultaneously shape and are shaped by—in other words, co-constitute—the condition of modernity."[15]

Infrastructures cause tensions in the "modernist settlement," or "the social contract to hold nature, society, and technology separate, as if they were ontologically independent of each other."[16] Modern epistemologies view nature as ontologically separate from human culture, and the infrastructures that sustain residential life support this viewpoint ideologically while simultaneously challenging it materially. Humans are dependent on one another, yet at the same time technologies of all kinds depend on and are inspired by nature. Modern infrastructures complicate our ontological demarcation of "nature" as distinct from human culture.

Infrastructures shape our individual daily lives as well as our individual and collective frames of mind; infrastructures operate as systems of provision that shape the context of possibilities for our thoughts and actions.[17] Society (comprised of individual actors, political and economic institutions and arrangements, and collective ways of thinking, knowing, and acting, all shaped by collective awareness or inattention to various social facts) also shapes the development of technological infrastructures. Yet one of the ironies of technology in modern society is that it is pervasive and extremely important, yet largely unnoticed—not always actually invisible, but rendered invisible through common acceptance and disregard. Perhaps the "most salient characteristic of technology in the modern (industrial and postindustrial) world is the degree to which most technology is *not* salient for most people, most of the time."[18] Most people "hardly ever stop to think about [residential dwelling technologies]; although they sustain our lives, they nonetheless remain myste-rious."[19] Technological infrastructures, as vast yet invisible systems, shape the social, political, and economic arrangement of society; "infrastructures consist not only of hardware, but also of legal, corporate, and political-economic ele-ments" and "infrastructures function for us, both conceptually and practically, as environment, as social setting, and as the invisible, unremarked basis of modernity itself."[20] This invisibility is a key feature of modern infrastructures; the dependence and isolation we experience as a consequence of residential technology use is also largely invisible, making dependence and isolation two hidden conditions of our daily lives.

Situating the Technology-Society Relationship

While technologies and technological systems may be unquestioned elements of life for most modern dwellers, scholars have spent significant time and energy attempting to explicate the relationship between technological devel-opment and the constitution of society. Two important ideas from the field of science and technology studies are, first, that technology and society co-constitute one another, with technologies shaping society and vice versa.[21] Second, this co-constitution is not mechanistic or deterministic. While Marx may have claimed, in a polemic moment, that the "hand-mill gives you soci-ety with the feudal lord; the steam-mill society with the industrial capitalist,"[22] the relationship between technology and society is much more complex than this simple one-way formulaic suggests. Technologies are not neutral—they are embedded with particular politics,[23] they help to construct and reinforce par-ticular forms of political, economic, and social arrangements,[24] and even shape particular practices, ways of being and ways of thinking.[25] Yet understanding their influence on social arrangements and individual practices does not require deterministic thinking and does not necessarily imply a conscious influence.

Many of the traits we come to associate with particular technologies are socially constructed; technological systems become viewed as "efficient" or "unrealistic" or "normal" or "alternative" based on social processes of inter-pretation and negotiation. Technologies can be viewed as "embodiments of human desires and ambitions, as solutions to complex problems, and as inter-acting networks and systems."[26] "The flip side of the claim that technology is socially shaped is the claim that *society is technologically shaped*, meaning that technologies shape their social contexts. . . . Technologies become part of the fabric of society, part of its social structure and culture, transforming it in the process. The idea of society as a network of social relations is false, because society is made up of sociotechnical networks, consisting of arrange-ments of linked human and nonhuman actors."[27] The language of "sociotech-nical networks" brings attention to the connections between both human and non-human agents within a network in which technology and society are mutually constituted. Technologies cannot be understood in isolation from their social context, and societies can only be understood in relationship with material and environmental systems.[28] Sociotechnology "consists of dynamic seamless webs of entities that are only labeled as technological or social after they have fully evolved."[29] The terms *sociotechnical networks* and *sociotechnology* emphasize how material systems are actively involved in shaping social systems and the ways we humans interact with those material systems, one another, and the natural world.

Technological infrastructures, aptly described as sociotechnical networks, are connected to social systems in a relationship that can be dubbed "co-constitution."[30] For me, there are two main consequences of acknowledging this relationship: one is recognizing that the technologies that support every-day residential life are not inevitable or set in stone, but are the outcome of processes of (power-laden) negotiation and interpretation. The other involves recognizing that technologies shape practice, or the habits and behaviors we engage in every day, actions that also demonstrate our understandings of what is normal, expected, and acceptable in terms of the organization of everyday life.

One way to further understand the relationship between the social and the technical elements of sociotechnical networks is to think about how technology "enframes" us, limiting and shaping our view as a frame does to a visual image.[31] This enframing can constrain the possibilities of thought, action, and social change. For some, all of modern technology is more than a compilation of co-constituted sociotechnical networks; instead, these critical theorists argue, modern technology represents a mind-set, or "technique," which promises progress through the rationalization and standardization of human behavior.[32] Technology, then, not only refers to something physical, but also something ideological. Technology represents rationality determined based on a means-end schema, and the sociotechnical networks that support modern life are assumed to be entirely neutral and simply the most rational means to a given end (whether this end be heating, lighting, or the ability to communicate across the world instantaneously). This faith in the neutrality and rationality of technology simultaneously shapes and reflects nonmaterial ideologies and practices because technologies and society are co-constitutive.[33]

The very definition of technology, then, encompasses more than material tools and systems; technology is seen as a particular mode of thinking. It is a mode of thinking that emphasizes the rational, the efficient, and the normal, thus shaping the categories we use to understand ourselves, other humans, our societies, and the natural world. Technology corresponds to technological or instrumental rationality, a mode of thinking that characterizes not only modern technologies but also modern thought and modern economic, political, and social arrangements.[34] However, technologies arguably also shape cognition and action in more acute ways.[35]

The work of Michel Foucault sheds light on how technology use shapes cognition through daily practice. For Foucault, individual practices serve a disciplinary function; our practices are disciplined by the technologies we use in our daily lives.[36] This type of disciplinary power is productive in that it produces a particular rendering of individual lifestyle and mentality. Foucault argues that "discipline 'makes' individuals" because individuals are both the objects and the instruments of discipline.[37] For Foucault, "technologies"—a term that applies to material arrangements such as the infamous Panopticon as well as a broad range of strategies of power and organization such as classroom routines, and that we can apply to seemingly innocuous things like electricity and water provision, waste removal systems, and food production—serve as instruments of discipline, shaping individual lives, bodies, and minds through daily practice.

There is a growing body of scholarship that recognizes the role of material systems in shaping daily behaviors at both the individual and social scale.[38] As reviewed in chapter 1, theories of practice argue that understandings of action cannot be isolated from the material systems that enable and constrain behavior.[39] In the co-constitution of technology and society, material systems shape shared understandings of what is normal and acceptable in terms of technology use as well as shared sets of social practices that contribute to and help to solidify conceptions of the normal in the sense of, in the particular case discussed here, normal residential life.

Further, technology use in modern society involves a constellation of practices that can be seen as forms of ritual.[40] When I flip a light switch, turn on a faucet, or flush a toilet, I am arguably engaged in a bodily ritual thoroughly saturated with social, political, economic, and technical relationships and meanings. The performance of these rituals works to "shape cognitions. The main objects or ideas that [are] the focus of attention during a successful ritual become loaded with emotional overtones. Those ideas or things become symbols; whatever else the ideas may refer to on the mundane level, there is also a deeper, Durkheimian level on which symbols invoke membership in the group that charged them up with ritual significance."[41]

Technology use is a ritual practice imbued with emotional significance. Flushing a toilet becomes a symbol of modernity, of normalcy, and these cognitive categories are reinforced through daily practice. Technologies shape ways of thinking about technology by engaging us, as residential dwellers, in daily rituals of the body.[42] They influence the ways we think about our neighbors and ourselves and the very societies in which we live as well as what we consider normal and abnormal or appropriate and inappropriate lifestyles. The dominant technological systems in a society both shape and reflect not only social arrangements and institutions but also ways of thinking and being.

The Sociotechnical Networks of Modern Dwelling

There are numerous exemplars in the field of social and cultural history that demonstrate how rigorous historical research can reveal the relationship between a given technology and particular social relations.[43] With a sweeping, broad-brush overview from the 1800s through the present, the brief review below borrows from this work and is intended to demonstrate historically how the technologies that support residential dwelling in the modern world have been shaped by and have simultaneously shaped social organizations and institutions, social relations and values, and individual ways of living, being, and knowing.

The technological infrastructures that support modern residential dwelling are sociotechnical networks, arguably social institutions in their own right. These institutions have histories, and as they develop, they promote some choices and possible actions while constraining others.[44] Examining the historical development of the technologies that support residential dwelling can shed light on how these technologies have come to shape our society, including the ritualized habits of behavior we consider normal when it comes to using technology in residential life.

The technologies used to support residential dwelling in America tell a story that is about more than material arrangements, including what types of political and economic arrangements and forms of organization are being reproduced and what kind of lives we have constructed as normal in the sense of both typical and normatively acceptable. As Lewis Mumford once wrote, "Throughout history, the treatment and arrangement of shelter have revealed more about a particular people than have any other products of the creative arts. . . . No society can be understood apart from the residences of its members."[45] The histories of the technologies that support modern dwelling, viewed as sociotechnical networks, demonstrate how technological development has supported material as well as discursive arrangements of dependence and isolation. This historical examination also highlights how technologies, as institutions, channel categories of human conduct and perception in particular ways, thus creating a categorical distinction called "normal" when it comes to residential life.

Electricity Provision

Thomas Edison's light bulb, invented in 1878, transformed many aspects of society, including residential dwelling. Electric technology was originally purchased as an isolated unit, generating power on site for use by individual households. In 1882, Edison constructed the first investor owned utility (IOU), a centralized electricity production facility on Pearl Street in New York City, which provided 110 volts of electrical current to fifty-nine customers.[46] This technological system transmitted centrally generated electricity to "all houses within a circle of half a mile."[47] The development of centralized electric generation and transmission over ever-increasing distances shaped the trajectory of the electricity technologies that would come to power much of the globe.

This is not to suggest that the history of modern technological systems was closed with the establishment of the centralized IOUs for electricity generation and transmission; in the "current wars" between direct current (DC) and alternating current (AC) technologies, Edison supported DC technologies, while George Westinghouse was a proponent of AC. In 1886, Edison wrote, "Just as certain as death Westinghouse will kill a customer within 6 months after he puts in a system of any size."[48] Edison conspired with Harold P. Brown, an electrical engineer outspoken about the dangers of AC technology, in attempt to thwart the expansion of Westinghouse's business.[49] However, Edison could not match the much longer transmission distances Westinghouse achieved through AC technology and, in the end, even Edison's companies adopted AC as their technology of choice.[50] AC electric technologies allowed transmission much farther distances than DC, and "eventually achieved such dramatic economies of generation that they became the foundation stone of electric power everywhere."[51] With the transition to AC electric technologies, existing DC stations, deemed inefficient, were converted to substations. These converters connected the older DC systems of the cities to the newer AC systems developing in the burgeoning suburbs.[52]

During the Great Depression of the 1930s, the Rural Electric Administration sought to provide electricity to rural areas, but the policy incentives of the New Deal specifically supported electrification based on centralized generation and transmission in pursuit of particular ideals regarding efficient, modernized populations.[53] Electric technologies and the appliances made possible by electrification were promoted as easing the drudgeries of farming and housework; a life of convenience and efficiency was the discursive promise of modern electric technologies.[54] Yet these ideals of efficiency did not necessarily match their actual consequences, particularly in terms of increasing time spent on "women's work" in the home.[55]

The regulatory regimes that worked to support electrification established a particular dominant technology based on centralized, fossil fuel–based generation transmitted over increasingly vast distances to the end-use consumer. These technologies require expert knowledge, are based on standardized codes of construction and conduct, and are characterized by two objective consequences: dependence and isolation.

The social arrangements that support modern electric technologies involve dependence on expert technicians and regulators as well as monthly payments for provision of light and in some cases heating, cooling, food refrigeration, and the ability to cook—all of which are, arguably, necessary for human subsistence and significant for human comfort. Modern electricity infrastructures simultaneously increase isolation, as we modern dwellers have light and temperate comfort regardless of the climate, at a significant cost to the natural world, as long as we have the ability to pay a monthly bill, every month for the rest of our lives. Furthermore, the structure of this sociotechnical network renders this dependence and isolation invisible, making the current structure of electricity provision and the associated social practices seem like the only possible option.[56] Other technological infrastructures, like those that provide water distribution, waste treatment, transportation, and food production, followed a similar historical trajectory of centralization and resulted in similar consequences in terms of dependent and isolated systems of practice.

Water Filtration and Distribution

The filtration of water and the distribution of this technologically clean water to residential homes across America is another technology that, like electricity, now silently pervades the lives and consciousness of modern dwellers to such an extent that it is rendered invisible. Until "the second half of the 19th century, most American urbanites depended for their water supplies on local surface sources such as ponds and streams, on rainwater cisterns, or wells and pumps drawing on groundwater. Under these supply conditions, water consumption per capita probably averaged between 3 and 5 gallons a day."[57] The first (capital-intensive, piped-in) waterworks system was constructed in Philadelphia in 1802, which led to increased water usage and the installation of "water closets" (bathrooms).[58] The development of water filtration and distribution systems corresponded to the construction of wastewater removal infrastructures and treatment facilities; "developments in water supply technologies have marched hand in hand with developments in sanitation and water treatment."[59]

There are two particularly notable points regarding the development of water filtration and distribution systems. First, the debates that took place in the early stages of technological development in both water filtration and wastewater removal revolved around themes of sanitation and health, promoted by a movement of "elites and professionals that aimed to change people's ideas about their personal habits of cleanliness, to create an enlarged role for government in areas related to health and sanitation, and to promote the construction of urban public works to achieve a healthful city."[60] While sanitation and health are laudable goals, there are multiple ways to arrange water usage and waste removal to achieve them, but centralized systems were presented as the only option. Second, water filtration was pursued upstream, meaning that communities filtered their water prior to distribution and consumption, but not after. The justification was that water treatment was a downstream issue—if every community filtered their water, no community had to deal with treating wastewater.

In the systems of water filtration, distribution, and treatment that provide water to most residential dwellers today, water is filtered away, somewhere; modern residential dwellers rarely know where, or how, or by whom, trusting that experts and standards keep best interests in mind and bodies safe and protected. Water is distributed to homes via a system of pipes that, again, you need know nothing about in terms of where they are, how they work, or who is responsible for their maintenance or repair. Just like electricity, water filtration and distribution systems encourage dependence and isolation, dependence on a network of material and nonmaterial systems accessible only to experts and isolation from other users (we are no longer gathering at the watering hole, if we ever did) and the natural world that provides this life-maintaining substance.

Recent water shortages and cases where water quality is clearly lacking are highlighting water consumption as a central issue for modern residential dwellers. Nonetheless, conversations about water scarcity seldom invoke any mention of rethinking the infrastructures that provide for residential water use. Modern residential dwellers have only experienced a world in which we need not think of the source of the water or its potential vulnerability; we trust experts to monitor and maintain a clean water supply that is infinitely abundant, providing water at the turn of the faucet ever more.[61] Further, when we're done with the water we use in our homes, we can remain comfortably ignorant about where it goes, how it's treated, or who may be using it again.

Waste Removal

In early American cities, wastewater was simply thrown on the ground or placed in a street gutter, a dry well, or a leeching cesspool. Human waste was put into a permeable cesspool or impervious privy vault. By the middle of the nineteenth century, a combination of demographic and technological factors (including urban population growth and the adoption of new household water fixtures such as the toilet made possible by the developing urban water supply system) began to "stress the cesspool-privy vault system, causing its eventual collapse and replacement."[62]

The first system used to address the need for waste removal was the water-carriage. The first municipal sewer systems—which were intended for human waste as well as storm water—were constructed in Brooklyn in 1855, Chicago in 1856, and Jersey City in 1859,[63] and "biological processes in man-made structures replaced land-treatment methods as the most popular approach to sewage treatment" during the first decades of the twentieth century. Yet "dumping raw sewage into waterways remained the cheapest and most popular means of disposal," and even by 1940, only half of sewage waste from those living in urban areas with sewage systems was being treated[64] (as mentioned above, the idea was that those downstream of the waste would treat it, as water was always filtered upstream of the community of users). Today, the idea of having a system in which we can flush away our waste without a thought to where it goes or how it is dealt with is so pervasive in American society that any other type of system seems unfathomable (and, due to regulations, often is).

Many Americans (an estimated 25 percent) use septic tanks for waste disposal, not a centralized waste treatment facility. Indeed, "the septic tank was a key element in the suburbanization of the United States."[65] The post–World War II housing boom that catalyzed suburban development in the United States took place before most land outside the city was equipped for sewage removal, and many suburban homes initially relied on septic systems. However, septic tank failures were (and still are) frequent. To address what became identified as a septic tank problem, septic tank installations became increasingly regulated during the 1950s, and attempts were made to expand centralized sewage lines and use septic tanks only when necessary.[66]

Septic tanks are technologies on the scale of the individual home; they are not centralized sociotechnical networks. Yet septic tanks rely on many of the same features that characterize centralized waste treatment facilities. They require experts to install, monitor, and maintain them. Some of the first suburbanites did not even ask about waste disposal when buying their homes, and "educated, sophisticated people have confessed that they did not know their home had a septic tank system until it overflowed."[67] Research suggests that this remains true today.[68] Septic tanks as a sociotechnical artifact demonstrate the overarching resistance to alternative technological systems that arguably stems from the normalizing influence of residential sociotechnical networks. For most Americans, there are two possible means of addressing human waste: a waste treatment facility or a septic tank. There are two potential options, and no more: outhouses or composting toilets are not "normal." Septic tanks demonstrate how options can be closed off through the very existence of options. Even options often correspond to the ideological systems in which we currently reside, where experts manage and maintain the systems we rely on to meet our residential needs and comforts.

Transportation and the Automobile

Henry Ford revolutionized factory work through the introduction of the assembly line, creating an affordable car and greatly contributing to the expansion of automobile use in America. While "it took the average worker twenty-two months to buy a Model T in 1909, by 1925 the same purchase would have required the labor of less than three months."[69] Sales soared in a very short time; vehicle registrations grew from 1 million in 1913 to 10 million in 1923. In 1923, Kansas had more cars than France or Germany and Michigan had more than Great Britain and Ireland combined. By 1927, there was one car for every five people in the country. This "rapid rise in motor-vehicle registrations created a blooming optimism, a national faith in technological progress."[70]

Combined private and government interests promulgated this national faith. The federal government promoted the development of automobile technology and the infrastructure of roads and service stations necessary for automobile travel as early as 1916, with the Federal Road Act. The act offered funds to states that organized highway departments. The Federal Road Act of 1921 designated 200,000 miles of highway as "primary" and thus eligible for federal funds on a 50/50 matching basis, and the government created a Bureau of Public Roads to plan a highway network to connect all cites with a population of 50,000 or more. Gasoline taxes developed rapidly, initiated by the state of Oregon in 1919 and instituted in all states by 1929.[71] By the 1920s, "a coalition of private-pressure groups, including tire manufacturers and dealers, parts suppliers, oil companies, service-station owners, road builders, and land developers were lobbying for new streets."[72] Supported financially and ideologically by the federal government, new street construction flourished, continuing even during the 1930s.

At the same time, "mass transportation was floundering because of government decisions that the streetcar represented private investment and should 'pay for itself.'"[73] In 1954, President Dwight Eisenhower appointed a committee "to 'study' the nation's highway requirements. Its conclusions were foregone, in part because the chairman was Lucius D. Clay, a member of the board of directors of General Motors. The committee considered no alternative to a massive highway system, and suggested a major redirection of national policy to benefit the car and the truck."[74]

Thus began our long road of dependence on the automobile and the sociotechnical networks necessary for its use. In the middle of the twentieth century, sociologists Robert and Helen Lynd conducted their now famous Middletown studies, and found that "in all six areas of social life in Muncie, Indiana—getting a living, making a home, raising the young, using leisure, engaging in religious practices, and participating in community activities— the private car played either a contributing or dominant role."[75]

Lewis Mumford wrote, "When the American people, through their Congress, voted . . . for a $26 billion highway program, the most charitable thing to assume is that they hadn't the faintest notion of what they were doing."[76] One of the most significant consequences of the rise of the automobile and our almost complete reliance on it for contemporary daily life is a substantial increase in our dependence. Most Americans are entirely dependent on the automobile; public transportation systems have floundered and the vast majority of Americans have no other option when it comes to daily transport.

There are many rich histories of rise of the automobile[77] and car culture[78] in America. The social historian Ruth Schwartz Cowan discusses this dependence using the language of "automobility,"[79] a concept that captures the consequences of automobile technology as an infrastructure and institution, recognizing that the private automobile dominates the American landscape and psyche to such an extent that, in many ways, to be without a private automobile is to be without full mobility. Automobiles are viewed as the epitome of personal freedom at the same time as they make us completely dependent on them and isolated from our neighbors, communities, and natural environments.[80] We are dependent on the automobile and the networks of associated infrastructure for our daily mobility and meeting our daily needs. Further, automobility serves as a tool of social inclusion and exclusion. Many people are unable to afford or access a reliable automobile, and automobility serves to perpetuate class distinctions and biases. Some individuals have disabilities that prevent them from having the freedom to privately transport themselves anywhere at anytime in their individually owned and operated automobile.[81] We are all dependent upon a vast array of experts and decision makers to maintain our vehicles and the roads necessary to drive them as well as all the standards placed on driving, from emission regulations to maintenance schedules to insurance rates. The automobile also increases our isolation from both the natural world and one another as we commute, shop, and travel in our own private metal boxes.

Food Production

The development of industrial food production has received widespread attention from both scholarly[82] and popular[83] sources. We can again turn to Ruth Schwartz Cowan for an excellent summary. Cowan reviews how a mother gets food for her child in hunter-gatherer societies, premodern agricultural societies, and industrial society, highlighting our increased dependence on innumerable technologies, organizations, institutions, and infrastructures in procuring the substance of life in contemporary society: we must now drive to a grocery store, where much of the available foodstuffs are grown by strangers using technologies we do not know or understand, prepackaged by strangers using similarly mysterious technologies, and transported across the nation or the globe using the sociotechnical networks of transportation.[84]

When it comes to food technologies, our societies have shifted from a dependence on nature to a social dependence: "People live longer and at a higher standard of living in industrial societies than in preindustrial ones, but they are not thereby rendered more independent (although advertising writers and politicians would like them to think they are) because, in the process of industrialization, one kind of dependency is traded for another: nature for technology."[85] This insight is applicable to all of the technologies of residential dwelling; while modern dwellers are less directly dependent on nature than we as a species once were, we are more dependent on a web of unknown and unidentifiable people, programs, and processes involved in the myriad sociotechnical networks upon which our lives depend. Many of the individuals in this web upon which the modern dweller depends are technical experts. As the popular writer Michael Pollan expresses, in modern society, we no longer even know how to eat well without the help of chemists, nutritionists, and government agencies.[86]

Yet our dependence on sociotechnical networks for food provision does not actually decrease our dependence on nature for subsistence; it simply veils that dependence.[87] Of course, the modern technologies of food production, transportation, sale, and consumption also increase our isolation from nature (as we are no longer directly dependent on it, thus we may no longer recognize our connection to it) as well as one another (as we no longer know how to grow food for ourselves, nor do we know those who are growing the food necessary for our survival).

Standardization of Technologies, Thoughts, and Practices

The histories of the technological infrastructures that meet the needs and comforts of modern residential dwellers are histories of standardization. In modern America, the vast majority of localities have adopted standards for how homes are wired for electricity and who is allowed to do the work. There are standards for who is allowed to construct home plumbing systems and what type of technology they are allowed to employ. There are standards for automobile maintenance, licensing, and insurance. There are food certification standards. By World War I, "following the negotiations between electricians, plumbers, employers, and the city's representative, . . . building codes and licensing requirements, along with instruction in the public schools, set the framework" that builders worked within when installing the systems that support dwelling in America.[88]

Standardization is a process that marks the history of many sociotechnical networks, revealing the common logic in the development of technologies as diverse as electricity, food production, and the organization of labor. Standardization explicitly institutionalizes limits to choice, indicating to the modern dweller what is acceptable and unacceptable, normal and alternative, when it comes to dwelling practices. Standards, in the forms of building codes and insurance rates and food-labeling requirements and monthly paychecks, set discursive limits on the possible, or at least on perceived possibility. Standardization is a process that shapes practice. It is because of standardization that modern American dwellers share the knowledge of how to find a light switch in a dark room, flush a toilet, and cook a meal in a modern kitchen. Standardization works to enframe technological practice, shaping habitual and unreflective behaviors as well as ideological (although not necessarily conscious) commitments to certain kinds of sociotechnical arrangements.

The imagery of the American suburb helps us to visualize this history of standardization. As Lewis Mumford wrote,

> In the mass movement into suburban areas a new kind of community was produced, which caricatured both the historic city and the archetypical suburban refuge: a multitude of uniform, unidentifiable houses, lined up inflexibly, at uniform distances, on uniform roads, in a treeless communal waste, inhabited by people of the same class, the same income, the same age group, witnessing the same television performances, eating the same tasteless pre-fabricated foods, from the same freezers, conforming in every outward and inward respect to the common mold, manufactured in the central metropolis. Thus, the ultimate effect of the suburban escape in our own time is, ironically, a low-grade uniform environment from which escape is impossible.[89]

The development of the American suburbs provides an historical caricature of our increasing dependence and isolation.[90] Suburbs have been popping up across the United States since the 1820s, but the biggest support for suburbanization, both materially and ideologically, occurred post–World War II with federal policies such as the GI Bill.[91] Government incentives encouraged returning GIs to move out to the 'burbs and purchase a newly constructed single-family home. These single-family homes have, over time, become much larger,[92] which means they require more natural resources (including fossil-fuel inputs) to build and maintain. They also require more economic resources because they cost more to light, heat, and cool, thus increasing dependence on a monthly paycheck to meet the needs and comforts of residential life. According to the environmental historian William Cronon, "This idealized domestic landscape [of the suburbs] became a refuge for the middle class, but also it would become an engine of sprawl, a high-energy-consuming economy, and a symbol of racial exclusion as well."[93] Suburbanization certainly had important social implications, including perpetuating racism and racial segregation through exclusionary neighborhood covenants that prohibited minorities from purchasing or renting a suburban home.[94]

In the suburbs, life-supporting technological systems are conspicuously invisible. The electricity supply is generated afar and transmitted to individual homes for power provision. Requirements for providing a clean water supply are imperceptible, as water arrives to individual homes somewhat mysteriously. Food production is completely removed from the visibility of daily life, with food products purchased at a centralized location; frequenting that location most likely requires automobile travel. Waste is taken "away" before being treated, processed, or deposited in a far-off landfill (which may not necessarily be so far away, but is likely unseen). The technological systems that support residential dwelling are constructed, monitored, and maintained by experts rather than users. Productive labor takes place outside the home, yet all necessary home functions are dependent upon it (as we pay for technological services). Economic transactions in the moneyed economy become increasingly necessary for survival, while increasingly taking place between isolated individuals.

Lewis Mumford described residential dwellings in the modern world as "stage-settings" that are, "on a strict psychological interpretation, cells: indeed, the addition of 'comforts' made them padded cells."[95] An ardent critic of modern technologies and particularly the suburbs, Mumford wrote, "the suburb served as an asylum for the preservation of illusion. Here domesticity could flourish, forgetful of the exploitation on which so much of it was based. Here individuality could prosper, oblivious of the pervasive regimentation beyond."[96]

The Consequences of Modern Residential Technologies

The suburbs represent an idealized vision of the standardization that pervades modern residential dwelling. Through the sociotechnical networks that support dwelling, individual lives are shaped by invisible yet necessary systems used in the privacy of one's home. Thinking about these networks that support residential life highlights our "dependence on interconnected technological systems," the centralized life-supporting technologies that provide electricity and water as well as the infrastructures that treat waste and provide food.[97] Modern residential dwellers in America experience multiple forms of isolation as a result of our technological dependence. Our work lives our isolated from our home lives. Through the organization of residential life and the technological infrastructures that support it, we are isolated from one another and from the natural resources and processes upon which our lives ultimately depend. Further, these technologies involve a normative component; the technologies that support residential life in America categorize and classify based on internalized concepts of typical, acceptable residential practice. The modern American dweller becomes defined through use of the technological systems that support residential dwelling, through the bodily practices and daily rituals in which we reenact our isolation and our dependence in residential life.[98]

The normalized technological systems supporting modern residential dwelling have particular characteristics as well as discernable consequences. First, the systems that meet the very basic needs of subsistence, providing food, heat, water, and electricity, have been taken out of the hands of individual citizens or localized communities. They have become increasingly centralized, distanced from the users themselves. This shapes the identity of modern dwellers as consumers, tacitly relying on technological systems to support daily life. Second, the technologies that support residential dwelling are increasingly specialized, constructed and maintained only by expert knowledge and decision making. This prevents democratic forms of technological control or policy making since only scientific, technical, and policy experts have the necessary power/knowledge.[99] Particular technologies are better suited to undemocratic forms of control and the general trend in technological development throughout the past century has been to create dependent and isolated citizen-consumers without the ability to meaningfully influence the forms of technology upon which modern dwelling depends.[100]

In a typical American dwelling, electricity technologies are based on centralized generation and transmission from fossil-fuel resources. The U.S. electricity transmission grid is the largest technological system in the world, yet it is conspicuously invisible in the modern home. When the normalized modern American consumer-citizen flips a light switch, he or she thinks little or nothing about the action, trusting in the reliability of the system. Yet we, as Americans, are both ignorant of and completely dependent on this system. The vast majority of us do not generate our own electricity or have the knowledge to do so, yet we are completely dependent on this technological system for both convenience and survival. Water provision technologies are also centralized, removed from the view of most American dwellers. Water is piped into our homes already cleaned and fit for drinking, or at least we trust that it is. Individuals need not be aware of the processes necessary to make that water suitable for consumption, but they depend upon the centralized water provision system for the substance of life itself.

Waste treatment technologies are also removed from the home as much as possible. Wastewater is transported to centralized facilities, where experts are expected to adequately treat our waste. Human waste is also transported away from the home. Consumer waste is taken to the nearest or most economically advantageous landfill. This landfill may be located in another residential dweller's backyard, but it is typically out of the sight and minds of the wealthiest waste producers.

Private automobiles are the dominant form of transportation throughout the United States; in many parts of the country, it is simply the only form of transportation available. The organization of our dwellings, in relation both to one another and to other aspects of the built environment, supports our dependence on private, individualized transportation technologies. Capturing our dependence on private automobiles and the ways car travel has shaped the entirety of our built environments, automobility highlights both the dependence and isolation of the modern American dweller.

Food technologies have been concentrated into the hands of a fistful of private corporations, and monocultures (where only one type of crop is grown on a large tract of land) have become the norm. The vast majority of American dwellings place no emphasis on individual or community food production. Overwhelmingly, food is purchased from a grocery store (typically accessible only by personal vehicle). Individual consumers have very little knowledge of or control over food production, further facilitating both the isolation and dependence of modern dwelling.

Building technologies have become increasingly standardized so that many homes across the nation are similar in building materials, home design, and aesthetics. Many of the materials we used to construct residential dwellings are the synthetic products of a modern industry, foreign to most individuals in their materials, their construction, and their dangers. Many modern building materials such as PVC, plywood, and carpeting are laden with dangerous chemicals; we depend upon the regulatory arms of the state to protect us from the materials approved for and expected in residential building. Further, building design, materials, and technologies standardize particular notions of residential dwelling that affect our individual daily practices.

Economic technologies (i.e., incomes) are held by and distributed within individuals or individual families. Economic resources are, for the most part, not directly distributed or shared outside of the dwelling unit. This increases the dependence of modern American dwelling, as every individual or individual family must obtain these necessary economic resources in order to dwell "normally."

Labor technologies (the material organization of work) are pursued individually, most likely in a location separate from the residential dwelling. This increases isolation, as our work communities become separate from our residential life. It increases dependence, as we work to create things we do not use and to buy things we use but do not create. It also creates a disjunction between work and home, a predominant feature of modern American life. In sum, modern life is largely characterized by dependence and isolation, objective features of modern life that are articulated through and perpetuated by the sociotechnical networks upon which residential dwelling depends.

A History of Dwelling in Resistance:
The Appropriate Technology Movement

The development of the technological systems that support modern residential dwelling did not occur without vocalized criticism and expressed alternatives. As Lewis Mumford once wrote, "Culture and technics, though intimately related to each other through the activities of living men, often lie like-noncomfortable strata in geology, and, so to say, weather differently. . . . The rift between the mechanization and humanization, between power bent on its own aggrandizement and power directed toward wider human fulfillment had already appeared: but its consequences had still to become fully visible."[101]

The appropriate technology movement emerged from critiques originally articulated by British economist E. F. Schumacher.[102] Schumacher was concerned by attempts of industrial nations, particularly the United States, to import their centralized, large-scale models of technological development to less industrialized nations. After World War II, the United States pursued a development strategy that emphasized rapid industrialization and the importing of monolithic technological systems to a so-called Third World context. For example, Harry S. Truman's Point Four program offered technical, scientific, and managerial assistance to promote the adoption of particular technologies and practices in order to increase productivity and efficiency in developing economies.[103] "To a significant degree, the American aid programs, and those of other developed nations, were captive to the notion that ideally all countries should follow the same pattern of industrialization."[104]

Schumacher wanted "to formulate an alternative to conventional strategies of economic development in the Third World."[105] Schumacher suggested another means of technological development, which he called "intermediate"—"one which fit between the primitive and poverty-reinforcing tools of much of the southern hemisphere and those large, powerful technological systems of the northern."[106] Appropriate technology "emerged as a popular case at a 1968 conference held in England on technological needs for lesser developed nations . . . to provide a possible solution to the problem of how to promote a more equitable distribution of wealth while avoiding the inherent environmental and social problems of industrialization."[107] Alternatives to large-scale technological development were given many names, from intermediate to soft to convivial, eventually coalescing on a single term—appropriate. A countercultural movement grew around ideas originally expressed in terms of international development, which were utilized to critique American development as well.

The word *appropriate* had many overlapping meanings for the appropriate technology movement: technologies were deemed appropriate "(1) with respect to limited natural resources, particularly fossil fuels; (2) with respect to the technology of production, in contrast to 'inappropriate technology' or forces of production; and (3) with respect to the nature of social harmony and social equity (peace), or the social relations of production."[108] Central principles advocated "self-reliance, local autonomy, and respect for Nature."[109] The movement promoted "land use and living patterns that improve the quality of life by reducing energy consumption and dependence on fossil fuels" and "self-sustaining living patterns that increase our awareness of the balance between the realities of Nature and the needs of Man."[110] Based on these values, appropriate technologies were defined as those "that are: (1) cheap enough to be accessible to nearly everyone, (2) simple enough to be easily maintained and repaired, (3) suitable for small-scale application, (4) compatible with man's needs for creativity, and (5) self-educative in environmental awareness."[111]

The appropriate technology movement recognized that human beings "do not want electricity or oil, nor such economic abstractions as 'residential services,' but rather comfortable rooms, light, vehicular motion, food, tables, and other real things," and that there are multiple ways to provide these real things.[112] It also recognized that each of these technologies came with embedded political and social implications and that the dominating, large-scale technological systems created conditions of isolation and dependence. As one supporter of appropriate technology wrote, "convivial tools are those which give each person who uses them the greatest opportunity to enrich the environment with the fruits of his or her vision. Industrial tools deny this possibility . . . and allow their designers to determine the meaning and expectations of others."[113] Appropriate technologies were defined as "small in scale, democratically controllable, low in capital commitment, environmentally sustainable, and adaptable to local cultural conditions."[114]

These design principles correspond to particular social values and a desire for particular forms of social organization. Recognizing the political nature of technological systems, the appropriate technology movement was located at an intersection of technological critique and political cause. The movement considered smaller-scale technologies more conducive to democratic control and viewed democratic control of technology as an important social value. Recognizing the politics of technological systems also contributed to the appropriate technology movement's emphasis on environmental sustainability. Arguably, dominant sociotechnical systems contribute to both government control of residential dwelling and government resistance to environmental protection. The appropriate technology movement, recognizing the role of modern technological systems as perpetrators of environmental degradation, looked to technologies that adjusted to and worked with, instead of fighting against, natural ecosystems. This search for ecologically considerate technologies corresponds to the appropriate technology movement's promotion of technological systems adaptable to local cultural conditions, as regional variation is often ignored by the one-size-fits-all technologies that produce the brunt of environmental damage; "key to intermediate technologies was to apply advances in science to specific local communities and ecosystems."[115]

The principles of the appropriate technology movement and the types of technologies the movement promoted were based on values of democratic control, individual freedom, and environmental responsibility, values similar to the broader counterculture movement of the 1960s. The movement had a collective vision of an ideal society, which is decentralized and based mainly on small communities rather than towns and cities. Each community is relatively self-sufficient, using local resources to grow much of its food, produce most of its energy, and make many of its manufactured goods in small craft-based workshops.... Organization is cooperative and based on participatory democracy. Tools are relatively small and simple, so that they can be readily controlled, used, maintained, understood and modified by people without specialist knowledge and with access only to small workshops.... As far as possible, only renewable energy flows are used, and environmental impacts are minimized.[116]

Challenging the assumption that technologies were necessarily destructive for both humans and the environment and proposing alternatives to the technologies most responsible for environmental degradation, the appropriate technology movement intersected with environmentalism, the back-to-the-land movement, libertarianism, technological optimism, and countercultural idealism to mold a particular amalgamation of social values and promote a collective idea regarding potential future sociotechnical arrangements, presenting a formidable critique of the "understanding of technology which operated as a hegemonic culture" and to dominant technological systems and the politics embedded within them.[117]

Resisting Normal

The cogency of the critique notwithstanding, by 1982 the solar panels installed by President Jimmy Carter had been taken off the White House by President Ronald Reagan and a "retrospective" was being published on the appropriate technology movement.[118] Despite the large gap between the movement's collective vision for society and the actual state of society today, "the debate over Appropriate Technology was rich in cultural meaning and ideological intent, as well as being a material and economic challenge to existing social interests."[119] The appropriate technology movement inspired "awareness of values not included in a mechanistic ideology—values derived, not from the machine, but from other provinces of life. Any just appreciation of the machine's contribution to civilization must reckon with these resistances and compensations."[120] The values and intentions of this movement are, to varying degrees, still present in the myriad ways that residential dwellers are seeking and adopting alternative technological systems in residential life. In the case studies that follow, we will see how communities of individuals are continuing to search for appropriate technologies, pursuing self-provisioning at the individual and community level by installing solar electric and solar thermal heating technology, buying alternative fuel vehicles and eschewing privately owned automobiles, growing food locally and organically, and choosing to live with technological systems that result in less isolation and dependence. These individuals are choosing to reshape their material relationship to the world through the adoption of alternative technological and material systems in ways that correspond to the ideas and goals of the appropriate technology movement, and—importantly—that involve an alternative set of residential dwelling practices.

As indicated in chapter 1, these choices are not always and not necessarily entirely conscious or consciously articulated, and they are not motivated by narrowly conceived environmental concerns or economic calculations. Instead, the residential dwellers I met throughout my explorations of alternative residential dwelling are acting in ways that correspond to general orientations. Rational considerations of means and ends do not exclusively drive action, and the adoption of alternative technologies is not exclusively or universally motivated by alternative values. By seeking alternatives to the technological systems that support mainstream residential dwelling in America, the people I met across these diverse case studies are dwelling in resistance, because they are resisting the dominant material technologies that support residential life, the dominant practices that correspond with the use of these technological systems, and the consequences of these systems in terms of dependence and isolation. They are resisting the culturally dominant conceptions of normal residential dwelling, demonstrating the potential to move away from the dependence and isolation that characterize dwelling in the modern world and the possibilities for reorganizing action, perception, and social relations in the process.

3

Custodians of the Earth, Witnesses to Transition

• •

The Story of The Farm

San Francisco, 1967: the summer of love. Thousands of young people gathered to experience and participate in the cultural upheaval, spiritual revolution, and the emerging possibilities for change. A man named Stephen Gaskin, an English teacher at San Francisco State College, began holding weekly meetings to share his ideas on life, spirituality, and the way forward, called "Monday Night Class."[1] The number of people attending these classes quickly grew, and in 1970, he and a couple hundred of his most enthusiastic participants traveled the country on a spiritually charged speaking tour organized by the American Academy of Religion. After traveling together, the group decided to purchase land and put their ideals regarding spiritual communion into practice.

The group bought land in rural Tennessee, formed an intentional community called The Farm, and devoted themselves to living communally with a shared set of principles that included nonviolence, veganism, voluntary poverty, and a recognition of the sanctity of pregnancy, children and natural, home childbirth under the care of The Farm midwives. Maryann, who still lives on The Farm today, told me that the community's history "is one long string stretching back to Monday Night Class," where together they discussed "love, sex, dope, God, gods, war, peace, enlightenment, mind-cop, free will and what-have-you."[2] Throughout more than four decades of life on The Farm, "the glue" that holds it all together is, according to Stephen Gaskin, "a belief in the moral imperative toward altruism that was implied by the telepathic spiritual communion . . . experienced together."[3]

Forty years after the community's founding, I drove from Wisconsin to visit The Farm, still located in rural Tennessee. On my way, I picked up a hitchhiker who told me about how he'd been stopped by the police because he had frightened some women by walking into a church where they were meeting; he had wanted to speak with the preacher. He seemed agitated that he hadn't been able to speak with someone, but he wasn't angry. He told me about his belief that children are the best teachers, said that they are actually teaching parents and other adults about themselves because they mirror back the actions and moods of parents and others around them. The conversation seemed serendipitous, since I was on my way to a community that originally formed based on a shared desire for spiritual leadership and has since its very beginnings viewed children as spiritual teachers and even the processes of pregnancy and childbirth as spiritually charged.[4]

The Farm is perhaps most well known for its midwives and history of midwifery. Ina May Gaskin is a famous, award-winning midwife. Her book, *Spiritual Midwifery*, shares how the values of positive communication, loving and connected relationships, spiritual manifestation, and community support that shaped life on The Farm also influenced (and continue to influence) the process and understandings of giving birth on The Farm.[5] During my visit to The Farm, a few women who were longtime members of the community reminisced that, in the community's early days, there were a lot of babies and young children in the community. The Farm midwives offered any pregnant woman the opportunity to come there, give birth, and leave their baby on The Farm (with the option always open to have their child back if they changed their mind) as an alternative to abortion. The Farm offered a supportive environment for single mothers, and Farm families connected having children to their spiritual understandings of natural birth, new life, and life sacraments.[6] The Farm also developed its own school to educate children using an alternative, participatory model of education.

Both the midwife clinic and The Farm School continue to operate today, although the community itself has changed substantially since its founding. Yet alternative forms of childbirth and alternative education are not the only unconventional technological systems on The Farm. Throughout the history of The Farm and the historic change from a full-fledged commune to an intentional community more in the style of a cooperative, the values of the community have survived through the decades and the massive transition, shaping the sociotechnical networks that support daily life.

The Farm relies on many forms of alternative technology, some of which are similar to those used by the other communities examined throughout this book, including solar electric power, solar hot water heating, a community well, local food production, and electric transportation (in the form of golf carts ridden around the community, some charged using solar electricity). The Farm was an early innovator in the use of passive solar design, and many of the buildings there are testament to the efficiency and longevity of buildings using passive design principles. The Farm was also an early innovator in developing other appropriate technologies like solar electric systems, soy cultivation and processing for the vegetarian diet, and in high tech ventures such as radiation detection equipment. Like the example of commercial success in manufacturing these Geiger counters, there are other alternative technologies used at The Farm—such as the alternative forms of reproductive technologies (natural home birth), education (The Farm School), and unconventional economic arrangements related to home financing, employment, and profit making—that are unique to this case. Nevertheless, these alternative material arrangements from home ownership to reproduction and education are forms of alternative technology adoption, examples of people changing their use of material systems in ways that change their behavior and how they interact with the natural world and with one another.

After over four decades and a significant transition, alternative sociotechnical arrangements related to care—care for the land and care for each other—are the most significant alternatives presented by The Farm. These alternative forms of technological organization are motivated by an underlying custodial ethic that has always been part of both spirituality and practice at The Farm. Seeing themselves as custodians, not owners, community members seek alternative arrangements for living based on their values. This custodial ethic has arguably helped The Farm survive many transitions, including the economic and community changes as well as population and generational changes. Yet it may also contribute to limiting the kinds of alternative technologies some members of The Farm are willing and able to pursue. Steeped in an ethic that orients them to offer custodial care, some residents are more concerned about protecting the land than making the land accessible for increased community membership. Attitudes that work to oppose things like composting toilets and animal husbandry, shaped by the same custodial ethic that motivates commitment to care for the land and for one another, potentially threaten the community's future longevity by limiting more proactive adoption of some kinds of technologies and practices.

The Story of The Farm

The Farm became an officially registered religious community in 1972, organized as a commune living as directed in the Book of Acts: "All that believed were together and had all things in common, and sold their possessions and goods, and parted them to all as every man had need."[8] The founding members originally purchased 1,050 acres, adding 700 adjoining acres to the community property through a later purchase. Today, approximately 5,000 acres is held in common management.[9]

People originally lived in shared community structures (large army tents or buildings constructed by the community), worked for the community, and shared all money, food, tools, and other goods in common. One member, recollecting on the community's history, told me that by 1975, there were 750 people on The Farm (250 of which were children); there were 1,100 by 1977 and 1,500 people (half children) by 1982. The Farm also formed satellite or sister communities in thirteen states and four countries, some on rural land like The Farm and others in cities like Washington, DC, so that by 1982 there were approximately 4,000 people living in communities associated with The Farm.

By 1982, the community's debt was substantial and their income meager. Concerned that the community would be lost, members entered a major transition (called the "change over" by community members and discussed in more detail below).[10] It was decided that the community simply could not survive based on its fully communitarian model; people had to become responsible for their own subsistence in order for the community to continue. In 1983, community dues were implemented—each member had to contribute a minimum of $35 per month to pay off the community's debts, in addition to being responsible for their own food, clothing, and shelter. Today, The Farm community is organized more like a cooperative than a commune; people are responsible for their own housing and means of subsistence. However, they cannot get a mortgage to build on the communally owned land and they share access to many resources, including the land.

The community has been working with alternative electricity generation since its founding, although the community is not entirely reliant on renewable energy systems. Some small PV systems were installed on homes decades ago, even as early as 1974. The Ecovillage Training Center (described below) uses solar systems to provide electricity to its Eco-Hostel and the homes located within the facilities' grounds.[11] In the 1980s, the community installed a small wind turbine. Recently, two Farm businesses and one Farm resident have installed large, visible solar PV systems, taking advantage of a short-lived regionally accessible grant program for solar energy adoption. Most members, however, live in homes connected to the local utility grid and pay monthly electric bills to the local rural electric cooperative utility provider (although several houses have signs outside indicating that they buy green power from their electric cooperative through an elective wind power purchasing program).

People are responsible for their own expenses, although there are still businesses on The Farm (some owned by the community, some by individual Farm members), and the folks I met estimated that about half of the people who currently live there (a population of approximately 200 adults) work on The Farm.

At the time of my visit, the minimum monthly dues requirement was $90 per month, which helps to pay for the communally maintained roads, the water system that is maintained by the community, the Front Gate and Visitor's Center, and the community center.[12] In addition to several businesses, there are several nonprofit organizations run by Farm members, and both the businesses and the other organizations on The Farm continue to be consistent with the spiritual and ethical values upon which The Farm was founded. A brochure from The Farm's Visitor Center describes the community in this way:

> We believe that there are non-material planes of being or levels of consciousness that everyone can experience, the highest of these being the spiritual plane. We believe that we are all one, that the material and spiritual are one, and the spirit is identical and one in all of creation. We believe that marriage, childbirth and death are sacraments of our church. We agree that child rearing and care of the elderly is a holy responsibility. We believe that being truthful and compassionate is instrumental to living together in peace and as a community. We agree to be honest and compassionate in our relationships with each other. We believe in nonviolence and pacifism and are conscientiously opposed to war. We agree to resolve conflicts or disagreements in a nonviolent manner. We agree to keep no weapons in the community. We believe that vegetarianism is the most ecologically sound and humane lifestyle for the planet, but that what a person eats does not dictate their spirituality. We agree that livestock, fish, or fowl will not be raised in the community for slaughter. We believe that the abuse of any substance is counterproductive to achieving a high consciousness. We agree to strive for a high level of consciousness in our daily lives. We believe the earth is sacred. We agree to be respectful of the forests, fields, streams and wildlife that are under our care. We agree that the community is a wildlife sanctuary with no hunting for sport or food. We believe that humanity must change to survive. We agree to participate in that change by accepting feedback about ourselves. We believe that we, individually and collectively, create our own life experience. We agree to accept personal responsibility for our actions. We believe that inner peace is the foundation for world peace.[13]

When I visited The Farm in the fall of 2011, I stayed with a host family, a couple who were both founding members of The Farm, although they were not a couple in those days. I got to see what life on The Farm is like by observing their lives, as well as talking with them directly about it and taking their suggestions on things to see and people to meet during my time there. I walked around the community, visiting with the people who live there and the organizations such as The Farm Store, the midwife clinic, the school, the Visitor's Center, and the Ecovillage Training Center, and participating in community events such as a wedding party and a fall harvest celebration.

One person I met at The Farm, a young woman who was not a member of the community but worked and lived at the Ecovillage Training Center, somewhat sarcastically described the community as a "hippie country club." In many ways, this humorous description fits: much of The Farm population these days is older, folks who were a part of its beginnings who have since aged. Many people drive around in electric golf carts, as the distances between far neighbors may seem too far to walk or bike.[14] There's even a disc golf course. There are also yoga classes, a weekly craft night, numerous meetings for community boards and nonprofit organizations, and a Sunday morning meditation. Community members can be involved in community administration, organizational meetings and events as much or as little as they choose. Yet the thing that struck me most about the community was the quiet peace at night—people's homes are not all lit up by outside security lights, there are no sirens or police, no traffic lights or road rage. When asked about the lack of outside home lighting, community members seemed to be hearing about something they'd never thought of. It seemed obvious to them that you would not waste electricity like that, and that they lived in a place where you do not need security lights or a patrolling police force to feel safe. The community today is in many ways organized primarily around their shared land, three square miles with even more land surrounding it placed in a land trust by a Farm-based nonprofit organization.[15] The Farm illustrates how a custodial ethic—a belief in caring for things that you do not own, even relinquishing the idea of ownership in order to more carefully tend to spiritually valuable places, practices, and relationships—can successfully organize a community and its technologies throughout a long history and significant transitions.

History: Spiritual Community with Charismatic Leader Moves Back to the Land

Stephen Gaskin was a student and then a college instructor in San Francisco throughout the 1960s. He was in the Marine Corps from 1952 to 1955, went to college on the GI Bill, and got "turned on" by the hippie counterculture he saw fomenting around him. He himself turned hippie, started tripping, and began exploring the spiritual undertones of the contemporary countercultural happenings.[16] In 1967, he was part of a group of a dozen people who started meeting on Monday nights to "compare notes with other trippers about tripping and the whole psychic and psychedelic world," a meeting that grew to be as many as a thousand or fifteen hundred people each week by 1970.[17] It was in 1970 that Stephen with his wife, Ina May, and a couple hundred other supporters, traveled to share their ideas on college campuses and public speaking halls throughout the country (now called "The Caravan" by those who participated).[18]

Stephen and those around him shared a set of spiritual and life principles. As Stephen put it: "I love the ethical teachings of almost all the religions, and I love the psychedelic testimony of their saints. But I do not believe in any of their dogmas. I think each one of us has a non-shirkable obligation to figure out the world on our own as best we can. The way we behave as a result of that investigation is our real and practiced religion."[19] A longtime member of The Farm put it this way: "It was, and is, a central tenet of The Farm that all things proceed from the Spiritual Plane to the Material Plane. . . . It was, and remains, our commonly held view that how you choose to live should be seamless with what you believe in."[20]

After months of traveling together, the Caravan crew returned to San Francisco, but not for long; they decided they wanted to put their spiritual ideas into practice on their own piece of land, forming their own community based on the Biblical teachings in the Book of Acts suggesting that all things should be shared to each from all. Stephen was the uncontested charismatic leader[21] of this group that came to settle in rural Tennessee and name their community The Farm. He was their spiritual guru, leading the Sunday Morning Service that came to augment and then replace Monday Night Class on the Farm (see figure 3.1) and marrying all Farm couples. He was also a practical leader in many ways. He was able to talk about the community as a church to locals, a culturally appropriate description for easing fears of the cultural clash in the rural south. He went to jail for the community in the early years; he was not present when the law enforcers arrived, but was punished for illegal marijuana cultivation on behalf of the community. His wife, Ina May, was the first and most experienced midwife on The Farm, another significant position of influence, perhaps even power, within this community striving for equality.

When the group of approximately 300 people arrived on their purchased piece of farmland, it had only a few old buildings and one source of utility-provided electricity. Although most of the founders of The Farm were young adults from the cities and suburbs, many had been exposed to the appropriate technology movement in the abstract sense (Stephen Gaskin and Buckminster Fuller actually met and conversed during, as one member described it, "some sort of hippie conference" in Vancouver, Canada, in the mid-1970s; see figure 3.2). The founders began to put some ideas from the appropriate technology movement into practice. One of the things these kids from the city learned to do when they moved to The Farm was to dismantle, move, and reconstruct old, unwanted buildings from neighboring lands, reusing old construction materials to make efficient passive solar buildings.[22] Some of the buildings still there today were moved there in the early years, including the midwife clinic, which was once an old church.

Mindy, one of my hosts during my visit to The Farm, moved to the community in 1974. She is now probably around my own mother's age, with long grey

FIG. 3.1 Stephan Gaskin leading the Sunday congregation at The Farm. Photo by David Frohman. Used with permission.

FIG. 3.2 Stephen Gaskin and Buckminster Fuller in Vancouver, mid-1970s. Photo by David Frohman. Used with permission.

hair. She works as a nurse outside the community. She tells me that her first ten years on The Farm were "basically like camping"—many families lived in the converted buses they'd driven there, or in buses that were even more modified once they arrived (see figure 3.3), or they lived communally in large army tents. There was no hot water and the only source of light was some DC light bulbs strung off car batteries that were trickle charged from their one main power

FIG. 3.3 A home on The Farm in the 1970s. Photo provided by The Farm Archive Library. Used with permission.

source (provided by the local rural electric utility, a bill paid collectively by the community from 1971 to 1983; see figure 3.4).

For the first twelve years, the community did a lot of farming in order to feed themselves. They learned to grow soybeans and make tofu. They also gave a lot of food away, as part of their spiritual practice, to relief programs and others in need. Mindy told me that in those days, their diets were good in the summer but winters were hard and "nutritionally, we were lacking," Yet eating simple vegan food and living without hot water or indoor plumbing were part of the community's spiritual practice. Each member took a vow of poverty in those days, and Mindy described it as "this sort of . . . mentality, that we were lucky to have running water at all because it was more than a lot of people had."[23]

According to The Farm's charismatic founder, "The spiritual revolution has got to be something that you do, rather than something that you just think."[24] Stephen and those drawn to his ideals sought to develop a community and a lifestyle based on this spiritual teaching. From the very beginning, the Buddhist principle of Right Livelihood—that work is seamless with and should not contradict spiritual practice, that hard work is both an act of love and a path to Enlightenment, and the idea of "putting our labor in the service of our dreams while forging high-quality relationships with people and resources"[25]—was an important part of understanding work on The Farm.

Thus, on The Farm, work was and continues to be viewed as spiritual practice—so, for example, farming is a way to practice Zen, dismantling old

FIG. 3.4 Early Farm Technologies. One way to trickle charge batteries to use as a DC power source on The Farm, applying physical work for communal benefit. Photo circa 1971. Photo provided by The Farm Archive Library. Used with permission.

buildings to construct new ones is enlightening, and being a midwife is a way to tap directly into spiritual experience. In the early years on The Farm, people accepted environmental values regarding the benefits of veganism, organic farming, and a low-impact lifestyle, but that was not the primary motivation to move back to the land and adopt low-tech and community based technological systems. Instead, The Farm saw direct experience as the only means to a spiritual existence and sought to live their lives in ways that improved the human experience as part of their spiritual values.

The Change Over

In 1982, there were 1,500 people living on The Farm, children made up half of the population, and the community was over $400,000 in debt. The Farm was very involved in relief work and community service, everything from running an ambulance service in the Bronx to teaching Guatemalans how to grow soy to make soy milk, tofu, and other products like ice cream and cookies to improve protein consumption. The Farm had become an extensive farming operation, cultivating much of their own as well as neighboring land, and had taken on heavy costs (and debts) for the seeds, heavy machinery, and necessary equipment. The Farm provided sanctuary for many people, some of whom could not or did not participate in the financial or subsistence ventures of the community. In September 1983, the entire economy of The Farm changed. Instead of providing for all needs communally, people became responsible for their own food, shelter, and necessities. Each member had to contribute monthly dues in order to help pay off the community's debts, keep the land, and maintain the community.

Maryann graduated from Bennington College in 1963; she then moved to France until she was "thrown out" during the 1969 student uprising. When she returned to the states, she heard about Stephen Gaskin. She came to visit his new community in 1973, and became a member. In those early years, she also spent time in a few of The Farm's satellite communities, returning to The Farm for good in 1979. She told me that the economic transition, what community members refer to as "the change over," was incredibly painful and embittering for the community. Yet Paul, a founding member of The Farm, told me that they likely would have lost the land without the economic transition.

Richard first saw Stephen Gaskin speak on a New England college campus during the Caravan tour. He moved to The Farm in the 1970s, but left after the change over and did not return for twenty years. He told me during my visit to The Farm that in his younger days, instead of fighting "the system," he wanted to build an alternative to it. He told me that he and the other folks on The Farm during that first decade learned firsthand how difficult that is, and they experienced a sadness and a sense of failure similar to those who watched the dismantling of the Communist bloc after hoping it was the beginning of worldwide revolution. Richard told me that similarly, The Farm had failed to become a communal utopia because it was stuck in the middle of a capitalist society. Like the early members of The Farm community, Karl Marx and Leon Trotsky hoped for worldwide economic revolution. Yet as analysts, Marx and Trotsky recognized the difficulty, or what they sometimes saw as the impossibility, of establishing a shared economic system within or while surrounded by economies based on private property, investment, and gain.[26]

The Farm Today

The Farm's community members now span several generations; founding members have gotten older, and some moved their parents to the community as they aged and died as part of a spiritual commitment to care for others through life's transitions. For example, after the change over Maryann and her family moved into one of the large homes that originally held as many as fifty people. She cared for her mother there, who died and was buried on The Farm.

Children who were born on The Farm during the community's first decades are now adults, and although the vast majority of them have left The Farm, some still live in the community. Some of these members now have children of their own. For instance, I met Garrick when he offered me a ride in his golf cart as I was wandering down the road. His wife grew up on what he called "the old

Farm," a reference to its years as a commune. They traveled back to The Farm for the delivery of their first child and had "been looking to get back for a long time," finally relocating permanently only eight months before my visit. Now, his children attend The Farm School and he works for SE International, one of the community-owned businesses on The Farm, a company that makes radiation detection equipment, holds million-dollar government contracts, and was incredibly busy after the nuclear disaster in Fukushima, Japan, that spring. When I met Garrick, he and his wife were staying in another community member's home while they constructed their own residence, which they hoped to finish within a year.

Today, The Farm is more like a small town than a stereotypical commune. People live in their own houses and are responsible for their own livelihoods. Yet in this small town, everybody knows one another. And, a lot of resources are still shared communally—The Farm has its own water source (a community well), its own system of roads, several businesses either owned by or housed within the community, several nonprofit organizations run by Farm members, a private school based on The Farm's educational and spiritual philosophy, a midwifery clinic and birthing center, and community structures such as a playground and outdoor music dome, a gazebo, and a community building for hosting events. The Farm's collectively purchased land is still owned collectively by the community.

The Farm continues to be involved in several alternative material forms and practices, technological arrangements, and ways of organizing social and economic life. In this unique community, a spiritually driven commitment to care and custodianship influences the way people build their homes and live their lives. A shared ethic shapes engagement with alternative sociotechnical networks and alternative social practices, how members organize material and social life in ways that have endured through a long history and significant transition.

Farm Technologies

Technologies on The Farm are, in some ways, fairly conventional. The community has its own well, and the community pays one of its own members to take care of the water system, testing water quality, handling the bureaucratic paperwork for the state's records, and dealing with all line maintenance and expansion. Jeff, the "water guy" when I visited, grew up on The Farm; he identified himself as "second-generation Farm family," a way to describe people who were born on The Farm. He received on-the-job training from the community member who was responsible for the water system before him.

Each home and office building has its own septic tank, which is typical in rural areas. The local utility provider, a rural electric association, provides electricity to the community, although (as mentioned above) two community businesses have invested in large solar PV arrays, some homes rely either entirely or in part on installed solar electric systems, and several homes advertise their purchase of wind power through the local utility's program. Homes are heated largely with wood; however, some use propane, which is typical of any home in the rural hills of middle Tennessee. A community member is paid to haul the community's trash once a week, as well as their separated recycling, which must go to a different facility forty minutes away.

The Farm no longer actually does any farming as a community. Many families have gardens near their homes, and several families shared in the expense of establishing a larger community garden with individual plots. The berry patches and apple orchards left over from the early years are fairly unattended, although community members sometimes organize community workdays to voluntarily tend to them.[27] The community has maintained its commitment to nonviolent dietary principles; many of the community members are still practicing vegans or vegetarians. Individuals may choose to eat whatever they wish, but notably, The Farm Store does not sell meat, and individual families may not raise animals for consumption.

There are many businesses owned by Farm members within the community, including a media services and consulting company that in its original form recorded The Farm Band. The Farm Store, which used to be the central distribution station of the commune, still operates as a community store. It was originally built in the shape of a mandala (see figure 3.5), a shape representing the universe in Buddhist imagery and the image adopted as the mascot of Monday Night Class and The Caravan. The Farm Store is owned and operated by a member of the community, and has changed ownership over the decades since 1983. The store is a popular place to grab lunch, use the Internet, or meet with other community members. During my visit, I saw community members congregate, eat, surf the web, and even hold a craft night at the large tables in the store, which are cluttered underneath their glass tops by photos of community members' weddings, babies, and events—the life transitions that are considered sacred sacraments in the community.

The community still commonly owns two large businesses. The Book Publishing Company first published Stephen and Ina May Gaskin's work and now publishes many other books that match the mission and vision of The Farm community such as vegetarian and vegan cook books, books by community members on environmental issues and spirituality, books on Native American spirituality, and of course, books on midwifery including Ina May's renowned *Spiritual Midwifery* and books on The Farm and its history.[28] The second community-owned business, SE International, is a high-tech industry that makes radiation detection equipment, yet it was originally born out of the low-tech practice of ham radio use.[29] The businesses and nonprofit organizations all continue to reflect the original spiritual and human values of The Farm community.

FIG. 3.5 The Farm Store Today. A community member's golf cart is parked out front and the community outdoor amphitheater in the background. Photo by author.

The Midwife Clinic and The Farm School have survived the decades and the changes on The Farm. Pregnant women come to live on The Farm before they give birth so that the midwives can attend to their vibes as well as their health, just like in the old days, and The Farm School operates as a fairly radical alternative educational environment officially registered as an institution of religious education. Both are still important parts of the community, although they both also involve people from outside The Farm.

The current principal of The Farm School, Preston, became a member of The Farm eight years ago.[30] He joked with me that he'd "moved to a commune after it underwent a capitalist revolution." The Farm is no longer purely communalist and egalitarian. Yet relations of ownership and profit making, aspects of the capitalist mode of production so vehemently criticized by Karl Marx,[31] are still different on The Farm. People who live on The Farm do not own any land individually, and they are not permitted to sell their homes or businesses for a profit. These alternatives to economic organization are, from my perspective, one of the most fundamental alternatives to the material organization of life presented by The Farm.

The Farm demonstrates the possibilities of steady-state economics,[32] even within a capitalist economy. When members of The Farm wish to sell their homes, they of course must sell only to community members. They also, importantly, cannot sell their homes for a profit—they can sell them for only the amount actually put into them.

For example, when Paul and Mindy finally built their own home after living in a mobile home for years because members cannot take out a mortgage to build or buy on communally owned land, they kept track of all costs, including materials and labor time (even their own), because, if they ever want to sell their home (which they can only sell to another fully accepted member of the community), they can only sell their home for its real production cost.

Similarly, when The Farm Store changes ownership (as it has several times), it can only be sold to another community member for its actual value, without any profit being incurred through the sale. This represents a radical alteration of a particular material relationship—the relations of profit making—and it has worked on The Farm for decades.

Farm Homes

Many of the homes on The Farm are similar to the typical American home. For example, Paul and Mindy's home, where I stayed while visiting The Farm, is beautiful—large and modern, equipped with a microwave and even a hot tub. Yet there are some notable and important differences between homes on The Farm and most modern dwellings. Since the land is owned communally, you cannot receive a mortgage loan to build or buy a home on The Farm. Thus, all community members live without the debt typically incurred by homeownership, which in itself introduces an economic freedom that most American dwellers never experience.

Further, the rules that shape the use of land around your home, what would typically be considered a private yard or lot, are clearly different at The Farm. There are no covenants restricting the use of clotheslines or the installation of solar panels. Paul and Mindy have a huge clothesline based on a pulley system across their yard. Most people on The Farm have large gardens near their home. Community members maintain gardens much more than they maintain pristine green lawns, valuing local and organic food production but not fertilizer-perfected mini fields of green. Another difference for homes on The Farm is the lack of outside lighting: when you know and trust your neighbors, care about minimizing electricity usage, and value the quiet peace of nature at night, security lighting becomes nothing more than an unnecessary nuisance.

Yet perhaps the biggest difference in terms of housing as well as all other material symbols of wealth or extravagance is one of attitude. On The Farm, there is what I can only think to call the opposite of "the Joneses." Some people have large and modern homes while others live in simple shelters such as mobile homes, the old communal buildings that have been converted into duplexes, or other small and inexpensive homes. Other folks I met on The Farm jokingly referred to the road where Paul and Mindy live as "mansion row," but the joke was made without envy or malice. Instead, it was a way for community members to indicate the lack of value placed on a large home with a two-car garage. The Farm community has its own cemetery, and this opposite of the Joneses attitude holds even there, where there are very few large or ostentatious tombstones and simple, handmade markers are much more common.

On The Farm, as Paul put it, individuals can "work as much as you can or desire to." Attitudes toward wealth and material demonstrations of it still correspond to the spiritual ideals upon which the community was originally founded. There is no value placed on excessive wealth, and attitudes toward homes as well as practices of home construction demonstrate the radical value of simplicity[33] that is still present on The Farm.

The Farm School

The K–12 school located on The Farm offers an alternative educational environment consistent with the values and practices of the community. It is technically a private religious institution, with students from both within and outside the community attending. Students from outside the community might attend because their parents value the open, participatory, and varied educational environment or because they struggle in public school. Parents on The Farm can send their kids to public school, but must pay $2,000 a year to enroll in the closest public school because of districting lines that make the closest school technically out of district. The Farm school costs approximately $3,000 a year, although parents can also do work trade to help cover the costs. Both of Paul's children attended The Farm School and went on to graduate college; he told me his kids "did well in college coming out of The Farm School."

The Farm School is small, with approximately thirty students at the time of my visit, about half from families within the community. There are only two full-time employees, although there are about ten teachers at the school, including two part-time employees and multiple parents who teach in trade for tuition. The school offers an incredibly flexible model of education, where many students attend part-time and are homeschooled for the other part of their education. Over 300 homeschooling students participate in The Farm's satellite campus program. As a state-recognized religious exempt institution, The Farm School requires its students to attend four hours a day, 180 days a year. Students are not segregated by standard grade categories, but are organized into three groups (corresponding to pre-K, elementary, and secondary education) based on not only age but also ability and interest. When I visited the school, I met an eleven-year-old boy in the high school group. Students at The Farm School study reading, writing, and arithmetic Monday through Thursday mornings, they have both a half-hour recess and a half-hour of "movement" each day (where they are not allowed to be on computers or reading), and their afternoon activities vary widely, with nature class being the only consistent afternoon activity every Thursday and Friday.

Preston, the school's current principal, describes it as an attempt to create a "true learning community." During our conversation about the school, he handed me a one-page sheet describing what such a community strives for: the sharing of both power and knowledge; honesty in all relationships; variation, because "diversity increases a community's adaptability and sustainability;"

integration with the larger community; a "responsive curriculum;" and a constant recognition that "[all] learners have gifts and special needs [and] all learners deserve an individualized educational program."

When I visited The Farm School, I had the opportunity to see this learning community in action. Preston told me that the school attempts to implement participatory decision making in a diverse community (part of having a "responsive curriculum") and I got to see the beauty and the challenge of doing this even with elementary school students. When some young students did not want to participate in nature class, instead of being reprimanded, they were asked what they would prefer to do instead. Through conversation, a deal was made: they could help in the school's garden for half the time and then take a guided nature walk. Especially with the younger students, education at The Farm School involves lots of outside time so that learning occurs in the context of everyday reality in communion with the natural world.

I also watched as the older students played a game of Risk during an afternoon class on governance with Preston, discussing as they took their turns how the game relates to politics in real life (zero sum games, strategic alliances, and the cumulative effects of power begetting power) and how they felt about taking from their friends, making others suffer, or the sense of supremacy they may feel when winning. My new eleven-year-old friend articulately told me about the congruencies between how he and the United States president may feel about being powerful and about the fragility of that power.

Preston has a PhD from a prestigious university and worked in a large metropolitan school district for over a decade before moving to The Farm. Frustrations with changes in public education (such as policies like No Child Left Behind, which he referred to as "all children left behind") and his inability to significantly impact the learning environment initially motivated him to find an alternative. He told me that he didn't move to The Farm as a revolutionary (although he acknowledged that living on The Farm has made him more radical in his views on education), but that he was sick of compromising and simply wanted to change himself in his lifetime. When talking with me, he relished in delight when describing the biggest changes from his life before moving here: his work commute no longer involves four hours in the car each day; instead, he takes a short walk next to a creek for his work commute. Further, at The Farm School, he is the principal as well as the janitor; he's the person who gets there at 6:30 to put wood in the furnace and who turns off the lights each afternoon. Contrasting the educational environment for students in public schools and The Farm School, he uses the words "diversity versus conformity; self-actualization versus complacency."

The children who attend The Farm School seem to benefit from this educational environment. The age segregation that typically occurs in school is nonexistent here. Students in different groups interact with one another regularly, so teenage boys learn to communicate with toddlers. During my time at the school, I met three teenage girls, all clearly good friends. Although they all walked around barefoot, they dressed completely differently—one looked like a punk rocker, one like a hippie, and one like a tomboy. They did not appear to feel the pressures of conformity experienced by most teenagers. After school, a very young boy who was playing outside with his mother approached me to tell me about his favorite tree; how many little boys get to have a favorite tree in their own safe haven of a three-square-mile playground, and feel comfortable enough to approach a stranger to talk about it?

Jackie, a community member whose children attend The Farm School, told me that enrollment has been low "since the change over." After over twenty-five years of low student enrollment, it is a challenge to cover costs and keep the school open. Preston told me that by trying to be as fluid as possible, allowing part-time enrollment and work exchange for parents, they are prioritizing values over finances and hurting themselves financially. Yet Preston also says enrollment will increase again as parents come to recognize the value of alternative education and seek out meaningful educational environments for their children. He told me that the school is "working uphill, because it's not what the system wants."

By valuing diversity and encouraging self-actualization, The Farm School views education as a means of true human development rather than simply a hurdle producing standardized documentation. The Farm School works to develop students as thoughtful humans rather than simply successful test takers. The Farm School is organized around principles that reflect the community's commitment to custodial care: together, the educational community supports the development of thoughtfulness and consideration, tending to students as whole human beings with their own spiritual and intellectual worth.

Spiritual Midwifery and The Farm Midwives Today

Women on The Farm became midwives for one primary reason: because women on The Farm (and before that, on The Caravan) were having babies. Wanting to continue their steadfast devotion to natural processes and good vibes, these women did not want to enter the male-dominated, sterile hospital environment to experience labor and childbirth, so they learned how to deliver babies themselves. Ina May Gaskin is an award-winning, world-renowned midwife known for the idea of spiritual midwifery, recognizing birth as a spiritual event as well as the importance of positive energy and physical touch in the sacred process of birth.

Certainly, spiritual midwifery has a gendered motivation: the professionalization and medicalization of childbirth has taken knowledge, power, and experience out of the hands of women. As midwives on The Farm, women are protecting other women through their knowledge and their care. Yet spiritual midwifery is about caring for both the women and men who participate in this sacramental life transition. The midwives on The Farm not only ensure that birth is physically safe; they also tend to the spiritual and emotional aspects of the birthing process. Women or couples come to stay on The Farm weeks before the birth so that the midwives can be physical care providers as well as counselors, checking in with a woman or couple's vibes and tending to emotional states to help create a successful and positive birthing experience.

Today, the Midwife Clinic on The Farm is located in an old church moved to The Farm decades ago. Six women work at the clinic, two founding members who were on The Caravan, two other older women from The Farm's early years, and two younger women in training. Pregnant women or couples travel to The Farm, stay with hosts in the community, and give birth in special birthing cabins or designated spaces in members' homes. Although they were much busier in the first years of The Farm, I was told during my visit that an average of four babies are born on The Farm each month (that's forty-eight births a year). The clinic also hosts workshops on midwifery and a midwife's assistant training program.

Spiritual midwifery recognizes birth as a sacred process, an opportunity for spiritual illumination and emotional elevation. A longtime member of The Farm told me that spirituality on The Farm is about recognizing life's transitions, including birth, as spiritual sacraments. To me, it felt as if the energy surrounding The Farm community was saturated with the history and energy of all the births that have happened there, of big pregnant bellies and new precious life; it felt like spring, in November.

Yet birth is not the only life transition given sacred meaning on The Farm. As member's parents have aged, and as they themselves have grown older, the idea of "midwifing death"[34] has become important to spiritual understandings of transition on The Farm. Some members of The Farm community consciously share their homes with the elderly and the dying, tending to those in their final stages of life. Maryann used the language "in spirit" to describe her deceased daughter and talked about the sacred nature of death and dying on The Farm. She told me, "If in our power to do so, we wouldn't let parents die in a nursing home attended by strangers." As the community itself has experienced the transitions of time, its spiritual ideals and practices have also come to embrace new sacraments of life, including death, and to tend to them with the same spiritual care that is given to the process of birth.

The Farm Ethic: Custodial Care and Transition Witnesses

The people who live on The Farm differ from one another many ways. Some live in bigger, newer homes, while others live in older structures originally built (mostly out of reused building materials) by the community. Some have financially successful careers off The Farm, while others work only part-time in businesses or organizations on The Farm. Some eat meat; others are vegan. Some drive old pickups; others drive new hybrids. Yet the members of The Farm community continue to share a spiritually based ethic, what I call a custodial orientation, which shapes their relationship to the land, material systems, and one another.

One woman who lives on The Farm put it this way: "Here you have a sense of the grandness of the land. You're part of that. You don't own it, and you get to be part of it."[35] Describing the original land purchase for The Farm, one member wrote, "In the Native American sense we felt a mystical relationship to this land we had bought, and our love for it bound us together."[36] Forty years later, Paul told me that he thinks the most radical thing about The Farm is still the common land ownership. He said, "Like Native Americans, we share a belief that it isn't 'ours'—that we're just passing through this plane temporarily, and don't own the land." Paul said the common land is now the core of the community and that owning land in common is the most radical idea that's left after all the years and changes. As one younger member put it, "Each individual has a different connection and relationship to the land, but it's some of the most sacred part of the community."[37]

Similarly, Richard told me that The Farm community "is about custodial. Which means not ownership, but use." He said, "Ultimately the amazing thing is that The Farm is still here and the land itself is still held collectively."[38] When Richard returned to The Farm after decades in the legal profession, he began working to solidify the legal status of The Farm so that the land cannot ever be subdivided or sold for a profit. He said that this legal work is his way of leaving a legacy through The Farm. He told me, "Ultimately, we're caretakers, we're custodians of the land." Given the tremendous changes that have taken place throughout The Farm's history, the community is no longer tied together by shared income, shared subsistence, or shared housing. Now, the land itself is the foundation of the community's connection, providing what one community member described as "something spiritual."

On The Farm today, there is a large centrally located prairie that used to be farmed by the community. Several people told me during my visit that this prairie is the only place in the state where one species of bird nests, because of loss of prairie elsewhere, some explicitly saying how important this is to The Farm community, others implying its importance by telling me about it.[39] After talking about the bird in the prairie, an older woman named Lyla told me that the most important thing my generation can do is buy land and put it in trusts, not to own it ourselves, but to protect it from private ownership by others, protecting land from the social and ecological consequences of private ownership.[40]

David moved to The Farm at age nineteen with his wife; they are now both nearing sixty. David is an avid gardener and cook; when I met him, he was wearing a tie-dyed button-down collared shirt under a tie-dyed apron, busy in the kitchen of his home, an old communal house converted into a duplex after the change over. David is soft spoken but articulate, with deep and intense eyes and a strong connection to what he calls "Farm spirituality." He told me that despite all the changes on The Farm since his arrival, "The land is still held in common. That's our strength and our union. No deeds; that involves a lot of trust. And a common understanding of some shared, basic values."

In my explorations of alternative technology adoption and alternative communities, I have met other people who share this custodial orientation. For example, Nancy built an Earthship (see chapter 6) in a subdivision that was once based on communal land ownership, where each homeowner owned only the land immediately surrounding their home and all other land in the neighborhood was owned in common. Describing the required legal change to standard, privately owned lots, Nancy told me, "I'm still kind of wondering about it. I hear what everybody tells me. They say, oh it's great that now you have a surveyed lot and you own it. But I'm telling you, I'm a white girl in northern New Mexico, I don't want to own land. This land has been here for so long, this is native people's land. This is not something I want to own." Yet at The Farm, this custodial orientation is commonly shared, and shapes the use of the sociotechnical networks involved in ownership of land and homes, choices in livelihood and business practice, as well as their practices, decisions, and interactions as a community. For members of The Farm, this understanding of land as something to be cared for without private ownership represents a radical departure from and a significant alternative future for normalized American dwelling.

On The Farm, this custodial orientation shapes not only views of the land; it also orients how people think about and act in relation to other people. What does it mean to have a custodial ethic when it comes to social interaction and individual practice? It means, simply, to care for people while recognizing their sovereignty. The Farm community has always held this ethic in common, and it continues to shape community practice today. Thus, there are two senses in which members of The Farm act as custodians—they are custodians of the land as well as caretakers of one another.

Longtime community member Sharon acknowledged this by saying, "It's not that you moved here because you liked the house, or the school was nearby. You don't choose to live here for those kinds of reasons. You choose to be here because you want to be in community and share your lives with people . . . because here people care about this property, they care about how we treat the land, they care about how we treat each other."[41] Similarly, Lyla told me that caring for one another has always been a part of The Farm's vision, especially caring for those less fortunate. Members of The Farm community have always had "a deeply understood commitment to creating a spiritual community and taking care of each other."[42] As Richard put it, "A lot of the stuff that reflects what The Farm was originally about, that higher idealism, looking out for somebody other than yourself, looking to better mankind or humanity, still exists here on The Farm."[43]

Richard's life demonstrates this custodial orientation when it comes to caring for both the earth and fellow humans. Richard installed a solar electric system on his home office building located on The Farm. Since the golf carts on The Farm are electric, he included a charging station right next to his solar panels during their installation, and anyone from the community can come over and charge their golf cart batteries with his solar system. David described this as "a nice gift" from Richard, and told me that his daughter—who lives on The Farm in a home near Richard's—uses the charging station regularly. David also uses it when he goes over to visit Richard; they play music together. Yet this solar charging station is more than a nice gift. It is just one example of how this community has successfully removed the mental, physical, and legal barriers that private ownership often puts on openly sharing both resources and opportunities.

The Farm's longest-lived nonprofit organization, Plenty International, also demonstrates how the custodial orientation can shape how humans treat other humans. Like much else on The Farm, Plenty was born serendipitously. Using ham radio in the early years on The Farm as a free means of communication, members of the community heard about the 1976 earthquake that devastated Guatemala and decided to help. Offering the skills they had learned out of necessity for life on The Farm—construction, medical care, agriculture, and food processing–volunteers were on the ground in Guatemala just two weeks after the disaster.[44] They learned that the Canadian embassy had sent building supplies but no builders, so they got in contact and got to work. Over 100 people from The Farm went to Guatemala in just four years, and over 100 people from Guatemala came to visit The Farm to learn about their processes of providing for themselves. The Farm helped establish soy production and soy-processing facilities in Guatemala, including a soy dairy that still exists today. Plenty also established an ambulance service in the Bronx for a time when the community most needed it, and has since worked on Pine Ridge Reservation, in Haiti, providing books to children post-Katrina, and in various other places and nations.

One community member told me that Plenty International was and is The Farm's national and foreign policy—looking at something outside itself with a bigger mission to serve. Paul told me that he thought this was one reason the community had survived. The Farm has never been an insular community; they have always been outward looking. Community members used to be encouraged to leave the community for "relativity," going away for "thirty-dayers" (months away), time on satellite communities, or time doing outreach work.

There is yet another way that this custodial orientation shapes social interaction on The Farm: the community members share a deep, spiritual appreciation for life's transitions and a commitment to care for one another through them. David told me that these life transitions—birth, marriage, even divorce and death—are the spiritual sacraments of the community. While most of the material arrangements in America have worked to remove us from these sacraments (with birth happening in a drug-induced haze in a hospital room, the aging segregated and isolated to nursing homes, and death being completely medicalized), members of The Farm seek to embrace these transitions as part of a holistic spirituality and to care for all those going through them.

Thus, members of The Farm are witnesses of transition in two important ways. Here, I use the word *witness* in the sense of "bearing witness," originally a "Quaker term for living life in a way that reflects fundamental truths. Bearing witness is about getting relationships right. The group of Quakers in the eighteenth century who built a movement to end slavery were bearing witness to the truth that slavery was wrong. Yet bearing witness to right relationships is not limited to Quakers. It is something done by inspired people of all faiths and cultures when they live life according to cherished values built on caring for other people and being stewards of the earth's gifts."[45] In terms of bearing witness to transition, the community itself has undergone profound change, yet survived this change and stands as an example of the potential to transform community without destroying it. Their custodial relationship to the land did not falter even though the structural organization of the community did. If a community can survive the transition from commune to cooperative, what kinds of transitions are possible in more mainstream communities today? In this example and many others, members of The Farm embrace life's transitions as spiritual experiences and seek to care for one another throughout these sacraments.

Yet in terms of technology use, The Farm community is no longer on the cutting edge of alternative material systems. The Farm as a community is fairly hesitant to embrace technological change, even the types of changes that would be good for the land they care about so deeply. As Richard put it, "When I came back, I tried to push the community, but now I've given up. . . . The Farm is one of the more conservative communities in America, not intolerant, but just a 'that's the way we do things' attitude that can really stifle change." Several different members told me the story of a family who wanted to move to The Farm and build a home without a septic tank, using a method of composting human waste instead. Some community members were skeptical, and one threatened to call the health department if the community allowed them to build. Instead of moving to The Farm, this family ended up buying land and building their home—composting toilet and all—just outside the community's boundaries, sending their kids to The Farm School. One person told me that there's "a whole crew" of people now living in this "just off-The-Farm bubble," young families who wanted to build radically environmental homes and were either turned off or turned away by The Farm's conservative attitude towards new technologies and new members.[46]

The most radical technological innovations being pursued on The Farm are located at the Ecovillage Training Center (ETC), an educational organization founded by a longtime community member and located on The Farm. A constant construction site, the ETC is comprised of a ramshackle collection of about ten structures that include straw-bale buildings, earthen plaster walls, rainwater catchment, a gray water wetland, solar electric power, solar hot-water heating, living roofs, outdoor earthen ovens, a geodesic dome that now serves as a storage shed, a chicken coop that I was told is the oldest cob structure east of the Mississippi, and even composting toilet outhouses. One of the interns at ETC during my visit told me that the buildings use alternatives to conventional insulation materials.[47] They grow bamboo, shiitake mushrooms, and other food using experimental permaculture systems. There is a small inn at ETC, where workshop participants and other visitors to The Farm can stay, built out of recycled barn wood and powered by solar electricity. The ETC hosts workshops on permaculture and edible forest design as well as natural building techniques and environmentally friendly technological systems, bringing in young and enthusiastic participants who are learning important, world-changing skills. Between eight and fifteen people work at ETC at any given time, including some Farm members, apprentices who come to live and work temporarily, and people who live near the community but are not community members—importantly, because they are interested in radical technological alternatives that are not fully embraced by members of The Farm.

The ETC is quite isolated from the larger community surrounding it; most community members have no connection to the ETC at all. Most importantly, the knowledge and skills developed at the ETC have no impact on The Farm as a whole. One community member said, "We're all frustrated that we're not further along in living with alternative technology."[48] Yet even my host, Paul, used the language of "return on investment" when talking about solar power, saying that the "financial incentives just aren't there." Paul also acknowledged that, "For some people, with the change over, it was like, well if The Farm can't work economically, what am I believing in? And some people lost their idealism or their optimism. I think the younger folks down at ETC have that the way we used to, but I don't think they're really seen or heard by the larger community, and I don't have any idea how to change that." With the community oriented so strongly to custodial care for the land and for one another, they have abandoned, ignored, or downright resisted some of the more radical, environmentally beneficial technological alternatives that others embrace when dwelling in resistance.

Brian is one relatively young, relatively new member of the community, who I think demonstrates the still present radical potential of life on The Farm. Brian and his partner moved to The Farm to have their first child, and have been community members for eight years. Brian works at the ETC (which he enthusiastically describes as "action-packed theory") and is incredibly passionate about radical, environmentally sound technology. He has been working on the construction of their family home, a cordwood and natural plaster building, since they moved to The Farm. He put a small solar panel directly onto his golf cart to keep the batteries charged. He charges the batteries for his power tools using a self-built, self-described perpetual motion device. He labels himself as "a mad scientist." One afternoon, Brian gave me a tour of his future home and the Quonset hut that currently serves as both their family home and his workshop. During our meeting, Brian told me, "We don't have an energy crisis. We have a crisis of scientific dogma at the institutional level right down to elementary school. . . . Thermodynamics assumes the system is closed, but everything has a frequency, some faster than we can perceive. The supply of energy in our society is the lever of control, the powers that be don't want to loose their grip on energy control, but it's about control of way more than money. . . . The question, though, is this: do we want technologies of life, or technologies of death?" In this last point, Brian sounded very much like the most well-known proponent of appropriate technologies, Buckminster Fuller, who inspired many of The Farm's original members. According to Fuller's expression of the appropriate technology movement, technologies could be user-friendly, small-scale, and require low capital commitment so that individuals gain control their own technology adoption and use. These aspects of the appropriate technology movement corresponded to the emphasis on self-expression and freedom from government control in the broader countercultural movement of the 1960s, the same counterculture that inspired the founding of The Farm. For those involved in the appropriate technology movement, "by staying home from the protest demonstration and modifying your toilet or building a geodesic dome or solar collector, you could make a more immediate and significant contribution to the effort to create an alternative future than through more conventional expressive politics."[49]

Buckminster Fuller wrote long ago that we could work with technologies that make less out of more, or more out of less; that it was a choice we as society could direct.[50] Talking about The Farm community making this choice, Brian said, "We have this great opportunity and this great challenge, to maintain this reservation for our tribe in a way that allows us to be even more independent from all of the different energy inputs it takes to have a town. So that's going to be a challenge for everybody everywhere, I feel like it's the same for us except that with the solidarity of everyone working with each other, we have an advantage I think, if we're smart about it."[51]

Technologies of Care as Technologies of More

People who live on The Farm certainly dwell differently than most contemporary Americans. They live in homes without mortgages, avoiding the unfreedom that inevitably accompanies large debts. They live on land that they do not privately own, indicating the potential to reshape neighborhoods and communities through sharing land in common. They live in a community where there are no streetlights or police, where people know one another and feel safe even without security lights at night. They live in a community that does not value material displays of wealth, valuing simplicity instead. Many of them drive private automobiles significantly less than most Americans because they can use electric-powered golf carts to go to work (if they work in the community), stop at the store, and visit a friend.

Begun as a spiritual community and still officially registered as a church, The Farm is held together by a common set of spiritual principles and values. As one female member said, "You don't have to argue about nonviolence on The Farm. And The Farm is such a school for nonviolence, with midwifery, nonviolence and birth, with the school, nonviolence and education, there's so many aspects of that . . . that's such a strong force for humanity."[52] The alternative forms of technological organization on The Farm—including steady-state economics,[53] common ownership, a shared water system, the blending of work and home life, alternative energy, alternative birth, alternative education, and alternative death—are motivated by an underlying custodial orientation that has always been part of both spirituality and practice at The Farm. Steady-state relations to profit mean both more community trust and arguably more community wealth. Technologies of care such as midwifing birth and death help to right both physical and spiritual relationships. Alternative forms of education can help students develop rather than teaching them to conform. Local and organic food production as well as vegetarianism and veganism are good for the resiliency of the body and the planet. Strong family and community connections provide resources in times of change and in times of need. The custodial orientation at The Farm, similar to Max Weber's "Protestant ethic,"[54] orients behavior in relation to a particular understanding of reason and rationality.

Seeing themselves as custodians, not owners, people on The Farm seek alternative arrangements for living based on their values. The alternative sociotechnical arrangements related to care—care for the land and care for each other—are the most significant alternatives presented by The Farm. Yet while this custodial ethic orients the values and practices of the community, it may also contribute to some of the more conservative attitudes toward new technology and new people that potentially threaten the community's future longevity.

New memberships on The Farm are rare, the population is certainly aging, and many of the younger folks on The Farm today were described to me by one longtime member as "second-generation kids with nowhere better to be." A couple older community members told me that these second-generation members (a way some members identify people who were born on The Farm) do not have the motivation or skills to keep the community going without their parents. Although Paul told me that he felt the community would thrive with more members, others are hesitant to increase their numbers. Further, some potential new members experience a sort of distancing from the more longstanding members in the community. An ETC apprentice told me about wanting to join a weekly dominoes game that takes place on The Farm and being told that she needed to live in the community for three years before she could attend. Other potential members choose to live just outside the community because they want to participant in The Farm School but are turned off by the community's reluctance to accept alternative technologies like composting toilets and animal husbandry. The community has embraced the focus on shared land and the sharing of human connection while largely turning away from the potential for more radical orientations to material arrangements, technological structures, and possibilities.

Many Americans today may be drawn to The Farm's nondogmatic and emotionally in-tune spirituality, and life on The Farm is not all that different from life in rural communities across America—except for the clear material, environmental, and social benefits of living without a mortgage, potentially being able to work where you live, and being closely connected to others who share the values of caring for the earth and one another through a custodial ethic, among many other differences. The Farm demonstrates that, collectively, a group of people who share a custodial ethic can successfully protect the land they hold dear, care for babies and children and the elderly in alternative ways that contribute to individual health and flourishing, live with fewer negative environmental impacts and greater social connection, and survive times of radical transition.

4

The Abundance
of the Commons

● ●

Twin Oaks and the
Plenitude Ethic

Welcome to Twin Oaks, "a world free of bills and bosses, where gourmet meals are presented at no cost and saving for retirement is a pointless exercise. Sounds pretty nice, right?"[1] Twin Oaks is an intentional community with over forty-five years of success as a radical alternative to mainstream dwelling.[2] The community is located on approximately 450 acres of land in rural Virginia and has an ever-evolving population of around 100 adult members and their children. Since the community's founding in 1967, hundreds of people have lived at Twin Oaks as members and thousands of other people have come to visit, some staying longer than others, to participate in this egalitarian, income-sharing community. The community is based on shared labor and shared income; every member is required to work the same number of hours per week (the weekly quota was forty-two hours a week when I visited in the fall of 2011) and receives the same spending cash allowance each month ($86 a month at the time of my visit; demonstrating much more success keeping wage growth paced with inflation than the United States overall, this allowance is $103 as of April 2016).

The community runs several small businesses, making tofu and hammocks as their main income-generating ventures; they also produce over half of the food they eat, including all their own meat, milk, yogurt, and cheese. They live in seven community structures organized into nine small living groups (called SLGs, see figure 4.1), living as some people may have only experienced during their days at the fraternity or sorority house, where eight to eighteen people share community living spaces, having only a small bedroom (some smaller than others) as personal, private space. Twin Oaks community members share access to community automobiles through a complicated yet efficient scheduling system for regular town runs and personal requests (the community owns seventeen vehicles, a number that has held steady for two decades; at the time of my visit, the fleet included hybrid and electric vehicles as well as large work trucks and vans). Community members have access to both a shared, community fleet of bikes and, if they desire, a personal bike provided by the community.[3] The community provides food, housing, medical and dental care, $50 for new shoes each year, approximately four weeks of vacation time each year, and a full wardrobe of clothing through their free-for-all consignment shop, "Commie Clothes." The community also provides various resource libraries for its members, including media libraries (books, music, movies), tools libraries, musical instruments, exercise equipment, and recreational equipment (including canoes, backpacks, a Ping-Pong table, and dozens of board games).

The most radical thing about life at Twin Oaks is the elaborate systems of sharing that organize the sociotechnical networks, material arrangements, and daily practices in the community. Community members share living spaces and automobiles and meals as well as musical instruments, public gathering spaces, bicycles, and clothing. Yet the sharing of common resources is not considered a

FIG. 4.1 An SLG at Twin Oaks powered by the sun. Photo by author.

form of sacrifice.[4] Instead, members of Twin Oaks talk about their lives as being more abundant, fuller and richer in myriad ways, because they share with others. In terms of access to material resources, which are all at their source about use of natural resources, Twin Oaks community members certainly use less—they have significantly fewer cars than the average group of 100 people in this country; the community claims that the carbon footprints of their members are about 20 percent that of the average American.[5] Yet they do not talk about their lessened consumption as an environmentally motivated sacrifice; instead, community members talk about their lives as motivated by a plenitude ethic, using the language of a politics of plenty, expressing that sharing communally can (and here, does) make residential life more abundant and more plentiful.

The individuals who choose to live at Twin Oaks are certainly less privately and monetarily wealthy than the average American. They live with very little cash income and no privately owned assets like cars, homes, or lawnmowers. Yet life at Twin Oaks is full of abundance. They are living with less, with fewer private resources and certainly less environmental impact, but without an ethic of guilt or sacrifice motivating their lifestyle choice or their explanation of it. Through a radically different organization of resources, members of Twin Oaks are living plentiful lives. I use the word *plentiful* here to intentionally invoke the idea of "plenitude" as put forth by Juliet Schor.[6] Life at Twin Oaks is an excellent example of a plenitude economy—as we will see in the pages that follow, people at Twin Oaks "work and spend less, create and connect more."[7] This unique community has made a plenitude economy, or what one Twin Oaks member calls a "radical sharing economy," work—and work well—for decades. The community is organized by a shared orientation suggesting that sharing material resources need not involve sacrifice, and that living by the logic of sufficiency can create abundance in the lives of individuals.[8]

Twin Oaks has made the structures of their shared systems work for over four decades, and community members embrace a particularly radical and positive attitude toward sharing systems. While they live with so much less of the stuff that most Americans seem to see as simultaneously empowering and cluttering, community members view their lifestyle as abundant and without sacrifice. While it may seem to an outsider like life at Twin Oaks involves an acute degree of limitation, the self-described "communards" who actually commit to the experience of life in this community describe it as just the opposite. They talk about and experience their lives as abundant, even at times when they are not consciously reflecting on the benefits of living in a sharing economy. At Twin Oaks, practices of plenitude also involve an ethic of plenitude, a belief in the abundance of the commons and an understanding of shared resource use as a politics of plenty. Members of the Twin Oaks community recognize that, as Schor writes, "True wealth can be attained by mobilizing and transforming the economies of time, creativity, community, and consumption."[9]

The replacement of private resources with public resources is seen as a practice of abundance, not scarcity. This ethic of plenitude offers the potential to radically reform how we think about common or public resources, focusing on the triumph rather than the tragedy.[10]

The History of Twin Oaks

The founding of Twin Oaks was directly inspired by *Walden Two*, a novel by B. F. Skinner, a famous psychologist most well known for the theory of behaviorism.[11] In *Walden Two*, Skinner tells the story of a Utopian community through the eyes of its founder, the protagonist, as an old acquaintance comes to visit. The founder of Skinner's utopia, named Frazier, tells the visitor, "Any group of people could secure economic self-sufficiency with the help of modern technology, and the psychological problems of group living could be solved with available principles of 'behavioral engineering.'"[12] Later in the book, the narrator posits, "The question is: Can men live in freedom and peace? And the answer is: Yes, if we can build a social structure which will satisfy the needs of everyone and in which everyone will want to observe the supporting code."[13]

The founding members used *Walden Two* as well as the back-to-the-land movement and counterculture of the 1960s to guide their creation of a new community. A shared and equal labor system based on labor credits is the core foundation of the community; at Twin Oaks, every hour of labor is worth one equal credit of labor time (with the exception of child care; one hour of child care is given one half hour of credit, with the intention of encouraging people to avoid the isolated clustering of nuclear families and to take their children to work with them, a specific deviation from fully equal labor exchange that corresponds to the community's values). People live, work, and eat communally, and are rewarded for their efforts with equality in provisions of sustenance and care. The governance system at Twin Oaks is also based on Skinner's utopia; Planners and Managers are responsible for decision making, and are given "sweeping authority."[14] However, their positions are temporary, come with no possible material benefit, and public commentary and feedback mechanisms are intentionally built in to decision-making processes. Further, as there are actually more manager positions than there are members, members often hold multiple manager-ships, so that many different people are involved in the multiple organizational aspects of community planning and engagement.[15]

Twin Oaks, unlike The Farm (see chapter 3), was not founded as a community based on the vision of a charismatic leader. Although one of the founding members has written about the Twin Oaks experience and was considered by many others as a charismatic figure in the community,[16] the founders followed Frazier's contention that any "society which functions for the good of all cannot tolerate the emergence of individual figures" because personal triumph is "not only unnecessary in cooperative culture, it's dangerous."[17] Today, many of the members haven't even read *Walden Two*. Twin Oaks is "expressly secular" and committed to egalitarianism; the community does not follow an individual leader of any sort—founder, psychologist, or divine. The community is registered as a 501-D religious organization, claiming that the core values and beliefs that drive behavior in the community are enough to gain them this classification.[18]

Some things have changed throughout the community's lifetime. The founding members have passed; the population has gone through a period of aging and then a more recent increase in younger adult members. Children at Twin Oaks used to be raised communally, living together and apart from their parents and under the care of many different adults, in an experimental style inspired by both *Walden Two* and kibbutzim, but this has been abandoned.[19] Mostly, the community has flourished, adding more land and more buildings, spawning an offshoot community located nearby[20] and currently working on the development of another neighboring community that is entirely off the grid.[21]

Many facets of life in the community have remained the same for over forty years. People live in dwelling structures designed for group living, sharing kitchens and community meals. Every adult in the community is expected to meet work quota, although the types of work available vary widely and are much more diverse than the average American's working life. Making community meals, cleaning the bathroom, and taking care of sick residents are given credit equal to time spent in the community's income-generating businesses (including hammock making, tofu production, a mail order seed business, a small book-indexing service, a contract services company, and an ornamental flower business) or sustenance activities (including, for example, the extensive gardening and animal husbandry operations, food preparation and processing, taking care of the community vehicles, maintaining the schedules of car usage, splitting wood and maintaining the wood burning stoves, and participating in community governance). The blending of working life and domestic life and prioritization of community scale self-provisioning over an exclusive focus on profit, the required commitment to public over private resources, and the systems of sharing that dominate life at Twin Oaks are all examples of alternative sociotechnical networks, inspired by and further motivating an orientation of abundance, an ethic of plenitude, shared and experienced by community members. This orientation to a politics of plenty, which supports an alternative organization of material and social life in which material resources are shared without a sense of sacrifice, is deeply tied to both the structure and the culture of life in this community, motivating and helping to explain the way people share and experience a sense of abundance in their daily lives at Twin Oaks.

Twin Oaks Today

Today, nine residential small living groups (SLGs) are scattered throughout the community, all within walking distance of one another and the community kitchens, work spaces, and gardens. Some SLGs are designated as children-friendly while others are expressly not for children. One SLG is for women only. One SLG was intentionally constructed as a single-level dwelling to house the aging population, and an addition to this structure was added to facilitate hospice care in the community.

Without police, streetlights, or even flashlights, members safely live, work, and play within the community, or as they say it, "on the farm." As one description of the community states: "We grow much of our own food, run several community-owned businesses and don't use money internally. . . . We enjoy a flexible work-schedule, a mix of ages living together sustainably by sharing resources, and an environment that is safe for all people."[22] In addition to the members of the community, Twin Oaks has a very active visitor's program and members can also host guests "on the farm," so there are typically between five and twenty nonmember adults visiting the community at any time. Visitors and guests visiting for more than a week are expected meet work quota, contributing to the community's activities.[23]

The community maintains full membership and has consistently had a waiting list for several years. People interested in joining the community must attend a three-week visitor's session, meet work quota while a visitor, and participate in a membership interview. Current members offer feedback to the membership committee about potential new members.[24] Member population is designated to remain within particular demographic formulations, with the gender disparity never being greater than 60/40, a rule requiring "pensioners" (people over fifty and thus receiving a reduced work quota) to go through an additional step if interested in requesting membership,[25] and a complicated formula for allowing families with children of particular ages to join the community.[26] During my time at Twin Oaks, I was told that typically fifteen to twenty members leave every year but that no one had left for six months prior to my arrival, an unusually long period of population stability. Unlike many other intentional communities, you do not have to pay a membership fee to join Twin Oaks.[27]

At Twin Oaks, the daily rhythms of work and home life are combined, and the resources available to all members are pretty much the same. People can walk or bike throughout the community, public bikes are plentifully available, and there are currently six golf carts available to those with limited mobility. Two community meals are prepared each day, and food is always available in the community kitchens and SLG kitchens. The community provides for the needs of life from clothing and health care to toothpaste.

Because the community provides the needs of subsistence (including food, water, housing, heat, electric light, clothing, and health care) to all community members, members can live rich and meaningful lives with very little interaction with moneyed transactions. As a self-identifying feminist community, Twin Oaks members strive for gender equality and nudity rules are gender neutral, so anyone may choose to garden topless or sauna naked, but all adults must wear a shirt to dinner. Twin Oaks is a thriving example of a plenitude society, where members engage in collective capitalism, coming together to run businesses and meet their human and social needs in a community dedicated to equality and simplicity as sources of abundance.

Technology at Twin Oaks

While the Twin Oaks community is certainly not self-sufficient or free of the use of fossil fuels, they have made many decisions throughout their history that correspond to the principles of appropriate technology. Technological choices have also been made based on necessity, economy, principles of radical sharing as a form of plenitude, and the policy contexts shaping the possibilities for sociotechnical alternatives.[28] Living in shared housing structures is an efficient way to organize dwelling, and the community benefits from lower electricity use and wood heating resource needs because of their communal dwelling style. Further, all the SLG structures were constructed with some renewable energy component—several of the buildings implement passive solar design for natural, cost-free heating, cooling and lighting, and some buildings have solar hot water heaters. One SLG is entirely powered by solar electricity and solar hot water heating. Most of the buildings at Twin Oaks were constructed using at least some reclaimed or recycled materials, like reusing wood from old barns.

The community buildings also serve as multiuse spaces, an efficient way to organize space and structurally contribute to the blending of work and home life. With the nine SLGs in seven different structures, multiple work buildings including the tofu production facility, the dairy barn, and greenhouses, two separate community kitchens, and a designated "smoking shack" nicknamed the "compost café" (smoking is not allowed in most community buildings or public community outdoor spaces), there are approximately twenty buildings at Twin Oaks. Most of these buildings serve multiple community purposes. The old community kitchen now serves as a community kitchen for the SLGs around it, which were built without individual kitchens, but it also serves as the milk-processing area for making cheese and yogurt. The new community building is used as an all-purpose meeting space, mailroom, and communication hub, and for big parties, in addition to providing space for two meals a day prepared in a large commercial kitchen. The women's only SLG holds a community library on its first floor, where anyone is welcome to borrow any book at any time. The second floor of the hammock-production facility is an SLG. The single-level SLG houses both the seed-packaging business and the indexing office. There is a large room for yoga and other group exercise activities on the first floor of another SLG, and a large music room full of instruments in another. The large mechanics shop has smaller office spaces in back used for the community mental health team and the sewing room. The multiuse buildings that dominate the structural organization of Twin Oaks contribute to its efficiency and to the sense of community among members.

In addition, the community has installed two large PV systems, producing a total of 23 KW of solar electricity (see figure 4.2), which provides for between 20 percent and 25 percent of the community's electricity usage, according to my tour guide (a member for over three decades) when I visited the community. The community took advantage of state and federal rebate programs, as well as selling their renewable energy credits, in order to install the system. The community was also actively switching from incandescent light bulbs to compact florescent (CFL) lighting, hoping that the increased efficiency would decrease their overall usage so that the community could meet a higher percentage of electricity needs with the solar system. During my time at Twin Oaks, an older gentleman who calls himself Wonka told me that the community still spends $10,000 a month for electricity (around $100 per member), but this includes the energy-intensive tofu production facility's usage as well as all other businesses and all residential usage for over 100 people. Because Twin Oaks also has its own well for water provisioning, treats its own gray water and human waste with an on-site treatment system, and meets over half of the community's food needs with on-site production, this electricity usage also encompasses many things (water supply, waste treatment, and agricultural production) that are not typically included in the electricity usage calculations for individual residential households.

One of the things that struck me while visiting Twin Oaks was that the community is decidedly not conservative with electricity usage. Lights are on in almost all the buildings almost all the time. This makes sense, given the community's inherent structural efficiency and their plenitude orientation (which is the very opposite of an environmental ethic of sacrifice).[29] Leaving the lights on can bolster community sentiment because people feel welcome in various spaces at any time of day or night, and it increases the flexibility and productivity of work because you can work late at night, if you choose, without working in the dark.

Religiously turning off the lights when you leave a room is not a part of community culture at Twin Oaks, but it does not need to be, and it makes sense that it would not be, given the community's orienting ethic. One community member, a man named Finn in his early thirties who has lived in the community for eight years, told me, "People know where the electricity comes

FIG. 4.2 The 10 kW solar electric array at Twin Oaks. Photo by author.

from, because it equates to how many nuts we buy this year." By this, Finn was explaining that community members are aware of the cost of electricity usage because they have a general sense of collective, community finances. Yet they need not have an orientation of sacrifice in order to live energy efficiently.

The primary heat source in the community is wood from their own land, and a "BTU Manager" in the community is responsible for making sure the wood-burning stoves are in working order and that enough wood is split and available for each building's usage. As another example of the community's ability to do more with less, the current BTU manager proudly told me that his SLG (a home for eighteen people) is heated with only four small logs per person per day, even on the coldest winter days.

Solar water heating is used as a supplemental water-heating source on many community buildings. The community has always had its own water source and maintains its own wells. The community also treats its own waste using a small-scale sewage treatment plant based on biological organisms and natural decomposition. In addition, there are several composting toilets in the community (see figure 4.3), which I was told, "some people use religiously."

With large gardens, several greenhouses, orchards, mushroom cultivation, managed beehives, chickens, and cows, the community produces over half of its own food, and it is all produced organically. Every year, the community invests over 10,000 labor hours into the garden and over 4,000 hours into food processing to preserve food grown in the community. Yet food at Twin

FIG. 4.3 A composting toilet at Twin Oaks. Photo by author.

Oaks also demonstrates the choices that face communities seeking to balance environmental integrity, personal health, and economic stability. The general annual food budget at Twin Oaks is approximately $52,000, which amounts to less than $10 per person per week for all members, kids, visitors, and guests. Much of the bulk food purchased by the community, such as flour, oats, rice, and cooking oil, comes from Sysco Corporation, a large food distribution company with little record of environmentally responsible practice.

The community does not restrict personal diets, and all food served in the community kitchen is clearly labeled so that vegetarians and vegans can navigate their meals. There are also labels for food that includes specific ingredients that members may be allergic to or simply not prefer (like peanuts, onions, garlic, and—I learned something new about vegetables during my visit with this one—nightshades) and labels for something "Dumpster dived," meaning it was picked up out of a Dumpster for free in one of the cities nearby. For example, during my visit, bagels labeled "Dumpster dived" were available in addition to the homemade bread from the community at the bread table. The community also works with a local organic food distribution company to "pre-Dumpster" food, getting produce unfit for sale before it is sent to the Dumpsters and sorting it to rescue what is still useful for human consumption, giving the rest to animals or to the compost pile.

The community owns and maintains a fleet of vehicles that can be used by community members, but no one has a personal vehicle at Twin Oaks. Vehicles are used for the regular town runs (several times a week to the closest towns, once a week to the nearest larger city) where community supplies are picked up. Individuals wishing to purchase something from town can write a request for the "trippers" who are driving to town, who do the shopping for the community (and get labor credits for their time). Individuals can also reserve cars for personal outings and trips.

Individual requests purchased by the trippers are paid for out of the individual's allowance. This allowance, which is tied to and changes based on the state of community finances, is saved up in an account managed by the community. Community members tasked with running into town for supplies for the community, who are earning labor credit hours (see below) to run errands and get supplies, can pick up things for other community members and charge it directly to their account. Thus, as one member told me, most community members are "not actually touching money often."

If you, as a member of this community, prefer, you can avoid automobiles and cash transactions entirely while living at Twin Oaks, because there's someone else to do your shopping and a personal account where your monthly stipends are deposited and accumulate. Yet a vehicle is available to use anytime you wish and you can always take cash out of your personal account, so members have the freedom and ability to drive away from the community and spend personal money whenever they choose. This is just one example of how the alternative sociotechnical networks that support life at Twin Oaks, instead of imposing limits on personal freedom, actually involve more freedom and flexibility than a typical residential American life, where both driving and moneyed transactions are for many people a required and unavoidable part of daily life.

The sociotechnical networks that support life at Twin Oaks are based on the concept of self-provisioning.[30] As Juliet Schor writes, "The long-term economic benefit of self-provisioning is that it expands a household's options with respect to employment choices, time use, and consumption. The more self-provisioning one can do, the less income one has to earn to reproduce a standard of living. In recent years, the technical feasibility of home production for basic needs such as food, energy, clothing, and housing has grown. Self-provisioning has become one leg of the stool for living smart and sustainably."[31] These benefits seem to multiply at the community scale. At Twin Oaks, community members grow much of their own food. They did much of the original building construction themselves, and continue to maintain and remodel these buildings themselves. The community produces some of its own electrical energy, uses on-site solar hot-water heating and wood fuel for heat, and treats its own waste. The community maintains, as much as possible, its own cars, trucks, farm machinery, and bicycles.[32] The community strives to meet its own needs with its own resources.

Further, the structures of sharing that organize life in the community are inherently efficient. People living together use less of the resources it takes to live comfortably. Working and living in the same spaces saves time, money, and material resources like the fossil fuels used to drive to work and the hot water used to get ready for work every day. These sociotechnical arrangements avoid the stress of commuting for people, cars, roads, and the planet. The technological organization of the community encourages efficiency. As one member notes, "The amount of waste we have is very minor. We do one dump run a week. . . . That's the waste of 100 people including several businesses."[33] Through sociotechnical networks organized around self-provisioning and to encourage the sharing of resources, Twin Oaks offers a radical technological alternative to mainstream dwelling in modern America.

The Twin Oaks Labor System

The labor-sharing system at Twin Oaks operates as an alterative sociotechnical arrangement, as it radically shifts the material and social organization of the community as well as relationships to time, tasks, and other people. At Twin Oaks, every member is required to work the same weekly quota and receives the same access to provisions and resources (including the same amount of monthly cash allowance) in exchange for their labor. A member named Mira, a mother of two young children who has lived at Twin Oaks for eleven years, told me that the shared-labor system is "probably why we're still around after forty-four years; we didn't spend five years arguing about who is going to do the dishes" (every member has to do one shift of washing dishes a week, as long as they are physically able to; dishes are the only required work task for every member). Mira said the shared-labor system is the primary community "value" that holds the community together. Another member told me that turning in your labor sheets and following the labor system is the "only active requirement" for living in the community.

The time schedule for filling out labor sheets sets a complicated weekly rhythm to life at Twin Oaks. On Monday, members fill in a labor sheet with their requests for time off blocked out, their general requests for work, and any specific labor activities already arranged. These are turned in to the labor assigners by 7 P.M. There are four labor assigners at all times, each responsible for the work one week per month, which takes between sixteen and twenty hours. After dinner on Wednesday until noon on Thursday, members can look at the labor sheets completed by the labor assigner and make revisions to them. The labor assigners then review the revisions, make the necessary changes, and the completed labor sheets are available for pick up in the central community building Thursday night. The labor week then starts on Friday. Completed labor sheets for the previous week must be returned by noon on Sunday, labor sheets with requests for the next week by Monday night, and the weekly scheduling begins again.

When I visited, the weekly quota at Twin Oaks was forty-two hours a week and was tied to hammock production. The Twin Oaks hammock business was its most successful income-generating activity, and it had mandatory production goals each week to meet demand. If the goal was not met, quota increased by 1.4 hours the next week. Time that counts toward meeting quota is measured in labor credit hours, and includes time spent in income-generating businesses like making hammocks or tofu, or subsistence activities like feeding chickens, milking cows, or working in the garden. It also includes making meals for the community, cleaning the public rooms in your SLG (all kitchens, living rooms, and bathrooms are public to everyone in the community), doing the shopping for the community, or taking care of children. If you work over the weekly quota, you can accumulate extra hours in a labor savings account for extra vacation time, although the community grants some vacation time (currently around four weeks) every year to all members. If you are ill, you can claim up to six hours a day of "sick hours" and your friend can claim "care hours" for bringing you meals or making you tea. Birthdays, the anniversary of your "join date," and several other holidays throughout the year are also "6 hours creditable." To accommodate aging and what we might think of as retirement, weekly quota decreases by one hour every year after your fiftieth birthday.

The Twin Oaks labor system involves careful planning, and several decisions with both financial and cultural consequences seem important to its success. First, the people who plan out the annual labor budget, tasked with figuring out how to balance the multiple needs and resources of the community as they vary across time and with temporal rhythms throughout the year, recognize that some work at Twin Oaks is seasonal and that conflicts in seasonal work can arise, and they balance their labor budgets accordingly. For example, most people buy hammocks in the summer, which is when work in the Twin Oaks garden is also at its peak. Thus, hammock-production goals are set in the winter to ensure the ability to meet summer demand and maintain a high productivity garden. The Twin Oaks community has had long-term financial success making hammocks. Recently, their tofu production business has become increasingly profitable, and the community seems to be making wise decisions to encourage this flourishing business. They've invested substantially in equipment and have shifted labor hours and community priorities to recognize the importance of this newly emerging income activity. As of this writing, the tofu business at Twin Oaks makes slightly more income than the hammock business and is, according to one community member, "still rising fast." Instead of getting stuck in the past, the community has successfully embraced this financial opportunity, shifting labor priorities and meeting record high demand, making five tons of tofu a week at the time of my visit.

Further, although there are managers of each specific organizational aspect of life at Twin Oaks, actual shifts that involve working with people do not include traditional bosses. "Honchos" are selected and indicated on each person's labor sheet, so everyone knows who's in charge of making dinner or packing tofu that day. But, honchos change with each shift, so you may be honcho for the dinner shift one evening but work with a group in which someone else is honcho another night that week. People only serve as honcho if they enjoy the organizational or facilitator role in a particular job, and different people choose to honcho different tasks. So, the person who came to make dinner with an idea in mind, already checked on all the ingredients and asked you to chop some carrots may, the next day, be asking you what needs to get done in the woodshop and checking in with you as they work. This honcho system allows people to serve as leaders only when they so desire, and only temporarily, with shifting positions attached to no differential reward. It reinforces the equality of community members while allowing them each to pursue their interests.

There are very few deviations from the commitment to equitable labor credit, where "an hour is an hour" and every hour counts equally, but these deviations serve the community well. One is that visitor labor does not come out of the annual labor budget, which determines how many hours will be dedicated to each aspect of the community overall an annual cycle. With three-week visitor programs almost every month of the year (ten or eleven out of twelve), Twin Oaks consistently boosts its productivity with the extra visitor hours.

The only derivation from the equal credit given to all labor is that child-care hours are the only form of labor given half credit, so an hour of child care is only worth half an hour toward the weekly quota. I was told this is intended to encourage parents to take their children to work as a form of practical education and participation, getting them involved in the community. Community members say this policy encourages child-care providers to take children to work with them, exposing them to new things and new skills, using the real world for education and integrating children into the labor system. Children at Twin Oaks are actually integrated into the labor system, as they are expected to begin working for the community for a very few number of hours each week at an early age, with quota increasing gradually as they get older. This policy toward child-care labor credits also encourages people to take care of one another's children and get children socializing together, as each child-care hour for each child counts as half an hour. This policy discourages people from being isolated in nuclear families when caring for children, corresponding to the community's values of integration and sharing resources.

More generally, the community uses the labor system to reward or encourage things the community values, like bringing the kids to help in the garden to teach them how to count with carrots or taking care of someone who's ill. When I asked Mira about "care hours" she told me, "We want people to stay away when they're sick, we live much closer than the average community, and sickness spreads. So it's nice of someone to do it, and we want to reward that. It's about what the community wants. That's what labor credits are about."

One of the things the community cares about is self-provisioning, and their labor system reflects that. A man named Sam who's lived at Twin Oaks for thirty years called it "vertical integration" when he told me, "In our businesses as in our community, we want to do as much of it as we can." Another community member put it this way: "We make lots of decisions you wouldn't make if you were running the organization like a business. For example, we have a huge organic garden. And we love the produce that comes from it, and we put a lot of hours into it. But if we were running it like a business, we'd close the garden down and we'd put everybody into the income areas."[34] Through an emphasis on community scale self-provisioning, the community pays "careful attention to multiple sources of wealth,"[35] encouraging activities that bring personal satisfaction and community cohesion.

From an outsider's point of view, the labor system at Twin Oaks can seem a bit limiting. There are particular tasks that need to be done in the community, like cooking and cleaning and gardening and maintenance and making profitable items like hammocks and tofu, and community members are the only people who do these tasks. When I visited, I couldn't help but wonder, what if a member of Twin Oaks wanted to spend time on a task that could potentially generate income, like making soaps or other crafts to sell at a farmer's market, or music, or writing? Within the limits of the annual labor budget process, the community seems to make space for these forms of freedoms as well.

Arris has been a member of Twin Oaks for fourteen years, and has spent the past seven years developing his skills as a woodworker and furniture maker. He makes items for the community and sells his items outside the community, through the Internet-based direct sales website Etsy (which specializes in handmade and vintage items)[36] as well as a local farmer's market and individual commissions. When I asked him about the opportunities to develop new skills and income generating activities at Twin Oaks, he said:

It's freeing and limiting at the same time. . . . If I make money on the farm, it all goes to the community and I get credit for it. One hour of work equals a credit no matter how well I'm paid. There are two ways of looking at this, at least. One is to be bummed I can't use the money to buy goodies for myself; I can't spend the money how I wish totally. I deposit the money and get compensated for my work. The other way is seeing it as I'm providing for our community and being a good breadwinner. Often, my work has been the highest dollar per hour here. . . . I can feel proud of being a contributor and strive for humility and service.

He went on to explain that, through the community's annual labor budgeting process, he has been able to ask for hours each year that are "creditable"— meaning he spends time woodworking and furniture making and gets labor credits for those hours. He's also free to spend additional time in his craft, and says that about half the time he currently spends making furniture is credited. He also identified key benefits to living at Twin Oaks, in terms of the ability to learn new skills and master a craft: "The community took a risk in . . . buying antique tools with the idea they work better. . . . It looked maybe like I was spending the community's money on what I wanted to do and disguising it as community work." He added, "The accounting makes it transparent." Thus, the community chose to invest in things Arris needed to develop his trade, and the transparency of accounting at Twin Oaks ensured that all community members were aware of the investment.

For Arris, the freedom to develop his craft is directly related to living in community, and he uses his craft to benefit the community in multiple ways. He said:

Much of it has been my gift to the community to make it a nicer place and run smoother and frankly, I wanted a better dustpan than our stupid snow shovel they used for sawdust, so I made one. . . . I think of it as the community bought me a workshop, a valuable tool collection, and lets me make choices I think are good and do what I want. . . . All in all, I'm beginning to make a good profit. . . . Many artists and builders and woodworkers hear me describe my set up and express incredulous jealousy. It's great in lots of ways. I have been able to learn skills and start a business on a steep learning curve and take risks that didn't involve choosing health care for my kid or tools.

Arris did, however, recognize that developing this valuable trade while living at Twin Oaks did include a trade-off in terms of monetary income, while being quick to recognize the nonmonetary abundance of life at Twin Oaks. He said:

Friends my age make maybe $60,000 a year. That's sixty times my income! They can buy anything they want, [but] they stress. My bills are paid, my commute is a very short walk, my daughter has about the best upbringing and care any rich parent might want. French, German, and Latin, math and art tutors, health and dental. All paid, and I just work forty-two hours a week. Quota is forty-two hours. But ten hours is child care for my own child! . . . I get credit for household things anyone would have to do. Mowing my own lawn, cleaning my own kitchen and bathrooms. I could make quota doing what anyone does on the weekend in their spare time.

The Twin Oaks economy, in many ways, successfully decouples enjoyable work from income generation. In mainstream America, when people ask, "What do you do?" they mean, "What do you do to earn income?" At Twin Oaks, people have the opportunity to explore what they like to do, rather than what they need to do to earn enough money to survive. In this way, members of the Twin Oaks community have opportunities to find the craftsperson within themselves.[37] The community's labor system has a built-in flexibility to allow people to pursue their interests, fostering the development of well-rounded individuals with many skills and experiences. Arris told me,

> I'm not certified in welding but I can weld. I can run tractors and implements.
> I've done HR type work. I have management skills, machinery mechanics,
> timber framing, rough and finish carpentry, furniture making of course, sewing,
> knitting and spinning, and I can mend any tool from a wheelbarrow to a shovel.
> Forestry and forest management. Wood splitting from felling trees to cutting
> to stove length or for sawmill and running a sawmill. Splitting, stacking and
> loading about 40 cords a year. I learned all of this in my fourteen years here. . . .
> When I moved here, I'd worked eight years in food service. . . . I figured I could
> do something well though but was unsure what would feel meaningful. . . .
> I didn't have a pot to piss in or a window to throw it out when I moved here. . . .
> I was poor and worked constantly and my commute was long. Now I don't work
> as much and I'm happier than many I know.

The labor system at Twin Oaks clearly demonstrates a new allocation of time, a key concept in plenitude economics, which recognizes that "all activities, whether they are monetized or not, have the potential to yield returns. We recognize wages and salaries as the returns to employment. But activities that do not earn dollars create returns as well. Doing work in one's own household, without a wage, is production. The cooked meal, the completed tax return, and the cared-for child all have economic value."[38]

Recognizing the abundance of living with the shared-labor system and the more diversified range of activities that count as worth counting, Eli, a member in his late thirties with long, braided hair, told me, "Some people think, forty-two hours, that's more than a full-time job, but here, so much is included.

Like, so much of just being productive in life that doesn't count as work when you're a salaried employee. And, there's no commuting!" Another member said, "One of the keys to the good life is having a flexible work schedule, having a pleasant work environment. If the day is beautiful outside, people can take the afternoon off and go for a walk in the woods, everybody has four or five different types of work that they do or more. The work that they really enjoy."[39]

For example, during my time at Twin Oaks, I helped to make several dinners, and these five-hour shifts were incredibly enjoyable social time. I enjoy cooking; admittedly, the task of planning a meal for over 100 people seems daunting, although I do think it would be fun to learn how, and some people get very good at planning meals for such a large group. Since it takes multiple hands to make such a big meal, dinner shifts meant spending five hours doing something I enjoyed doing, and these shifts allowed me to spend time with enjoyable people. Making dinner is something I would need to do anyway, because everybody's got to eat sometime, but the task of doing it for myself and others counted as labor credit. These work shifts didn't feel like work at all, but they were creditable hours based on the community's labor system. At Twin Oaks, work and life become flexibly intertwined. People can do things they enjoy and can spend time with people they enjoy, while working for the good of the community in a multitude of ways. The resources that support life in the community, including but not at all limited to the resources of monetary income, are created and maintained through communal effort and become decidedly more public.[40]

How Policies Shape Practice: Life at Twin Oaks

Twin Oaks demonstrates how policies, the formal and informal rules and regulations that govern land and technology use in their local community, work to shape social organization and behavioral practice.[41] I was told during my visit to Twin Oaks that communal living in SLGs is only possible because of the area's agricultural history; zoning laws there still allow multiple unrelated people to live together in the same domicile. This regulation remains on the books from a time when slaves or agricultural laborers lived together on the farm. There are many locations throughout the United States that prohibit dwelling communally through restrictions on shared dwelling space among nonrelated inhabitants. This demonstrates just one way that local policies, in the form of zoning regulation, can encourage or limit the potential to reshape residential life through alternative sociotechnical arrangements that arguably benefit both the natural environment and human communities.[42] Zoning also affects the ability of people to use their home residence as a business, a limitation faced by many Americans, but one that does not hinder the environmental efficiencies and social benefits of the alternative organization of life at Twin

Oaks, where members live where they work. Policies at the county level also make it possible for the Twin Oaks community to maintain its own well and wastewater treatment facility. As will also be explored in the cases of Dancing Rabbit (chapter 5) and Earthships (chapter 6), local policies are extremely important for either allowing or prohibiting changes to the organization of residential life and the material systems that support it.

Both formal and informal policies within the Twin Oaks community are also integral to the community's success. The community has very clear delineations of the kinds of behaviors it does and does not regulate. You cannot have a personal vehicle or access personal economic assets while living at Twin Oaks, but the community does not dictate diet or whether you choose to use the composting or flushing toilets. The community does have strict guidelines for the proportion of children of various ages the community will support at any time, and even has a formal application procedure and rules for those requesting permission to procreate; but the community has never required someone to leave because of a pregnancy, and will support people's healthcare needs throughout their entire lives. Perhaps connected to its foundation in behavioralism, Twin Oaks encourages and discourages behavior through internal policies, both formal (like limiting where people can smoke cigarettes and use cell phones) and informal, unwritten but well understood among community members.

At Twin Oaks, people walk well-worn and familiar trails every day, to travel from home to various jobs, and to meals, and to see friends, and back home. One informal norm that struck me during my visit is that community members do not universally look up to exchange eye contact, smiles, or hellos when they pass one another on these trails. This was very different behavior than I expected in coming to visit a small, successful intentional community! When I asked, a member explained to me that this lacking expectation of a passing exchange is not considered rudeness at Twin Oaks; in a place where people have so little personal space, this norm allows people to be alone with their thoughts, if they choose. These kinds of informal norms, internal and unwritten policies of engagement, make this community work, so that individuals feel a sense of abundance rather than sacrifice in the radical practices of sharing involved in life at Twin Oaks.

Making the Private Public: Shared Resources at Twin Oaks

At Twin Oaks, many of the resources typically owned and used by a single individual or household in mainstream society are public goods shared by the entire community. Buildings themselves are public, multiuse spaces. Vehicles, tools, dishes, and silverware are all community resources. All members share community assets such as musical instruments, golf carts, bicycles, and the

FIG. 4.4 The sauna at Twin Oaks. Photo by author.

sauna (see figure 4.4). Personal ownership is limited to the property that can fit in your room.[43] With anything else, I was told, "you should be willing to practice nonattachment," as it becomes community property.

The sharing of resources is key to what makes Twin Oaks such an interesting model of community success. One community member described Twin Oaks as: "a model of sustainability. This is how we're able to live a middle-class or upper-middle-class lifestyle with significantly less income. We don't have individual washing machines, we don't have a mower for each person, but we share them across the whole group. We have seventeen cars that we share amongst 100 people, so we're getting by with about forty cars less than the average group of Americans of that size. And we do that through a fairly complicated set of carpooling, ride sharing, that sort of thing."[44]

The Twin Oaks community works to achieve a balance between private and publicly shared resources. As a community, Twin Oaks has worked out the scheduling issues that inevitably arise from car sharing because they value this system of sharing and allocate time to organize it effectively. As one community member told me, "Work and shopping are the number one and number two reasons for driving, but that doesn't happen on an individual basis here. Our system is so efficient, I could request an item from town by 9 A.M. and have it at 1 P.M., with all the accounting done, if I wanted to." This is a prime example of how a "shift toward plenitude, which economizes on materials and is rich in time, enhances the value of sharing."[45] The transaction costs involved in shared ownership and communal use might be high, but Twin Oaks values a sharing economy and thus designates labor hours to maintaining the communal, shared systems.[46]

Bikes at Twin Oaks are organized based on what one member named Arrow described as an "augmented anarchist" system. Bikes are generally public, but people do have personal bikes (and the community will provide individuals with a personal bike); you can label your bike with your name if you want to, if perhaps you brought it with you when you moved into the community or if you are a child, an odd size, have a special need that makes only a particular bike suitable for you, or just like having the same bike available to you wherever you last left it.[47] Clothing, too, can be labeled if you're particularly attached to it. During my visit, I was warned that someone might walk off with my coat or umbrella, mistaking it for a community item, without a label. These types of sharing systems provide community members with bikes to use, clothes to wear, a free and public lending library full of books to read, access to a room full of computers, and musical instruments of all kinds to pick up and learn.

The sharing systems work successfully at Twin Oaks partially because, as Arrow explained to me:

> The problem with free stuff, or at least what people think is the problem with free stuff, is hoarding. But here, we have so little private space, that's not really an issue. And, it's really a dual system. For example, you can take all the clothes you want from commie clothes, but you're responsible for your own laundry. You can read all the books in the library, but you put them back in the library when you're done instead of hiding it away in your room. You can use the guitars, but they should stay in the music room so other people can do the same. It's sharing, but sharing with some agreed upon principles to govern it.

As another example of this way of inventively organized shared and communal resources: the community maintains the fleet of public bikes, but people with personal bikes are expected to maintain their own bikes (or trade labor credit hours to the person who does the bike maintenance, as a way of accounting for the time without it being the community's responsibility to maintain personal property). These kinds of "augmented anarchist" systems seem to work well for organizing shared resource use at Twin Oaks to support the politics of plenty.

Shared community resources like those at Twin Oaks and in other community housing groups can "include guest rooms, larger spaces for entertaining, gyms, or even media rooms, pools, and other amenities that sit empty much of the time in McMansions. By combining resources, owners can obtain the benefits of large homes, but at a fraction of the cost."[48] A member named

Mason told me, "I am particularly fond of the public woodshop at Harmony [the name of one particular SLG] for any project or craft. Not too many McMansions have that." At Twin Oaks, members can enjoy the benefits of accessing significantly more material resources than most individuals have access to, because these resources are shared. As Arrow told me, "There are two possible responses when resources are scarce: reduce or share. In the United States, 95 percent of people's shit sits idle 95 percent of the time. But here, we choose to share."

The Twin Oaks community is able to successfully share resources without a sense of sacrifice or a dutiful guilt to reduce.[49] Members of the Twin Oaks community consume significantly less, but they have access to significantly more than their monetary incomes would suggest possible, and to more different kinds of resources (like woodshops, Latin tutors, and the ability to learn new trades without monetary investment or risk) than the vast majority of Americans across most incomes. This is because they share community assets and resources from buildings to cars to clothes. The community also invests time in this shared economy through labor credits dedicated to organizing the car sharing, running community errands, and buying community clothing. The community values these sharing systems and encourages them through the labor system. Broader community values regarding anticonsumerism and nonattachment also help to make the sharing economy of Twin Oaks a success.[50] The community has formal and informal mechanisms for supporting the sense that sharing is a form of abundance, with both organizational and cultural practices motivated by a common orientation to plenitude. The community's orienting ethic, a plenitude ethic, motivates a sense among community members that sharing is a means of abundance rather than a required sacrifice.

The Plenitude Ethic

The concept of plenitude brings "attention to the inherent bounty of nature that we need to recover. It directs us to the chance to be rich in the things that matter to us most, and the wealth that is available in our relations with one another. . . . It puts ecological and social functioning at its core, but it is not a paradigm of sacrifice."[51] The idea of living in a plenitude economy also reminds us, "We will not arrest ecological decline or regain financial health without also introducing a different rhythm of work, consumption, and daily life."[52] These kinds of changes in sociotechnical arrangements shift the ecological impacts of human life and experiences of social isolation and dependence in positive, and some would argue absolutely essential, ways.[53]

Twin Oaks demonstrates one model for plenitude living, where shared resource use, shared labor, and communal living can provide "security to have real rather than illusory freedom of choice."[54] The community has expanded what counts as work, bringing value to the things that matter and allowing community members to diversify the allocation of their time. The community prioritizes community scale self-provisioning, meeting many but certainly not all the community's needs through gardening and animal husbandry, maintaining their own vehicles, constructing their own buildings, and using technologies that are sustained by community members themselves such as bicycles, wood heat, solar power, composting, and the biological treatment of waste. The community has a successful sharing economy, what Arrow called a "less is more economy," with many resources that are typically owned privately becoming part of the public good. This means the community needs fewer cars, washing machines, bathrooms, and guitars—and they also use less electricity, water, gasoline, and plastic—than the typical group of 100 Americans. They are living proof that there are alternatives to "principles of ownership in everyday life that have broader applications at other levels."[55]

Yet it is not only that they are living in a plenitude economy, with shared access to material resources and an alternative relationship to work, time, and money; their orientation to life more generally is based on an ethic of plenitude. The people living at Twin Oaks are not motivated to share by a sense of sacrifice.[56] Instead, they recognize that sharing can be a source of abundance. As one member put it, "Our per capita income is substantially less than the per capita American standard, which means we live by Uncle Sam's standards in poverty, but we actually enjoy all the amenities of contemporary life. We have the Internet; we have computers. I have a whole fleet of cars and trucks and minivans at my disposal. And I have people who go out and do my shopping for me, and clean the kitchen for me."[57]

During my time at Twin Oaks, I heard things like this a lot. During a casual conversation about life at Twin Oaks, Finn offhandedly commented, "I have a sauna. I have seventeen cars." In a different conversation he told me, "I have, like, eight guitars." In yet another conversation, a member named Ezekiel said, "There's a tax manager here, so you even have someone to do your taxes for you." These men know that these things belong to the community, of course, but they feel no loss of satisfaction in that. They experience a sense of richness and abundance in life completely unrelated to their personal economic status. They live with a sense that their lives are fuller because they live in a sharing society. Arris also pointed out the sense of abundance that stems from the shared labor system at Twin Oaks: "Someone cooks almost every meal I eat here for me. I've cooked maybe ten meals living at Twin Oaks in fourteen years. A fact my mother often remarks on. I'm not a great cook so it's fitting to let others do it if it suits their fancy. My fancy is with tool use and I'm afforded that privilege."

Members of Twin Oaks really do live more abundantly because of their successful sharing community. Members have rightful access to multiple pieces of property that many Americans simply cannot afford—a sauna, a music room full of instruments, an industrial kitchen and several smaller kitchens, a yoga space, an expansive library of books and movies, an adult-sized playground with a huge hammock and swings, several common living rooms, a creek and canoes, a pond, a piano, a shack for smoking cigarettes where friends gather to listen (and sing along) to music and socialize, and a particularly designated shack for watching movies that can be reserved by anyone at anytime for collective, organized movie viewing. Twin Oaks even has a retreat cabin in the woods that can be reserved anytime by any member; many members told me they used it often, either for some time alone or to host a date (see figure 4.5). People at Twin Oaks never have to worry about not having enough money to pay the bills and they never have to worry about getting fired.

FIG. 4.5 The retreat cabin at Twin Oaks. Photo by author.

Twin Oaks provides a living example of a plenitude society, where alternative material and social organization supports alternatives to "the scale of production, how knowledge is accessed, skill diffusion, the ownership of natural assets, and mechanisms for generating employment. These questions move beyond the prescriptions of conventional economics, to a deeper reconceptualization of how to organize an economy when natural resources are valuable . . . and equity matters."[58] The Twin Oaks community has radically reshaped the organization of labor and the structure of dwelling so that members live communal, egalitarian, abundant lives. The people living at Twin Oaks eschew roles like "employee" and "boss" and "consumer" and "owner" to live more plentiful lives through radical sharing of material resources and radical alternatives in technological, economic, and social organization.

By dramatically changing the dwelling technologies and organizational structures that shape life, community members live with more freedom over their time, equal credit given to all tasks, priority given to self-provisioning, and equality among themselves. Demonstrating the guiding principles of the appropriate technology movement, as articulated by Buckminster Fuller, people at Twin Oaks do not "struggle for survival on a 'you' or 'me' basis" and are therefore "able to trust one another and be free to co-operate in spontaneous and logical ways."[59] Twin Oaks has successfully done what Lewis Mumford called for when he argued for normalizing "a vital standard":

> less in the mechanical apparatus, more in the organic fulfillment. When we have such a norm, our success in life will not be judged by the size of the rubbish heaps we have produced: it will be judged by the immaterial and non-consumable goods we have learned to enjoy, and by our biological fulfillment as lovers, mates, parents and by our personal fulfillment as thinking, feeling men and women. Distinction and individuality will reside in the personality, where it belongs, not in the size of the house we live in, in the expense of our trappings, or in the amount of labor we can arbitrarily command. Handsome bodies, fine minds, plain living, high thinking, keen perceptions, sensitive emotional responses, and a group life keyed to make these things possible and to enhance them—these are some of the objectives of a normalized standard.[60]

Twin Oaks is not an isolated or insular community. They have a very active visitors program; they allocate labor credit hours for members to engage in political activism outside the community (another example of using their economic system to support community values in ways that are not possible without a shared-labor system). They host a national conference for intentional communities every year, and receive both national and international media attention.[61] Popular media stories seem to fixate on the community's child-rearing practices, with kids often being homeschooled, having regular interaction with adults beyond their nuclear parents, having work requirements, and otherwise being described as "free-range." Popular media accounts sometimes tell outright lies; Twin Oaks is not nor has the community ever been "off-the-grid," as some headlines claim.[62] Most importantly, these media stories seem to miss the most important part of life at Twin Oaks: how the alternative material organization supports a shift in social organization, a shift that allows members more freedom, flexibility, and ultimately well-being because of their commitment to living with technological and material systems that make sense within the context of a shared plenitude ethic, an orientation to sharing as a source of abundance.

People who live at Twin Oaks do share access to many of the resources that meet their needs and comforts, resources that many Americans own and access as individuals or nuclear families. By "sharing everything"[63]—which is still quite an exaggeration—people at Twin Oaks minimize the environmental impacts of their resource consumption, as well as their economic needs. It is more efficient, both environmentally and economically, for people to share large housing spaces and especially the resource intensive infrastructures of bathrooms and kitchens. It is also socially beneficial for many individuals. It is something that can be either supported or hindered through local policies. The alternative sociotechnical arrangements that support life at Twin Oaks, from car sharing, to organic gardening and animal husbandry, to spatially combining work and home life, to structurally combining multiuse spaces, share these general elements: environmentally, economically, and socially beneficial, and supported or hindered through local policy contexts.

Yet what makes life at Twin Oaks such interesting study—and what arguably makes the community so successful—is that people there share an orientation suggesting that sharing itself is a form of abundance. Rather than feeling a motivating sense of the need for sacrifice for reasons related to either the natural environment or global social equity, people are motivated to give up individual cars and private spaces because they recognize that their lives will be more abundant, not more full of sacrifice, if they choose to live at Twin Oaks. As a community, Twin Oaks has a carefully crafted set of policies and a community culture that encourages people to live based on this plenitude ethic, which shapes their individual and communal practices.

The shared orientation, this plenitude ethic, where the values and practices of the plenitude economy are internalized and orient decisions about and attitudes toward behavior, shapes initial motivation, but it also shapes ongoing practice. The community is successfully balancing shared space and resources with individual needs, based on a common ethic that supports values related to sharing as a form of abundance. In doing so, community members are living full, fulfilling lives, with much less negative impact on the natural world and much greater connection to it.

Twin Oaks, like the other communities I've visited and write about here, does not claim to be the only way to live a successful, fulfilling alternative lifestyle through alternative technology and alternative forms of community: "We are not perfect, nor appropriate for everyone. We never claim to be. . . . What we do claim is that our living situation is far better than most."[64] The communities described in this book all identify as experiments, options, alternatives—not the one and only answer. Twin Oaks demonstrates that there are options in terms of reorganizing work and home life, and for sharing access to infrastructures, technologies, systems, and resources, that can minimize environmental impacts and the need for individual economic wealth while providing a life that is abundant, plentiful, and fulfilling materially and socially.[65]

The Twin Oaks community demonstrates the potential for living with an ethic of plenitude. Sharing resources, increasing the kinds and accessibility of public and common goods, does not have to feel like sacrifice. Sharing communally can be part of a politics of plenty. People at Twin Oaks talk about their lives in this radical sharing economy as involving more freedom, not less, and the replacement of private ownership with public resources is seen as a practice of abundance, not scarcity. This ethic of plenitude offers the potential to transform how we think about common or public resources, giving us hope to contemplate their potential in communities of various forms.

5

Individualism
and Symbiosis

• •

The Dance at Dancing Rabbit

I visited Dancing Rabbit Ecovillage during the hottest part of the summer, and the Missouri prairie was hot and muggy. It rained quite a bit, too, and the walking paths around the community were muddy and slick, while the tomatoes still growing on the vines looked desperately oversaturated. Many of the people who live in the community, including approximately fifty adult self-identified "rabbits" and their children who are members of Dancing Rabbit Ecovillage[1] as well as the temporary visitors, residents and "WEXers" (work exchange interns), walked around barefoot with muddy feet, while others wore tall rain boots. My time there felt a bit like summer camp, with people exchanging friendly greetings as they passed one another on the roads and trails, casually moving from place to place under the hot sun. At Dancing Rabbit (abbreviated below as DR, a shorthand members often use, even in speech), there is no rush hour traffic, there are no office hours, and people are—for the most part—friendly and sociable with one another and with temporary visitors.

Some longtime community members assured me that it did not feel much like summer camp to them, as their lives were busy with generating income, making meals, and navigating daily life with family and friends, and they knew from experience that summer was busy with gardening and food preservation to prepare for the winter months. Yet for me, the energy of the community felt reminiscent of something I imagine most American adults would associate with summer camp—where there are no cars, you can walk around barefoot, use the nearby swimming hole to cool off every day, and spend your days outside enjoying physical activity, recreation, and the company of people who seem genuinely interested to see you.

There are no personal vehicles in the community, and the physical layout is intentionally designed to create a walk-able, sociable village. Community members travel around the village by foot and share access to a fleet of cars organized in a community cooperative, owned and maintained by all community members who use the vehicles. Community members live in individual homes (located on land leased for a home site, as described below), but there is a large common house with shared facilities for bathing, laundry, meals, Internet use, playing piano, playing a board game, or just hanging out; the central courtyard next to the common house provides an outdoor space for gathering at picnic tables, baking bread in the large sun oven, and enjoying the company of others.

Dancing Rabbit was designed to meet two goals simultaneously: to enhance social connection and to address resource sustainability. As such, the community aims, as member Greg describes it, "to represent positive aspects of engagement with each other in the built environment." For example, when a new member joins the community and wants to build a home, they must adhere to the community's ecological covenants (described below) but must also consider the social elements of the community in the siting and orientation of their building.

Dancing Rabbit proactively attends to ecological and social aims simultaneously. Members describe interpersonal growth and social connection as the least talked about, but perhaps most important, elements of living sustainably. As a community, DR has big dreams to grow, aiming to someday be home to a population of 500 to 1,000 people, organized as a village where a larger community infrastructure supports sub-communities.[2] In many ways (like vehicle ownership and the use of building materials), daily life for members of DR is radically different from mainstream residential dwelling in America. In other ways (like economic activity, schooling, diets, and religion), the community does not expressly limit or restrict members' participation in many aspects of conventional American life. The material arrangements and social practices at DR are organized based on a dance between two set of values: one that recognizes the symbiotic relationship between humans and the natural world as well as humans with one another, on the one hand, and one that values independence, on the other. Members recognize that "human actions are inescapably entwined with a larger web of life forms (human and nonhuman), natural formations, technologies, and built environments."[3] Yet they also embrace freedom of choice and freedom of practice, corresponding to values regarding individual liberty, based on a shared orienting ethic of symbiotic independence. Both elements—connection and relationality, on the one hand, and independence and liberty, on the other—are important for community members, and the dance between these two elements shapes daily life and practice in the community.[4]

The Story of Dancing Rabbit

Dancing Rabbit was founded in 1997, when a group of friends from California moved to the northeastern corner of Missouri in search of land on which to begin an intentional community. For the seven founders, it was important to find a place where building codes and zoning regulations would not limit the kinds of experimental buildings and infrastructures they hoped to employ. Thus, independence from the standardization that enforces normalized aspects of conventional residential dwelling in America was important to the community from the very beginning.

When I asked members of the community about the environmental motivations of the founders, they acknowledged that climate change was not the central issue organizing environmental concern like it is for many environmentalists today. Some longtime members remembered discussions of climate change (then called global warming) and more frequent discussion of peak oil among members in the late 1990s, but the motivation for pursuing a more environmentally sustainable lifestyle through alternative technologies in alternative community was described more pragmatically. Keith and his partner have been members since 1999, joining just four days in to their first visit. As Keith explained, "Really, it's just logic."[5] Other community members expressed similar ideas when asked why humans should care about their environmental impact, focusing on the pragmatic nature of environmental responsibility, recognizing the inextricable connection between human beings and the natural world in terms of the impact of human activities on both ecosystem and human well-being.[6]

Members describe the model for DR as a "community of communities," where groups of individuals share access to material resources like kitchens and gardening space, where there is room for multiple and overlapping groups of sharing systems based on interest and availability, and where individual freedom to design and build material and economic structures of support prevails within the context of community covenants that dictate possible variations in material practice. For example, if you were to become a DR member, you would have the option of renting a living space (if there's a space available; see figure 5.1 as an example of a home that was being rented by a new member at the time of my visit) or building your own residential dwelling. If you chose to build, you could build a single-family home or a unit for multiple families to share, but you would have to use local and/or sustainable sources of lumber

FIG. 5.1 A "Rabbit Warren" – Grain bin converted to duplex residence. Photo by Savannah Fox, used with permission.

and live without a septic tank, instead using the community's system for composting human waste. You could build a self-contained home with space for bathing and cooking, or a home without a bathroom or a kitchen, instead using the community's shared bathing and cooking facilities.

In many ways, DR represents what Dolores Hayden described in 1984 as the future for a redesigned American dream "in which household practices, as well as appliances for cooking, cleaning, and maintenance, can be shared collectively, thus lessening the time and financial burden on any single individual or household."[7] By sharing resources, members are less reliant on monetary income and have more time to do things they enjoy. Member Vance explained that, "Here, I have the highest quality of life for the lowest cost of living. Members live on between $250 and $500 a month here, with an incredibly high quality of life."[8] Another community member, Greg, grows food, does natural building, sells yogurt, and participates in community governance for his income, supporting his family of three; he said, "My principal income strategy is to avoid the need for income."

At Dancing Rabbit, an orienting ethic of symbiotic independence motivates life and practice for community members, balancing a commitment to honoring the relationship of mutualism among humans and between humans and nature with a commitment to respecting individualism. The community covenants and sharing systems that govern sociotechnical arrangements and practices do not overshadow the individuality of community members and the shared emphasis on individualism. During my visit, this individualism was apparent in the way different people dressed, spoke, carried themselves, interacted with others, and talked about their motivations for living in the community and their memories of the community over time.[9] Dancing Rabbit is very much a community of individuals and a community that values individuality.

Evan has been a DR member for four years. His small cob home, which he built himself, is beautiful, with rounded corners and embellishments of colored glass, and chaotic, with a disheveled feel you might expect in a home built by a young man who lives alone. He spent several summer breaks from college working on a cob home on family land, a project later abandoned to join DR because the life in the community provides "more freedom to do what I wanted to do anyway, like build a weird house. This is something you can't really do in other places, because of money, like my own lack of money, and because of codes and what not that limit the kinds of things you can build. Here, I have liberty in pursuing a life I want to live."

The cooperative systems and the overarching ecological covenants at DR may seem like limitations on the personal freedoms that residential dwellers in America often associate with residential life. Yet members talked about the benefits of saving time and money by using shared systems, recognizing the efficiency and freedom that comes from sharing access to costly and environmentally impactful material goods, systems, and infrastructures.[10] Furthermore, members balance the limitations on personal choices imposed by the community co-ops and covenants with a strong culture of individualism. It is the dance between valuing individuality while also recognizing connectivity (and thus limiting the environmental harm caused by the material systems designed and used by humans that are inextricably connected to the natural world, and creating a community and culture that honors the ways that humans are themselves connected to one another) that characterizes life at in this community. This dance is dynamic, and Dancing Rabbit Ecovillage has certainly evolved since 1997. The dance is also promising. Thinking beyond the bounds of this one community in Missouri (a part of the mission of DR's educational nonprofit, The Center for Sustainable and Cooperative Culture[11]), this case study demonstrates that changes to residential dwelling through the use of sharing systems for the social and environmental benefits they bring does not have to contradict values related to individuality; instead, shared systems can support individual freedom in perhaps unexpected ways.[12]

Technology at Dancing Rabbit

At DR, the use of sociotechnical systems is guided by six ecological covenants. These covenants currently read:

1. Dancing Rabbit members will not use personal motorized vehicles, or store them on Dancing Rabbit property.

2. At Dancing Rabbit, fossil fuels will not be applied to the following uses: powering vehicles,[13] space-heating and -cooling, refrigeration, and heating domestic water.[14]

3. All gardening, landscaping, horticulture, silviculture, and agriculture conducted on Dancing Rabbit property must conform to the standards as set by OCIA for organic procedures and processing. In addition, no petrochemical biocides may be used or stored on DR property for household or other purposes.

4. All electricity produced at Dancing Rabbit shall be from sustainable sources. Any electricity imported from off-site shall be balanced by Dancing Rabbit exporting enough on site, sustainably generated electricity, to offset the imported electricity.

5. Lumber used for construction at Dancing Rabbit shall be either reused/reclaimed, locally harvested, or certified as sustainably harvested.

6. Waste disposal systems at Dancing Rabbit shall reclaim organic and recyclable materials.[15]

Karen, a DR member for six years, describes the covenants "sacred agreements." She said the covenants are "not meant to be like heavy-handed HOA covenants that determine all of six colors you're allowed to paint your house. We have a lot of autonomy and freedom of choice here, except in regards to the covenants. They identify the things we hold sacred and dear—we can question them, but we must hold to them at a fundamental level." All members and residents must adhere to the covenants; they are part of a legal document that residents and members sign. These covenants indicate the community's commitment to alternative residential dwelling; they also demonstrate the community's reflexivity about what it means to live sustainably, their internal flexibility and commitment to individual autonomy.

The covenants have changed throughout the community's history. The first change altered the language of Covenant 2, which initially prohibited all fossil-fuel usage, thus community members used wood year-round for cooking, even in the hot summers when they did not need the heat. In thinking about their impact on forest ecosystems and local air quality through particulate matter, the community decided to allow propane usage for cooking. The covenant now stipulates a list of activities that could but are not allowed at DR to be powered by fossil fuels, with cooking absent from the list to accommodate the use of propane for cooking. The community also chose to alter Covenant 5, which dictates the source of lumber used in the community, when certified sustainably harvest wood became readily available in the market.[16] This, community members said, made building a home in the community more accessible as it eliminated the limitation of locally available lumber.

A more recent and controversial change to the covenants involved Covenant 4. The community was entirely off the grid, disconnected from the surrounding utility infrastructure, until 2012. The community made the decision to connect to the local electric utility grid, but has maintained its commitment to electricity generated from renewable energy sources by stipulating that the community must produce twice as much electricity from renewable energy as is used from the grid annually.

While community members are committed to living by these covenants that dictate use of sociotechnical arrangements, they also have the ability to rethink and reshape the community's use of sociotechnical systems through revisions to the covenants. Also, note the many things not included in the community's covenants. While community members cannot use chemical fertilizers in their gardens, they do not have to garden at all; the covenants do not dictate food practices in any way. Further, community members can build working lives, personal relationships, and religious practices in any way that fulfills them; the covenants allow for personal freedom, while guiding usage of sociotechnical systems in ways that recognize the fundamental symbiosis among humans and between humans and the natural world.

Karen told me, "The covenants come from our values, and it's challenging to find shared and aligned values. It's hard to make something work for everyone, because we have a huge spectrum of what sustainability means, or what the 'eco' in eco-village means, among our membership." Karen is on a committee reviewing the covenants for potential changes, so she knows firsthand how diverse community members' perspectives really are.

Karen also recognized that the covenants are meant to encourage rather than prohibit behaviors. She said, "People don't respond well to thou shalt not, in our experience, it's not our culture. Our culture is not one of heavy hands." Despite covenants that might seem limiting to an outsider observer, the community's culture is not about limiting individuality or individual choice. Instead, the covenants are meant to reflect best practices for environmental and social sustainability. As one visitor put it, "In some ways the covenants make it easier for you, because some things are decided for you. You know, we have so many choices, and that can be really stressful too. Here, they've worked out best practices, and members can avoid the stress of so many choices."

One afternoon, I heard a member named Alice giving a tour to a small group of visitors who had stopped to stay in a bed and breakfast located in the community.[17] She told the group, "It's kind of like a condo association—we have some rules that we agree to in exchange for services provided by the community." In her explanation, she seemed to be searching for a way to explain the community in terms the visitor's might understand—people agree to live by the covenants, and in exchange they get access to many shared resources. One difference, of course, is that there are many more shared resources available at

DR. As one visitor, after being introduced to the community covenants, put it: "As humans, we have the gift of organization. As social beings, it's what we do. Why not use it for good? It's like our shared super power!" At DR, members have harnessed the power of social organization to articulate and work toward shared goals that recognize the fundamental interconnections between humans and natural systems with flexibility to evolve community practices as new understandings of sustainability and new challenges emerge.

Arguably because the community emphasizes social connectivity as well as environmental sustainability, members are willing to have real conversations about their behaviors. For example, online shopping increases the community's total usage of fossil fuels through production and transportation externalities. When I asked about seeing the UPS truck several days in a row, Karen told me, "we have to recognize all our externalities, and we are not environmental angels. We have a long way to go. The prevalence of online shopping and long-distance shipping is a challenge that the community's founders could not really have known or anticipated. But we as a community are talking about it."

The community is also actively involved in an Eco-Audit process, where resource use is continually measured and compared with average American usage.[18] Karen tells me that participation in this process is at least in part related to the educational and outreach goals of the community; she says, "When we speak about ourselves, what story do we tell? We claim to use 10 percent of the average American's fossil-fuel usage, and it's important that we really look into how and why we're able to do that." The community's Eco-Audit demonstrates that DR does consume significantly less. In 2013, community members threw away just 7 percent of the average American's per capita annual nonrecyclable landfill waste, and the community recycled 80 percent of their waste, representing 229 percent more recycling than the average American. Using a fleet of shared vehicles, community members drove only 10 percent of America's average vehicle miles/person/year in 2015, and because of the efficiency of their vehicles and the emphasis on shared car trips, used only 5.5 percent of the U.S. per capita average annual fuel consumption.[19]

During my summer visit, DR was a very busy place. There were at least three active construction sites, with people working on new homes and community kitchens. People were outside a lot, gardening, feeding chickens, and (in the sub-community called "Critters") taking care of goats, which are organized on a cooperative system where people collectively take turns caring for the goats and share access to milk and meat. There were organized activities almost every day, from evening song circles to morning games of ultimate Frisbee. The daily happy hours (from 4 to 6 P.M.) and Thursday pizza nights at the café were lively and well attended by both community members and local community residents. Community members were coming and going every day, using community cars to run errands or just go to town for ice cream.

Despite reliance on shared systems, I never saw anyone impatiently waiting to use a bathroom, shower, or washer. Community members clearly had freedom in how to spend their days and seemed to enjoy the dense social environment that the community provides. The sociotechnical networks that shape social practice at DR provide for rich, fulfilling, connected but independent lives, by valuing both symbiosis and individualism to support both environmental and social sustainability.

Rabbit Warrens

The majority of Dancing Rabbit Land Trust's (DRLT) 280 acres are in a federal conservation reserve program (CRP), meaning that the community is paid a federal credit to keep the land undeveloped. This program helped to pay off the mortgage for the land, and matches the values of community members, who are interested in, as Karen put it, "making room for life to do its thing."[20] In 2012, the community decided to take some land out of CRP to increase space for orchards, vineyards, larger permaculture gardens, and animal husbandry.[21] Yet most land is still undeveloped in CRP, the community has planted over 15,000 trees, and residences and community buildings are all built in a small part of the community's land in a densely populated village.

Because the land is owned by a community land trust, no individual can own land (a common theme among the case studies here).[22] Individual members lease what they call "warrens" from DRLT, and houses are built on leased warren land. The community advises that new members lease 2,500 square feet per person living on that warren, and lease fees are currently set at a penny per square foot per month.[23] Individuals can build any kind of building they choose on their warren (see figures 5.1–5.3 for examples), as long as it abides by the community covenants described above and is able to pass a full, transparent, and sometimes contentious group consensus process.

The residences throughout DR are identified by whimsical names like "Woodhenge" and "Skyhouse" and "Lobelia" and "Robinia." Members identify their homes using these names, saying "Lobelia" more frequently than "my house" or "her house" as reference. This, I was told by one member, dissociates built spaces from individual identity, allowing the names to stay with spaces as the residents inside change.

Homes at DR are small compared to the average American domicile; a tour guide told me that 400 to 600 square feet per person is the typical range for homes in the community, while another member suggested that 230 square feet per person is the average amount of indoor space (many homes include extensive outside space, some even have outside kitchens for warm weather food preparation and cooking, see figure 5.4). Some people live in much smaller spaces, some even less than 100 square feet. Members can build homes

FIG. 5.2 Rabbit Warren. Photo by Savannah Fox, used with permission.

FIG. 5.3 Rabbit Warren. Photo by Savannah Fox, used with permission.

without the space or expense requirements of kitchens and bathrooms and participate in community cooperatives for these dwelling needs.

Moose (a name he adopted when moving to the community because there were already several men with his common given name) has been involved with building several of the residential and community structures at DR. He has concerns about people living in spaces that are too small and the ways that affects personal, mental, and physical health. The first home he built for his

FIG. 5.4 Outdoor dish washing station in an outdoor kitchen. Photo by Savannah Fox, used with permission.

own family was 400 square feet for three people; his current home is approximately 900 square feet for three people. Karen also talked about the physical and mental health impacts of living in a small space, and said she hadn't even realized how much her health was suffering until she and her partner moved out of their small temporary space into their comfortable 800-square-foot home.[24] This is another example of the community's reflexivity and openness to changes in their own experiment. While their houses are still nowhere near the average size in America, they are cognizant of balancing personal health and environmental footprint.

The community considers both individual and community needs when thinking about the citing and orientation of new buildings. Several members mentioned the influence of *A Pattern Language*, a book about radically reconceptualizing community design and architecture, for shaping their thoughts about home building and village planning.[25] Inspired by this revolutionary architectural treatise, the community adopts "a fundamental view" that recognizes "that when you build a thing you cannot merely build that thing in isolation, but must also repair the world around it, and within it, so that the larger world at that one place becomes more coherent, and more whole; and the thing which you make takes its places in the web of nature, as you make it."[26] As Greg put it, "We are always looking at developing, organizing, and maintaining spaces to use for a happy life, good social contact, balancing privacy and proximity. For example, we have the central courtyard and shared spaces like the common house. And when someone wants to build a new house, we as a community talk about its orientation and how it will fit in socially as well as in the built environment, aware of things like spaces between buildings and orientations of front doors." Greg says the community's development plans are about "balancing personal needs and impacts." As one concrete example: the community is strongly supportive of passive solar design. As Moose said, "Passive solar works with ecology, and we try to work against it as little as possible." Yet a community where every house is built strictly to passive design principles is a community where all houses are oriented to the south, so every house's front faces another house's back, which hinders social connection through building design. Thus, the community aims to balance ecologically responsible design principles like passive solar with attention to how the built environment shapes patterns of social engagement. Warren leases are also spatially determined by a variety of factors, including the availability of open land and an individual's desired location in terms of the sub-communities that operate as clustered micro-neighborhoods within the village.

Most residences at DR involve some variation of natural building; straw-bale homes are particularly popular. These homes are certainly more time-intensive to build, as they require hand application of several coats of plaster (with adequate time in between coats to dry) to protect the bale walls. Use of natural building materials corresponds to the community's values and its identity; during my visit, one visitor talked about a lifelong desire to build "a house that is part of the organic landscape" and another visitor used the metaphor of chicken noodle soup: while the ingredients in the can might be the same ones you use, it's always better when you make it yourself, because "there's more love in it when you know the ingredients and you know the people who helped make them."

Homes vary in whether or not they are built with access to running water; some members use the phrase "walking water" to describe living without indoor plumbing, as they still access community supplies of potable water (via walking) and bring it to their domiciles. Members use two sources of water: rainwater is filtered and used for drinking water as well as for washing and watering gardens in many homes, but some members choose to pay an additional water co-op fee to access the several taps to county water supply pipes scattered throughout the community. Based on an estimate combining county water usage and total potential rainwater usage (calculated using rainfall totals and catchment system space), the Eco-Audit estimates that DR members used 19 gallons of water per person per day (7.4 gallons of county water and 11.6 gallons of rainwater) in 2013. While this is likely largely overestimated, since rainwater catchments systems are very rarely used to capacity, this number is still significantly less than the approximately 90 gallons per person per day used by the average American.[27] In 2014, calculated water usage represented just 7.6 percent of the average American's municipally supplied water.[28]

While many members join eating cooperatives, some members choose to make meals at home, and the means of doing so vary widely. One member, who rents a grain bin converted to a two-story domicile, uses a microwave. Other members of the community still cook exclusively with wood, and one member who hosted me for dinner during my visit cooked our delicious meal outside on a wood-fired rocket stove. Two members talked to me at length about experimentation with cooking and diet to minimize environmental impact and maximize well-being; one talked about her outdoor biochar stove, while the other talked about using an electric crockpot. This points to the dance of symbiotic individualism: while those who use wood might prefer that everyone avoid fossil-fuel usage, they also recognize the negative environmental externalities of wood as a cook fuel (such as lumber consumption and localized air pollution), and thus people make their own choices within a spectrum that values both individuality and the connectivity of human behaviors in impacting both the earth and other human beings.

Dancing Rabbit Co-ops

Access to many sociotechnical networks at DR is organized based on a system of cooperatives. The land on which people build their homes is shared, as are the cars they drive, and members have the potential to access shared systems for electricity, showers, bathrooms, laundry, kitchens, meals, Internet, and other community spaces and provisions.

All users of the systems pay the same amount to access the resources of that cooperative. With the humanure waste composting system (members call it "humey") cooperative, every member pays a co-op fee but there are different rates depending on whether they have their own bucket on their warren or use the bathroom stalls in the common house. There's the DRVC car-sharing co-op that every member joins in order to access a vehicle, the BEDR electricity co-op for those who do not want to install a self-sufficient electrical system, and multiple eating cooperatives in which people come together to share in preparation and enjoyment of meals.[29]

The community's choice to organize car use cooperatively is about much more than the environmental impact of gasoline powered automobiles—the community also recognizes the social implications of private vehicle ownership and the way cars force organization of social space for their own

convenience, with loud, dangerous, and isolating highways, privatized parking lots and garages, and the entire social organization of work, home, and leisure based on vehicle transportation. The lack of personal vehicles is another
example of how the community considers both ecology and community in its
decisions. DR is not car-centric, as so many of our modern communities are.
The community, thus, is moving away from the negative social consequences
of automobility in terms of dependence and isolation:[30]

> One the one hand, automobility generates a tremendously flexible—and often
> appealing—source of mobility, privacy, and independence. On the other hand,
> it can be understood as a coercive practice that consumes massive amounts of
> space, requires lengthy commutes, increases dependence among youth, elderly,
> and others unable to drive or without access to a vehicle, relies upon exten-
> sive state surveillance, harms or eliminates environmental options for future
> generations, and structures patterns of living, working, and playing that often
> preclude many from more than nominal use of alternatives even where these are
> available.[31]

Dancing Rabbit members get the physical benefits of a walkable village,
the social benefits of seeing one another outside rather than being confined to
their private metal boxes, and the personal benefits of living in a space without
the noise of traffic and without the personal economic investments necessary
to own, maintain, and insure a personal car. As one visitor put it, "It's about a
commitment to using less fossil fuels, but also about avoiding the infrastructure built entirely around the use of cars, and the anxiety that comes with it."
In response, a member named Ethan said, "Yes, cars are butterfly killers. That
said, we find use for the utility of a pickup truck, and cars to get to the dentist."
Dancing Rabbit members recognize the environmental and social impacts
of vehicle usage while also valuing independence, because people can and do
access cars or, conversely, can choose not to pay into the co-op and never personally access a community co-op vehicle.[32]

Karen said the vehicle covenant works as "somewhat of a filter" for potential new members. She said, "To really answer the question—Are you willing
to shift your culture to car-less? Note I did not say carless." While members
have access to vehicles, they do have to shift their mind-set and their practices
after spending a lifetime in a culture that equates personal vehicles with the
epitome of freedom.

Greg, a member of the community for over eight years, said of the lack
of personal vehicles and the resulting physical and social organization of the
community: "being able to work where you live is very powerful for me. . . .
Lifestyle change is about choices, basic choices we make every day about
where we live, where we work." Living in a community where people do not

drive to work made life more fulfilling for Greg; it did not involve sacrifice of individual freedom but instead provided individual quality of life by avoiding traffic jams, traffic noise, and your own required commute. As Evan said, when describing DRVC, "some people describe this as a sacrifice, but with this sacrifice, other sacrifices, like the soul-sucking time spent commuting and in traffic jams, are abandoned." Dancing Rabbit's car-sharing system recognizes the choice to rely on private automobiles as political, and that "[w]hen political choices lead to a different communal structure, one that facilitates proximity as much or more than movement over vast spaces, then the constraints imposed and the opportunities enabled for our freedom by the present system of automobility will become more evident. These choices not only make it more convenient to walk, bike, and use public transportation, they can make it more feasible and appealing to choose *not* to take a trip at all."[33]

BEDR, the electricity co-op at DR, was established after the community connected to the utility grid. The decision to move from off-grid to utility grid-tie was contentious, and members provided several explanations for the decision. Some said being off-grid was a barrier to new members, who had to invest thousands of dollars in independent renewable energy generation for new homes. Others said the grid connection allowed the community to avoid the toxicity and planned obsolescence of battery storage systems; the decision to connect to the grid was made when a large battery system was due for replacement. A few talked about off-grid living as a barrier to economic growth because of unreliable Internet access, and said they wanted to rectify the offsetting of electricity usage when members went into the nearby town to use Internet during times of low power in the community. A few others talked about connecting to the grid as a way to increase food security, as the community can now store more food using freezer storage.

Some mentioned that connecting to the grid allowed the community to increase their positive impact on the world because DR contributes electricity produced from renewable energy sources to the local grid. As Karen said, one way to think about the decision is that "the grid connection was making the world a better place by feeding our power into the grid. But we [she and her partner] felt really strongly about the 2:1 rule [that the community must produce twice as much renewably sourced electricity as the power it uses from the grid annually]." Another comment from Karen (who continues to live without a grid-tied electrical connection) again demonstrates the balance between individual needs and a desire to live in symbiotic relations: "I was really troubled by the decision, I wanted to be off-grid, and I didn't want to use any fossil fuels at all. Here, I have that freedom, I can still live that way, without having to choose for other people how they live."

Very few warrens remain entirely off the grid, and most people rely at least partially on BEDR for electricity. Evan has a BEDR connection that he says is incredibly useful for tools and appliances, although he still uses a small off-grid solar system for most personal use. Many homes have solar panels and/or wind turbines, but without battery storage. The consistency of electricity from the grid is important to many community members; I heard stories of frustration about the color-coded scheme the community once used to communicate the level of power supply available (ranging from green, which meant use anything, to black, which meant everything's off). Yet most of these stories were told with a twinge of nostalgia, as if people were remembering the good old days of more principled—albeit more electrically simplified—living. I also heard many people talk about the benefits of being off-grid, specifically having power when the grid goes out. People who continue to be off-grid seem delighted by their continued independence (and their ability to provide lighting and refrigerator space to others when they lose their grid-tied power), while those who need the reliability of grid-interconnected power seem pleased that they can always use the Internet without having to go into town on cloudy days like they used to.

The grid connection is having an impact on the community's electricity usage and culture of conservation. In 2012, the community consumed slightly more than 10 percent of the average U.S. per capita electricity; in 2013, this increased to 14 percent, and in 2014, the community members consumed approximately 20 percent of the average American per capita electricity.[34] During the community meeting where results from the most recent community Eco-Audit were presented, community members loudly vocalized their disappointment in themselves (as they aim to consume 10 percent or less of the average American in all resources). They immediately started interrupting the presenter to ask why the increase in consumption was happening. In response, Ethan yelled from the back of the room, "Because it's like crack!" In a private conversation, Ethan said:

> I visited an off-grid community, and moved to a grid-tied community. It rained for two weeks during my visitor period, and they held the Q&A session by candlelight, and the fridge was turned off. Connecting to the grid was a big watershed community decision. It was about economics, but it seems like it was also about the inconvenience and hardship of folks, especially for people who need an external connection for work. The community was kind of offsetting the needs of members, because people traveled into town to use electricity, because they needed phones and computers for work. We were also at the point of needing new batteries. Now, we have this two-to-one ratio in the covenants, but we're not compliant. We need to remain focused and aware of conservation. We have people here who had dwelled in an off-grid world for a long time, and so were very tuned in to energy conservation. But now we have people here who have never been off-grid, and they just aren't as aware. They leave lights on, or open

fridge doors to linger and think about what they want, or don't think about when they shower or charge their laptops, and so they're using power at night unnecessarily. BEDR energy co-op is like a sixteen-year-old getting a car—like, wow, I can plug anything in anytime of day. Suddenly, people see tools they'd never seen before, because they have all this potential power available to them.

As one visitor put it, "the grid tie dilutes the feedback, increases the convenience, and now the community is looking to increase responsible use of what's available." Some community members are working to foster more electricity conservation. For example, Ethan hung signs in the common house explaining the benefits of showering and doing laundry during sunny days, when the community can draw on its own generation. The community made the decision during my visit to change how they schedule use of their all-electric car so that it can be charged during the day, rather than at night. The community actively deliberates about electricity usage, considering how to balance individual needs and comforts with their commitment to live as sustainably as possible in recognition of the reverberating impact of their behaviors. As Karen said, "It gets back to the fact that there's not an easy distinction between what's eco and not eco. Electric induction cooking is very efficient, versus wood and propane, but we still use wood to avoid all fossil fuels. The electric car is great, unless it's being charged with coal-based fuel. We as a community continually grapple with these questions."

In addition to homes, cars, and electrical energy, members also engage in different material practices related to food that reflect their commitment to both mutualism and individuality. The community covenants make no restrictions on diet, although the community's culture is one in which many people grow food and eat a plant-based diet. As Greg said, "People in the community meet a lot of their veggie needs, not grains though. I'm really curious about it, our lack of sufficiency in terms of meeting the bulk of our calorie needs is something I'd like to see change, but given our commitment to reduced fossil-fuel usage, there's a built-in ambivalence about tractors and tractor tillage." He also reminded me: "We're all independent operators in this thing, we're not a commune."

While members are responsible for the monetary requirements of meeting their own food needs, they have the option of cooking, eating, and enjoying food more communally than most American dwellers. Organized based on shared dietary choices, eating patterns, social preferences, and locational convenience, there are several eating cooperatives, in individual homes or in separate kitchen spaces. The cost for joining an eating co-op ranges from "pay what you can" to $11 per day, and for this, members have full access to the food and shared kitchen space, within the rules stipulated by each co-op, and members take turns preparing meals.

Some people choose not to join a co-op, eating alone or within a nuclear family. In choosing how to eat, members are able to balance their multiple interests. As Moose said, "It is good for my family's health to have our own eating scene, but the downside is having to make every meal. At the co-ops, you cook one or two times a week. At our house, we're talking at lunch about what we should make for dinner." The flexible "eating scene" means that people can come together with others to benefit from the material and temporal efficiencies and the sense of community provided by sharing meals, or they can choose to eat alone or in nuclear family units, and these choices can change throughout time to match the needs of the individual or family as children grow and families transition. While most intentional communities have a single sociotechnical arrangement for meals, so that members either must eat together or must arrange for meals individually, the structure of food-provisioning arrangements at DR suits the individualism and preference for choice among its members.

While people make individual choices regarding food, there is a shared culture of food, as most people prefer whole foods to processed convenience foods. As one member, Hannah, explained to me: "Food is actually the thing that sometimes reminds me of how different we are here. My son plays baseball on a local team, and at a recent potluck, we were the only family to bring vegetables, and the only family to eat them." In a recent community survey, members indicated that while it's not part of the community's mission to grow their own food, some people do want that, and most members think that food security—defined as providing your own food or knowing your food sources—is important to the community.[35] In terms of meat consumption, some members are completely vegan while others engage in animal husbandry. Some members think raising and eating meat is an appropriate form of food security in this particular landscape (prairie with significant winters), while others think that veganism is "necessary to be ecological." In this too, the community honors individual freedom while encouraging food habits that recognize the impacts of providing food sustenance on both the environmental and social systems that support human life.

Rabbit Economy and Labor Technology

Dancing Rabbit is not an income-sharing community like Twin Oaks; here, members are responsible for their own income to support subsistence needs and comforts.[36] However, the community is actively thinking about and working toward alternative forms of economic organization. The community houses a nonprofit educational and outreach organization, The Center for Sustainable and Cooperative Culture.[37] The headquarters of the Fellowship for Intentional Community is also located at DR.[38] A small number of members work for these organizations. Many community members said they are able to spend more time working and volunteering for causes they care about because they need less money to support their ideal quality of life at DR.[39] Further, as the community looks to become an economically viable and sustainable village, community members work for one another—at the inn and café located in the community,[40] selling eggs, cheese, meat, and produce, supplying firewood, providing child care, and building houses.

The community has established a (suggested, not required or enforced) minimum wage of $9.50 per hour. This was once a standard wage, treating all work as equally valuable, but the community recently made a decision to allow wage inequality in order to more highly value skilled labor. Moose said, "Some members want to value all labor equally. I'm a fan of an expert labor workforce. Do you want an experienced builder building you a house, or someone who's willing to work but has no experience? You'll end up paying for a lot more hours of labor, and getting a lower-quality house in the end. Not valuing our labor artificially deflates our economic system." Other members also talked about building the community's own economy by paying one another for the work people exchange, including work on village committees.

Dancing Rabbit also has its own alternative currency system, ELMs (which stands for "exchange local money"). Members can pay one another and their co-op fees using ELMs. One member told me, "I use ELM almost entirely for all of my living expenses. Almost all of my income is in ELMs. . . . I pay for nearly everything with ELMs, including transportation, lease fees, my Mercantile tab, all of my co-ops and utilities, etc. . . . There is no one here that won't accept ELMs for anything that I know of. It is also routinely used in transactions with folks from the other communities and even some locals who don't live in one of the official communities."[41] This alternative currency system allows people to provide for much of their subsistence without needing to participate in the conventional money economy.[42] Although the structure of economic organization at DR is very different from that of Twin Oaks (see chapter 4), both communities have sociotechnical networks related to work and economic exchange that allow people the freedom to engage in alternative practices when it comes to interactions with minted money. Claire told me, "People get less attached to making money after living here. They get to explore how much they really need to be comfortable, instead of living like they always need more."

Rabbit Communication

Dancing Rabbit's structures of communication focus on personal growth, interpersonal connection, and nonviolent communication processes. Dancing Rabbit embraces personal growth and interpersonal connection as key pillars of sustainability. The educational sessions organized for visitors include a session on "inner sustainability," addressing ways of meeting personal needs and ways of communicating effectively in community.[43] Sasha, one of the session facilitators, said, "Here, you learn to know yourself better. You have to learn to communicate effectively about your emotional landscape."

Karen said, "We embrace conflict here; we don't shy away from it. We see conflict as an opportunity to get to know each other better." You cannot simply ignore people or decisions you don't like at DR. Further, the community recognizes that personal well-being and social connection are necessary for environmental sustainability. One member calls cooperation "the mother of all sustainability skills" because "we can't really get down to 10 percent of the average American's consumption without sharing resources in a significant way."

The Dancing Rabbit Ethic: Symbiotic Individualism

On a walking tour of the community, Moose told a group of visitors: "People who come here interested in natural building don't want to be told how to build. . . . The beauty of Dancing Rabbit is people decide for themselves." In a separate conversation, Corey said, "I live on ninety percent less than what I needed to live prior to coming here. That creates so much freedom." Dancing Rabbit allows people to experience the freedom and individuality that they value.

Yet members also talked about intentionally seeking out intentional community—finding DR after searching online for communities or after failed attempts to find or build community elsewhere. As Claire told me, "I love being able to connect with friends so much easier, to see them more often and know them more deeply. I don't like driving, I don't feel an urge to get out. I love just taking a walk and seeing other people, and here we have a lot of support from one another when times are rocky." Claire also said,

> It's important to me that here, we focus on noticing how much what we as individuals do impacts other people. We're used to having our own domains, but one of the biggest changes living here is needing to check in with people before doing anything, really. We live very close to one another and what we do does impact one another; we're just admitting that that's true. . . . People do have an impact on one another . . . this is also true in cities and other communities, but we kind of learn to ignore our impacts on one another. Here, we are very conscious about it, and see it as important to honor the ways our lives impact one another.[44]

The alternative sociotechnical networks that shape residential practice at DR are organized based on a dynamic dance, recognizing and appreciating individual freedom while honoring the symbiotic relationality among human beings and between humans and the natural world. Members of the community acknowledge that human activities impact the earth as well as other humans, recognizing this as a simple, pragmatic truth. Yet the community also values individual freedom.

Decisions about how to organize eating co-ops, how to exchange labor, goods, and services, and how to cooperate in any shared venture from gardening to goats, are up to individually organized groups. Co-ops and sub-communities form, evolve, and sometimes fade over time. Within the context of the community covenants, there is autonomy to make decisions about sharing incomes, child care, food-provision responsibilities, and many other things. As Karen said, "Our culture is one of letting people make their own choices." As one visitor noted, "I've been observing the didactic here between individuation and connection, which seems really resilient here. Community is a very fragile thing, but a part of what makes this community strong is the space made for individuality."

It might seem counterintuitive to describe a community where people cannot use private vehicles as one that values individuality. Members express that private automobiles limit freedoms, as much or more than they facilitate them, but they also recognize that some limitations on personal freedom are just logically necessary if we recognize our symbiotic relationship with the natural world and seek to sustain human life. As Moose said, "Yeah, people here do all agree on the environmental stuff. And do they agree on why to care? Sure, we know that if we don't, we're all fucked. Nature will be fine. We need to take care of nature for our own sake." Greg also expressed a pragmatic (and individualistic) understanding of environmental concern when he said, "Why care about the environment? Well, it's just practical. And, I feel good when I live responsibly and connected."

Certainly, there are members who recognize the natural world as a sacred space. The community allows for various perspectives on the natural world. Yet the community's overarching orientation suggests that human beings should care about the natural world and seek to minimize negative impacts on it because it is necessary for human well-being. It is seen as logical and pragmatic to recognize the symbiotic relationship between humans and nature, the fact that humans need natural resources to survive.

Community members also care about one another, and many expressed that the community's culture of caring and connection has helped them through a time of adversity. Thus, recognizing that how humans organize their societies is related to how humans relate to one another can also support individuality and individual well-being. As Greg said, "The principles that we share, the principles that hold us together, will be there through adversity, whether that adversity is related to climatic changes or resources scarcity or social growing pains. The larger American society doesn't have that, doesn't have a shared set of principles to help them hold together when times get tough." Another member, Chris, also said: "When things get to be unstable, this is the place to be, with a lot of people who care deeply about one another and a lot of resources to share."

Dancing Rabbit members care about the natural world and about one another, arguably because they recognize that this caring contributes to their own individual well-being. Members also share; they share access to economically and resource intensive material systems like cars, kitchens, and land, and they talk about sharing as something that contributes to rather than limiting individual freedoms. Bathrooms, vehicles, washing machines, and kitchen stoves sit idle the vast majority of the time, and sharing them can be an efficient way to use material resources. Sharing systems are also time efficient, freeing up time for people to do other things they individually enjoy.[45]

As Chris put it, "sharing systems are really, really efficient. Life is easier, and you can do more, with a shared group. I guess the phrase 'live simply so others can simply live' captures my values. The U.S. poverty level is still in the top 5 percent of global wealth. My family lives well below the poverty line, but we don't take any assistance. We don't need to, because here we live rich and abundant lives and we need a lot less. I'm happy to have the time to do things I enjoy, I'll trade that for any additional wealth."

Dancing Rabbit balances use of shared systems with valuing private freedoms and individuality through both community organization and social norms, similar to Twin Oaks (see chapter 4). At DR, there is a gradient of public to private space; the courtyard and common house at the community's entrance are both very public, but many people live in individual residences, and the community has additional building spaces (a meditation space and a yoga/dancing/community room) available for members who need extra space for temporary privacy.[46] Through both the built environment and cultural norms, DR balances shared systems with valuing individuality, and shared systems themselves can contribute to individual freedoms by creating more time to pursue individually valuable activities.

Community members recognize that they are continuously dancing to find a balance in shared systems and individual freedoms to meet the dual goals of environmental sustainability and personal sustainability. As Claire told me,

> Sharing can be hard. The changes to infrastructure use were definitely not the biggest changes for me in moving here. . . . I was nervous about not having running water, but it's totally fine. It's actually great, because one benefit of no running water is there's less to worry about, because there's less to break. And in terms of the environmental benefits of sharing, that environmental piece, it just makes economic sense too. I bought a house here for $28,000, and the house is comfortable for my family, and I don't owe a bank for that house.[47] I also live

without a car payment. There are only two showers in the common house, but I practically never have to wait at all. The shower in your individual house is empty most of the time. We have a baby grand piano, and even though it isn't played all that much here either, it's played more than it would be sitting in someone's individual house.

Sharing systems, then, are beneficial for individuals, because they do not have to work to pay for individual ownership of systems they hardly ever use. While there are also clear environmental benefits, it is the benefit for individuals that DR members articulate (just as at Twin Oaks) when talking about why they share. Sharing systems, it seems, can contribute to, rather than limiting, human freedoms.

Dancing Rabbit Ethic in Practice

Life at DR is different than mainstream residential life in a multitude of ways. As a visitor, I assumed that people would talk about the lack of flush toilets as one of the biggest differences; not so! When I asked Claire about her use of technology at DR, she told me about using the Internet to create an online personal coaching business and develop a series of workshops on relationships and self-growth. She focused on something about technology that was new to her—an online business—but did not mention that her house has no kitchen, bathroom, or running water. This was universally true among members and residents.[48] When asked about the biggest changes involved in living at DR, they talked about the climate (both geophysical and social) of the midwestern United States and about the social connection that comes from living in intentional community. A few talked about the benefits of not spending so much time alone on the computer. No one talked about the lack of flush toilets.

This indicates an important point: alternative forms of practice very quickly become new norms in practice. One visitor put it this way: "In the past two days I've been wondering, why have I flushed my poop down a toilet filled with water my whole life? I don't know anymore!" New habits of practice become unthinking patterns very quickly (as Claire put it, the infrastructural changes were not the biggest changes in moving to the community),[49] which means there is a lot of room for change in the practices of residential dwelling that may seem unusual simply because we aren't doing them yet.[50] Alternative material systems correspond to alternative residential practices, but people who live these new practices quickly accept them as the new normal. As Claire told me, "People change when they move here. They change their standards of cleanliness and presentation, and start experimenting with what they are comfortable with, in the context of less social pressure to look a certain way."[51]

Alternative technologies support alternative practices, and abandoning socially normative understandings of material practice makes space for experimentation with practice; arguably, residential dwellers have a "shared inability to imagine an alternative" to current configurations of residential life, because "the alternative . . . cannot be formulated a priori but only emerges out of the entanglements of material practices themselves."[52]

Practice is also important for understanding DR culture in a second sense. Members talked about being able to learn through firsthand experience and bodily practice in a way that is not supported by mainstream residential dwelling. As Greg said, there is "a very accessible knowledge base" in the community. People have experience building homes, growing and preserving food, and facilitating community growth, and are willing to share knowledge and experience openly with one another. Because members live in a dense village and have shared goals and shared culture, they can exchange ideas about how to pursue valued practices; they also have more time to do so, because economic pressures on time use are lessened.

Greg also said, "Probably half of my personal knowledge is based on just experimenting. I learn best through firsthand bodily knowledge. I think a lot of people around here do." Ethan said, "I need to be able to work with my hands, learning through doing and really doing it with my own hands; that suits my personality. I'm a tactile learner. When you use gas for heat or cooking, you don't know how much is there, it doesn't feel any different even when it's empty. When I cook on my woodstove, I have to get the sticks, and chop the wood, and I like that." The sociotechnical networks at DR encourage the kind of knowledge that comes from physical, bodily doing, a kind of knowledge that mainstream residential dwelling denies to residential dwellers.[53]

Why "I Heart DR"

During the weekly "WIP"—which stands for week in preview, a weekly Sunday meeting to discuss the week ahead, including schedules for shared vehicles, visitors to the community, and community events—the last part of the meeting is devoted to "I heart DR," a time when members can publicly recognize one another for appreciation. In concluding this chapter, I'd like to highlight some of the things I heart about DR, while also acknowledging the tensions in the dance that orients their practice, summarized clearly by Dave, who has lived at DR since 2002. Over morning coffee at the café, he told me, "A lot has changed since I moved here. The biggest change is the energy consciousness, some people use lights unthinkingly, they are people who have come since BEDR was established, and they never lived off-grid. I miss that consciousness. . . . Overall, you know, we're a bunch of nonconformists coming together in community, and it's not always easy. But you know, I've changed, we all change, and the community is constantly experimenting with new things and evolving, just as we as individuals can experiment and change here, too."

Dancing Rabbit is very self-aware of their role as a space for experimentation in sociotechnical systems (as is true for the other communities described here, see in particular the Earthship community, chapter 6) and how to organize socially. For example, while the community was not meeting their goal of producing twice as much electricity as they consume at the time of my visit, there was a lot of conversation about behavioral slippage, the lack of institutional memory from those who moved to the community after the grid connection, and what the community could do about it. Further, they are not hiding their deficiencies: the information about electricity production and consumption was very publicly displaced in the community common house, a space open to visitors. The level of reflexivity, the ability to critically reflect and work toward change, is one of the strengths of the community; their "activities are the embodiment of an ongoing dialogue about what sort of life it is good to live and what sort of household practices might help facilitate that life."[54]

The sociotechnical networks at DR address many of the tendencies of isolation and dependence that characterize mainstream residential dwelling. Even if members choose to connect to BEDR and pay a monthly utility bill and continue to eat as individuals or nuclear families rather than in shared food co-ops, the people who live at DR are paying their own community for their power, cannot avoid conversations about electricity consumption, must reserve a DRVC car to go grocery shopping, and must change their practices of using water and dealing with human waste. Members are also given the freedom and are culturally encouraged to change their material practices in ways that increase their connectivity while simultaneously fostering independence. As Greg put it, "The biggest piece of alienation in modern societies is having no way to know how many degrees of distance I am from the things I'm relying on. Here, I am able to connect the dots a lot more easily, and can change the forms of dependence that I'm not comfortable with." Greg was very aware of the community's ability to contribute to his own sense of self-sufficiency and resilience; he said, "I have a very different relationship to food than I did growing up. Now, I know where my food comes from and I grow a lot of it myself. Mostly, that's about gaining knowledge. You don't need this knowledge, you don't need to know, if you think the system will continue to function as it does. Here, I think a lot of us think it's to our benefit to know and to have control over the systems we need to the extent possible." As one author recently put it, "neither the preferences of individual 'Rabbits,' nor the strong community environmental rules, nor the materials employed can—as isolated units of analysis—explain the ecovillage's impressive achievements."[55] Instead, understanding the possibilities represented by DR requires attentiveness to orientations that are broader than individual values, beyond environmental values, and to how shifts in practice create new understandings of a life to be desired.

Since 1997, Dancing Rabbit Ecovillage has been experimenting with ways to radically reorganize the material and social foundations of residential life. Members share an orientation that values both symbiotic relationality and individual freedom, and it is the dance between these values that characterizes life in this community. The community's character will likely continue to evolve. However, recognizing the potential for a relationship of symbiotic mutualism among humans and between humans and the natural world, while at the same time valuing individual freedom, represents an orientation that corresponds to evidence regarding the human impact on the natural world while also corresponding to widely held values regarding human liberty. While the community may seem from an outsider's perspective like a limit on personal freedom, members talk about how the community actually contributes to their freedom.[56] As Greg put it, "The future will likely look like downsizing to some, but it's really about figuring out what we need versus what we want. Here, we get to pursue more of what we want."

6

Self-Sufficiency
as Social Justice

• •

The Case of Earthship
Biotecture

What comes to mind when you read the word Earthship? Perhaps you think of a spaceship, and you wouldn't be alone; over the years, I've had many people ask me about the spaceship houses I study. Yet Earthships are not designed to carry us away from this planet. Instead, they invoke many of the principles expressed by the "spaceship earth" metaphor coined decades ago: they are buildings designed to provide for the needs of dwellers through the systems and resources of the dwelling itself.[1]

To a very large extent, these buildings operate as self-sufficient systems, with heating and cooling, water collection and filtration, waste treatment, and electricity generation from renewable resources without connection to grid infrastructure, providing for many of the needs and comforts of residential dwelling through radically efficient design. In an Earthship, the "building you live in looks after you and cares for your needs. Ecological living [in an Earthship] is not about privation but about an improvement of the quality of life for its inhabitants and their descendents."[2] Earthships operate based on the principles of what Buckminster Fuller called ephemeralization, a radical kind of efficiency that does more with less and meets multiple aims through parsimonious system design; they offer a "holistic version of sustainability" in residential dwelling.[3] Earthships are a synergistic system; they efficiently do more with less, meeting the needs and comforts of dwellers through less resource-intensive and less infrastructure-dependent technologies.[4] Earthships articulate many of the same criticisms and potential solutions once offered by the appropriate technology movement, attempting to move away from technological systems that create economic dependency and social isolation.

Architect Michael Reynolds started conceptualizing, designing, and building Earthships in the 1970s.[5] His company, Earthship Biotecture, is a mix of architectural firm, nonprofit development organization, educational facility, subculture, and social movement. Although Reynolds himself is the charismatic leader of Earthship Biotecture, it is, in many ways, an organization without authority.[6] A key component of Earthship design is that they are simple and low-tech; anyone can, in principle, build their own Earthship home. Further, individuals must be self-motivated to participate in the Earthship community, and people come from across the country and around the world to participate as interns or volunteer on builds to learn about Earthship construction.[7]

A distinct feature of Earthships is that these residential dwellings operate on the imagery of an organic system: they require the interaction and participation of the dweller. When it comes to the bodily practices that mediate and shape technology usage, Earthship dwellers are actually doing differently through interactions with their home and the material systems that support life within it. It is through active participation in the systems that support dwelling in an Earthship, organic systems that require attentiveness and interaction, that the individual becomes transformed. As active participants in their own technologically derived comfort, these people are no longer isolated by and dependent upon the technological infrastructure that typically supports residential dwelling. Just as the dominating technological systems that support residential dwelling shape individuals in both thought and action,[8] the technological systems of an Earthship reshape individual thought and action by changing technological practice.

To be clear, Earthships are not at all an intentional community in the sense of the case studies in the previous chapters. Community here means simply people who share an interest or experience with Earthships, and those I met only loosely identify as being part of a kind of community. Yet those involved in the Earthship community (here meaning those living in Earthships or participating in Earthship Biotecture through employment, volunteering, or educational opportunities) are inspired by a shared orienting ethic, an orientation of self-sufficiency as social justice, seeing self-sufficiency as a human responsibility as well as a catalyst for and the ultimate aim of full human potential.

As Lewis Mumford once wrote, "The living organism demands a life-sustaining environment."[9] According to the ethic orienting Earthship dwelling, human beings can provide for their own needs through the resources of their own dwelling in ways that do not impinge upon the needs of other human beings, other species, or the natural world. For those involved with Earthships, every human has the right to sustainable dwelling that provides for needs and comforts without dependency on larger economic and technological structures, and social justice means allowing and providing the means for individuals to build and exist in their dwelling without dependency and isolation, where people take responsibility to live within the limits of the natural world and where self-sufficiency is viewed as social justice.

Earthship Dwelling

Earthships are made from tires filled with rammed earth (stacked tires filled with dirt provide the structural stability of the building as well as thermal mass), aluminum cans, glass bottles, packed mud, and earthen plaster.[10] They utilize passive solar design and earth berm construction for natural heating and cooling; they collect and filter their own water and treat their own waste; power is provided by small renewable energy systems and battery storage.[11] Architect Michael Reynolds coined the term "Earthship biotecture" to describe the fundamental principles of Earthships construction, highlighting their integration with natural, biological systems (in contrast to mainstream architecture).

Water is collected on site using a rainwater collection system. It is first filtered and used in sinks and showers. It is then filtered again using internal plant beds capable of producing food for home consumption, and then used in toilets. It is filtered a final time using an external plant bed. The internal greenhouse is built along the south-facing side of the building[12] and glass panels are angled to maximize sunlight in the winter and optimize sunlight while minimizing heat in the summer. The only cost associated with living in a typical Earthship, after its construction, is propane for cooking (see figure 6.1).

Earthship design (and the foundational principles behind the design) is certainly different than a typical modern dwelling. Even from the outside, Earthships don't look like a conventional home, with walls made of trash—tires and bottles—packed with dirt and covered with mud. By reusing the waste materials of society (Mike Reynolds calls tires an "abundant natural resource"), Earthship design articulates a different understanding of the relationships among human, nature, and technology. Earthship design presents a "passionate argument against the concept of building shells that are almost wholly uninhabitable without services being piped in" and "a very plausible alternative to the slavish reliance on centralized solutions that characterizes our present housing."[13] The self-sufficiency of Earthships is part of their contribution to resisting modern forms of power that operate by shaping us as particular kinds of consumers, dwellers, and spenders.[14]

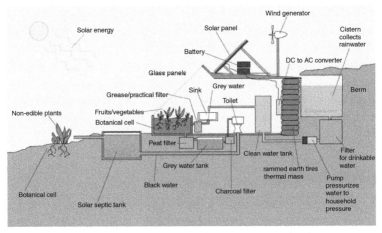

FIG. 6.1 Earthship Design. Image by Earthship Biotecture, used with permission.

Earthship Biotecture: Company, Subculture, and Social Movement

Earthship Biotecture is located in Taos, New Mexico, and there are three neighborhoods of exclusively Earthship homes located outside Taos. Not easy labeled or categorized, the founder of Earthship Biotecture can perhaps best be described as a charismatic leader;[15] he operates less as a typical businessman and more as a "wizard"[16] who is deeply emotionally driven and speaks to the emotional concerns and motivations of individuals to educate and inspire.[17] As such, Mike does not run a typical business. Although Earthship Biotecture as a company does charge fees for architectural designs, construction plans, and the building of houses as would any other architectural firm, it also operates as a large educational facility with a visitor center, internship and academy programs, and volunteer opportunities. Earthship Biotecture engages in public speaking events around the world and runs a nonprofit organization for global humanitarian building projects. During my own participation in the internship program, I stayed in internship housing at the Greater World Community (see figures 6.2 and 6.3) and got hands-on experience with the types of work done by the company and community.

Earthship Biotecture is more than just a company. It is also a loose subculture of individuals who live in Earthships and/or work for or volunteer with Earthship Biotecture. Although the company claims that there is at least one Earthship in every American state and many throughout the world, Taos is certainly the hub of Earthship culture. The first piece of land Mike purchased still serves as a largely experimental site; some of the oldest original designs still stand there. Two other pieces of land are organized as Mike originally

FIG. 6.2 The Hive, Internship Housing at the Greater World Community. Described as a 'phantasmagorical palace' by my advisor (Dr. Mustafa Emirbayer), this is where interns live during their stay at the Greater World Community. Photo by author.

FIG. 6.3 The Hive, Internship Housing at the Greater World Community. Photo by author.

FIG. 6.4 The Greater World Community. A view of the community from the visitor center; the community itself is private. Photo by author.

intended them: homeowners purchase only a piece of land big enough for the home they are going to build and the rest of the land remains communally owned space.[18] A third, named the Greater World Community and located in the high desert mesa just outside Taos, was forced by county administration to become an official subdivision with legally designated plots of land. The visitor center and internship program are based at the Greater World Community (see figure 6.4).

Some of Mike's employees have been working with him for fifteen years or more, and are deeply connected to the principles and goals of Earthship construction. Some people come to participate in the internship program or academy and then stay on as employees, volunteers, or as the internship coordinator lovingly calls them, "lingerers."[19] All the people I met were inspired; they were motivated to participate, to stick around, and to work physically hard in pursuit of learning about and constructing Earthships. Employees and interns have a unique language and culture based on the foundational ideas behind Earthship design and the hard physical work involved in building Earthships (see figure 6.5). The subculture of Earthships is shared with hundreds of people every year through the internship and academy programs, volunteer participation with building construction across the globe, as well as Mike's public speaking engagements.

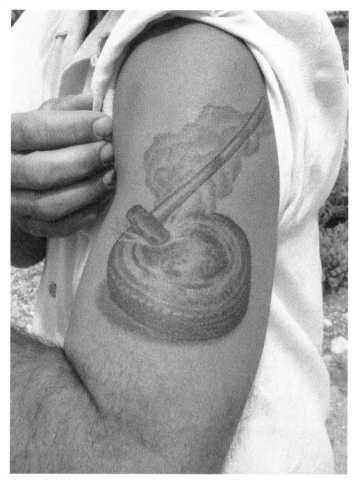

FIG. 6.5 The Earthship Subculture: Shared Symbols. An employee's tattoo of a tire being pounded. The tattoo even includes the size written on the tire, which the employee explained was his favorite size tire to use for home construction. Photo by author.

The Appropriate Technology Movement and Its Continuing Relevance

Critique of technological structures and promotion of alternative technological practices have a long history. While the hippies of the 1960s counterculture were moving back to the land and trying to stick it to the man (including Mike Reynolds himself, who I was told moved to Taos as a draft dodger), scientists and other academics were thinking about new ways to understand the world and the role of humans and technology in it.[20] The fusion of philosophy and practice created a movement for what is broadly called appropriate technology.[21]

The appropriate technology movement considered the environmental damages and economic inequities caused by the sociotechnical networks of modern technology to be, simply put, inappropriate. Appropriate technologies, advocates argue, are aligned rather than in conflict with natural resources and rhythms; are participatory and small-scale rather than alienating and monstrous; make practical sense in terms of efficiency; and concur with social values regarding relationships among humans (related to justice and freedom) and between humans and the natural world. Thus,

> "Appropriate technology" is a term that represents a particular view of society and technology. It suggests that technology is neither neutral nor does it evolve along a single path . . . that the only wise technologies are those which seek to accommodate themselves to the biological environment within which they are used. It assumes that the purpose of economically productive activity is to produce what is determined by need, in an enjoyable, creative process; not what is determined by endless greed, in an alienating, repetitive production process. . . . And it presumes that the only development that makes sense is the development of the people and their skills, by the people and for the people.[22]

The appropriate technology movement explicitly acknowledged that the sociotechnical networks that provide for human needs and comforts reflect and reinforce particular values, shaping social structures and relationships. Among the people, ideas, and things associated with the appropriate technology movement, "the two themes that pervade are (1) decentralization and localism and (2) voluntary compliance with humanistic, proenvironmental behavior."[23]

Earthships as Appropriate Technology

The appropriate technology movement promoted ephemeralization (radical efficiency) and do-it-yourself, localized participation[24] in decentralized, distributed technologies and soft energy paths.[25] These ideas continue to inspire alternative building practices like those found in an Earthship. The similarities are visually apparent when we compare the Dymaxion House (Buckminster Fuller's design for affordable, adaptable, and nonresource intensive housing) and a modern Earthship dwelling (see figure 6.6). The similarities in these structures—the passive solar design to naturally provide light, heating, and cooling as well as the use of cheap, readily available, and easy to work with materials—reflect the similar normative values that drove their design.

Sandy has a degree in architecture; she moved to Taos to work for Earthship Biotecture, and she designed and built her own Earthship where she now lives with her husband and two young children. She acknowledged that people have been using solar technology and water catchment systems

FIG. 6.6 A Modern Earthship. In Taos, NM. Photo by author.

since at least the 1970s, but she said that Earthships are unique because Mike Reynolds "put all the pieces together" in the design of Earthships. She told me, "There are multiple reasons for every architectural move. So it's built out of tires, which are recycled, but those are your structure, they're also your heating and cooling system. There's not a lot of façade or superficial—I mean, you go into most buildings and you have no idea what kind of heating or cooling system; you just know that it's hot or cold." The overlapping efficiencies of ephemeralization are clear in her statements about having multiple reasons for every architectural move, and efficiently using waste materials (tires) to create an inexpensive, structurally sound, naturally heated and cooled home.

Shiloh, a graduate student who came to Taos for an internship at Earthship Biotecture, said, "The Earthship lifestyle is acknowledging a lot of different environmental problems and seeking one way to address those things." He told me that the environmental problems related to how humans deal with water, electricity, and waste are often considered in isolation, but Earthships address multiple environmental concerns within a single alternative dwelling system. In contrast, an Earthship homeowner named Jane said that the dominating technological systems and architectural designs are "all built on single premises that aren't thought out, and also because there are always greedy people who want to sell you something." Earthships are designed to be comfortable dwellings without reliance on outside resources like water, electricity, or the money it takes to pay monthly bills for these things in conventional residential life.

The appropriate technology movement and Earthship Biotecture share a common emphasis on do-it-yourself, distributed, and participatory technologies. The appropriate technology movement sought to promote technologies that were "not controlled by dominant institutions . . . and accessible to amateurs,"[26] and advocates like Buckminster Fuller were "touting this necessity in 1969 as a paradigm shift in civilization to benefit everyone."[27] Instead of technologies that cultivate isolation and dependency (such as nuclear power, which necessitates elite access to knowledge and control,[28] and hard energy paths such as modern electricity generation and transmission, water filtration and waste treatment, which are largely out of the control of the people who rely on them[29]), the appropriate technology movement advocated for sociotechnical networks designed to promote participation and empowerment, systems that users construct and control.

Earthships are a complete dwelling system designed based on these principles. Sandy told me, "I think it's really important, the low-tech aspects of it. Even though it's high-tech in some ways, it's designed for people to do it by themselves, for people who've never done any construction. I mean, we'd never built a house before, and . . . we built a huge percentage of the house with hand tools. Anyone can do it, and that appealed to me."

Tasha is from India, where she completed a degree in architecture. She came to participate in an Earthship internship because she was frustrated that she received her degree without doing any actual building. After spending a month working on Earthships, she said, "You don't have to have a background in architecture to do this . . . which actually just makes architecture just pointless. . . . If you can't give people what they want and need, then what's the point of it." Earthships are designed based on the principle that people are capable of and ought to be able to design and build a residential structure that meets their needs and comforts on-site with their own skills and resources. Matthew, one of the intern "lingerers" at Earthship Biotecture, told me, "I think everyone here could agree that the systems should serve to serve us, so that we can all participate in it."

The appropriate technology movement recognized how technologies cultivate and promote particular values, relationships, and practices, and how dominant technologies were degrading both environmental and social systems as well as individual lives. Earthship Biotecture is one contemporary expression of similar values and concerns. Riley is an Australian man who had visited Earthship Biotecture twice for both professional and personal reasons; he's working on a PhD in architecture and he's building an Earthship. He put it this way: "The Earthship's really just a big bundle of technology put together to make more technology, isn't it? It's like a really sensible way of using our technology to achieve our ultimate goal, which is to live and be able to continue living."

Both Then and Now: Why Rethinking (and Reshaping) Systems Matters

Certainly, environmental concern is one reason to move away from dominant, fossil fuel–based technologies,[30] and Earthships offer a means of living with significantly less environmental impact. They recycle materials and use fewer new resources for building, use substantially less water, and harness renewable energy sources. Nancy told me that by living in an Earthship, "You aren't using up precious resources, god knows, you know, there are enough people using up tons of resources. . . . So we aren't using so many precious resources, we aren't using nonrenewable resources for sure. I love the fact that you have your own electricity, we have a wind generator as well, so the sun and the wind are giving us our electricity. How cool is that? I just think that's cooler than shit."

Most of the people I met in the Earthship community (but, perhaps surprisingly, not all) were concerned about the environmental degradation caused by the dominating technological systems, which at least partly inspired them to be involved with Earthships.[31] Audrey, a grandmother from Alabama who came to participate in an internship during a summer off from teaching, told me, "When I say I'm here because I want my grandson to have a place a live, it is the environment that I'm thinking of, we're running out of oil, obviously, and water, and it's about that." Jen, a young employee at Earthship Biotecture, said, "Centralization is huge in the environmental aspect too, I mean waste treatment plants and taking water from places far away and stuff like that, this is about just reducing that kind of need."

Yet as an intern named Kevin put it, "The environment is sort of a given now. The environment is definitely something we have to be conscious about, and these houses are certainly good in terms of that, but I think we're all here for that as well as for other reasons." That July evening, with the sunset pouring in the expansive glass windows that comprise the south side of internship housing in The Hive, surrounded by the greenery of the internal plant bed, the other interns all agreed. Another reason to rethink and reshape our technological systems, expressed by both the appropriate technology movement and the Earthship community, is the danger of dependency and the ways modern technological systems require it. Dependency is dangerous because, as an intern named Nate said, "When centralization goes away, you're left with a big gaping hole that no one knows how to fill."

The winter prior to my first visit to Taos, northern New Mexico experienced a natural gas outage, and most grid-tied homes had no means of heating or cooking for days. Every single Earthship homeowner I met that summer told me the story of the outage. Some talked about how lucky they had felt then; it was days before they even heard about the outage because their homes are not dependent on this system. Others talked about what a tragedy it is that this poor, isolated county in New Mexico was dependent on a system controlled so far away as Texas for such basic human needs. As Nancy told me, "Here, I'm not relying on a power company. Power outage—oh really, didn't know, I don't care. . . . The fact that I'm not relying on these giant grids of electricity, or water, or anything, and you know, come on, there's so many people on this planet, everything's going to get scarce. Did you hear about the natural gas outage last winter? God, it's terrible, just awful. And that part of it is pretty amazing, that you aren't dependent on any gigantic system. You just aren't." According to Jane, another homeowner, "To be in a building that's completely dependent on external systems, it's just a bad way to be."

For the Earthship homeowners as well as the Earthship Biotecture employees, interns and volunteers I met, large, centralized technological systems are dangerous because they make us dependent upon them; without experts to construct and control these systems, most Americans would be without light, heat, or water. Further, these systems make us dependent upon a money economy because we are required to pay a bill each and every month our entire adult lives just to access the things upon which our lives depend. As an intern named Crysta told me, "We're interested in having communities that aren't based on a central industrial structure, you know, or some sort of business that's controlling us all." Rethinking and reshaping the sociotechnical networks that provide for the needs and comforts of human dwelling through the construction of Earthships is about minimizing negative environmental impacts, but also about much more than that. Minimizing dependency on the same large systems that cause so much environmental degradation also means minimizing arguably dangerous economic and social dependencies.

An intern named Chris told me, "I think that here, people recognize that sustainability means really asking what you actually need and reassessing that." Similarly, a homeowner named Jane said, "It has to do with, how do you interact with your environment, and what is the most minimal thing you can do and still maintain the things you want." For others, Earthships were about choosing to live sustainably as a means of improving the quality of one's life. Brianna, a summer intern, said, "We've been convinced in American culture that what we have makes our quality of life better, if it's more luxurious then our quality of life is better, but it's not." As Earthship homeowner Sandy told me, "To me, sustainability is—I don't know—I think I'm more interested in, I think for me it's been about a higher quality of life with less impact." Resisting the values embedded in the sociotechnical networks that currently dominate residential life, Earthships offer an alternative means of living an improved quality of life. One homeowner named Maggie put it this way: "This is about alternatives to what we're forced into, really. I mean, we have to have a car to get to work so that we can pay for our car. How can we get out of that cycle?"

Earthship Dwelling, Earthship Being:
Motivations and Experiences

The people I met at Earthship Biotecture are a diverse group in many ways. Most were concerned about environmental issues, but not everyone was. Employees and interns talked about shopping at Walmart and eating meat as illustrations that they aren't environmental purists. Most of the homeowners, employees, and interns were consciously interested in moving away from dependency on technological systems, but they did not share a particular political orientation that could be classified based on America's dichotomous political system. In fact, many talked about how Earthships can appeal to both liberal and conservative individuals in America, for different reasons. Yet while they did not share an environmental ethic or political viewpoint, there are common motivations and experiences that tie this diverse group of people together.

First, all the people I met at Earthship Biotecture—interns, employees, and seminar participants as well as homeowners—came to Taos specifically for Earthships. Some came from as close as Albuquerque, others from as far away as India, but all traveled explicitly to learn about Earthships, live in an Earthship, or work for Earthship Biotecture. Second, many (but certainly not all) of the people I met were already living what may be classified as an alternative lifestyle prior to coming to Taos. Stephanie showed up in Taos in her VW bus, which she lived in at the time. Nancy and her partner lived in an RV when they arrived at the Greater World Community. Jane worked outside the United States for twenty years before retiring, so she felt like, "I've always been separate from this culture. I've worked in places where you can't buy coffee, instead of having four aisles of potato chips. I also don't have a TV, so I don't have to be exposed to commercials." One intern hitchhiked to Taos to participate in the internship program, and another intern lived in a converted school bus back home. I met a man who rode his bicycle from Canada to Taos to participate in the academy program. Many of the people I met were already living, or were at least willing to think, "outside the box" so to speak.

Even among those who weren't hitchhikers or nomads and who didn't already identify as outside or separate from mainstream culture, everyone shared a common belief in the possibility of change. This may seem simple, but it is really quite significant when we think about how power works—through shaping our actions and our thoughts about what's considered normal action—and about how we can resist it.[32] Believing change is possible is an essential first step in making change of any kind.

Sandy first came to Taos to work for Earthship Biotecture. When she arrived, she was asked if she would manage a rental home in the REACH community in return for a free place to live. Describing this experience, she told me, "When you're really warm with no utilities, no fossil fuels, and you're totally with the solar power, and all the systems totally function beautifully . . . for me it was really, it kind of fundamentally changed my view and how I understand what buildings can be." Later in our conversation she told me, "No matter what the individual motivation, people here share a desire to see system change and think that Earthships contribute to that system change." As an architect, Sandy has always been interested in alternatives to conventional building construction and she actively sought out more sensible building concepts that were better for both the planet and the people on it. As an individual, she believes something different is possible, and she went looking for it at Earthship Biotecture. Similarly, Margaret told me that the only commonality among the people who own and live in Earthships is that "they are all people looking for something different."

A building contractor named Jacob who came from British Columbia to participate in the internship program told me, "It's like the boiling water: you put a frog in cold water and boil it, the frog stays there and dies while it heats up. I'd say that's what the masses are pretty much doing. It seems to me that the people here are the type of people who are interested in jumping out of that boiling pot, you know." During that same conversation, Mitch—a professional actor—told the group, "I think the biggest thing for me is just the fact that it exists, that it's just an example of a better way to live. What it is of itself is interesting. While I would love to build one, for me it's more a metaphor of the way that we're moving. I want to be around something so inspirational." To this, an intern named Matthew added, "Yeah, it's like Walden, Walden Pond,[33] just going out and doing something completely different that's positive, reclaiming and rethinking our lives and reclaiming our own freedom." Later in our conversation, Riley similarly said, "I think we're all rebels just by default. By choosing this, we are rebelling against current, you know, economic paradigms, manufacturing paradigms, we're voting for solar power, and that is rebelling."

Believing that it is possible to jump out of the boiling pot and choose an alternative way to dwell may require rejection of dominant technologies or paradigms, but, not necessarily. This group (nine interns and two employees participated in this conversation) did not all agree that they were rebels, and some were quite uncomfortable with this classification. Yet they did all agree that they were looking for change, or as Crysta put it, "looking for hope." However, the people I met acknowledged that Earthships are not the only answer for pursuing system change. Some said that while an Earthship could be designed to function in any climate, it would be difficult to utilize the design in urban or densely populated areas (although many talked about the possibility of having communal water collection or waste treatment in these situations, still on a smaller and more personalized scale than the current systems). Some talked about the other types of sustainable building (such as cordwood or straw-bale) that could be used as an alternative to rammed earth tires. Several said that passive solar design is a good idea for any building.[34] Many said that Earthships are an experiment without definitive findings; the designs are constantly evolving and improving. Earthships are "as much as anything a provocation."[35] Believing that change is possible does not require having the one and only answer for what that change should be. Instead, and more interestingly, it involves acknowledging the multitude of possibilities and the importance of experimentation that could allow humans to live more sustainably for the sake of themselves, the planet, and one another.

A final thread that brings together this diverse group of people is the common experiences related to bodily learning and bodily practice. Earthships are a very bodily experience—human bodies build them and living in one requires different bodily actions than a conventional home, so they reshape bodily practice through both construction and daily dwelling. From pounding dirt into tires to maintaining the energy, water, and waste systems, Earthships involve participatory bodily practice, requiring the dweller to interact with their dwelling space in a radically new way, demonstrating the relationships among technologies of dwelling, dwelling practice, and bodily-codified conceptions of normal practice.

Bodily Dwelling, Bodily Knowing

Our individual bodies, body imageries, and bodily practices are socially embedded, constructed, and normalized.[36] Michel Foucault focused on how strategies of discipline from the structures of prisons to educational techniques to medical technologies shape bodily practice and the very ways that humans perform and define normal action.[37] Marcel Mauss examined how bodily techniques like the simple acts of walking and talking are learned through and shaped by the societies in which we live,[38] concluding that, "What emerges very clearly . . . is the fact that we are everywhere faced with physio-psycho-sociological assemblages of series of action."[39] The sociotechnical networks that support human life also shape bodily practice and how those practices relate to conceptions of normal ways of living.[40]

When electricity comes from some unknown and mysterious system, it is arguably harder to tangibly reshape how we use it. In contrast, when you know your electricity comes from the sun, you can look outside to gauge how much power is available to you that day, which reshapes how you use it. Environmentalists concerned about the future availability of clean water may espouse the value of short showers. Some Earthships are built with a valve in the shower so that you can turn off the water but have it come out at your preferred temperature when you turn it on again; the material world of a shower head reshapes the possibility of new bodily practices. With new technologies come new practices and new conceptions of normal practice.

Earthships are very bodily technological systems; they require the active participation of the dweller. They are built using active human labor. Dwellers must be conscious of their water and energy usage, paying attention to the availability of resources and changing their behavior accordingly. Earthships require bodily participation through the maintenance of the systems; the batteries used for energy storage need water added to them approximately once a month, the filter for the grey water system must be cleaned; the windows and skylights used to maintain a comfortable temperature must be opened and closed. The experiences of Earthship participants demonstrate how material systems shape bodily practices and how bodily practices reshape codified understandings of normal.[41]

Earthships were intentionally designed so that individuals without building experience could learn about and build their own; this is part of the do-it-yourself culture. Sandy told me, "The people who are motivated to build their own, work for the company, be interns or volunteer for builds have more in common in terms of motivation than those who just buy an Earthship but don't participate in the build. There's something about the hard work, the sweat equity, that is unique about Earthships." She referenced a quote from a current employee and friend used in one of Earthship Biotecture's recent books: "You know how the house works because you put your body into it; you don't pound tires with your mind."[42] For "owner-builders" (people who live in an Earthship that they built), the "sweat equity" (as Sandy called it) is an important part of forming a knowledge of and relationship with your dwelling.

People who voluntarily participate in internships or Earthship builds around the world are also involved in bodily learning and active bodily participation. By actual doing (building), they are learning about the possibilities for construction as well as their own capabilities. Phillip, a charismatic and talkative intern who lives in Sweden and earns a living as a sustainable business leadership consultant, said, "It's quite amazing, this hands-on experience. There's intelligence in your body. Also, I think all of us have done things that we didn't think we could do. And there's just that explosion when you expand your idea of what you're capable of and what you can achieve. I mean, I for one, have never done anything more than put up a bookcase, so this is way over my competence levels, but I'm learning and you see results."

Brianna, an eager young woman who had graduated from an East Coast college just before her summer internship, told me that Mike Reynolds told her, "I don't want you to help. I want you to help yourself." She took this to mean that the internship was about learning to do things for yourself, things like building but also basic skills like cutting firewood or maintaining solar battery systems, so that you could utilize these skills later in your own life. Similarly, the entire group of summer interns told me about being inspired by a talk that Mike had given them about "being the cancer"—that they had come here to learn by doing so that they could spread that knowledge through active doing elsewhere. This is one of the key principles for the Earthship community: active, bodily engagement and learning to do things you did not think you could is empowering, and that empowerment is important.[43]

The House That Hugs You: Earthships as Organic Systems

One of the commonalities among the Earthship homeowners I met was how they talked about living in their homes. They used words like hug, embrace, and hold to describe their relationship with their dwelling space. Sandy told me, "My home is so quiet, and the mass of the building, it really kind of embraces you. . . . When we go to a hotel, we wake up a lot because it's hot and then cold and hot and then cold and then you hear the mechanical system at night. Now I'm really much more in tune with those things, and I love that the house doesn't need any of that."

There are several reasons these homeowners talk about their homes this way. First, the earthen walls and rounded corners of Earthships make the homes look and feel differently than a conventional dwelling (see the discussion of aesthetics below). Second, the process of building your own home creates a different kind of bodily relationship with your space ("sweat equity," see above). However, the most common explanation I got for why people experience Earthships in this way was—in the words of Noel, a middle-aged British woman who lives in Maine and came to Taos to participate in the December internship program—"because it's a living system."

An Earthship is like a living system in many ways. The systems are there to take care of the dweller, which in itself creates a bodily relationship with your home, because it takes care of you in a way that is unusual for modern residential dwellings. Sarah, who built her own Earthship but has since left the Greater World Community, told me, "Just to run on solar energy, it was a trip, it was a high. Taos would have a blackout and you wouldn't even know it." A "high" is a bodily sensation, and Sarah's description of experiencing a high while living off the grid in an Earthship demonstrates how a dwelling that takes care of you can create a sense of bodily engagement and bodily excitement.

Yet while Earthships are designed to take care of the dweller, the dweller must also take care of the home. As Maggie, a homeowner, said, "It's a living

FIG. 6.7 Earthship Aesthetics. An Earthship front entrance. Photo by author.

dynamic system and must be lived in. Take the example of the solar batteries.
You can't just walk away from the house for most of the year and expect it to
function like the living system it is." There is a bodily relationship that forms
between dwelling and dweller when they exist in this sort of dynamic, recipro-
cal system.

One of the things that many people love about Earthships is their charismatic aesthetic beauty. Full of rounded corners and a unique, colorful play-
fulness from the bottle walls, they are if nothing else an enjoyable visual
experience (see figures 6.7 and 6.8). All the homeowners I met talked about
how much they prefer the aesthetic of an Earthship to that of a conventional
home. The finished earthen walls and rounded corners make the homes look
and feel differently than a conventional dwelling.

FIG. 6.8 Earthship Aesthetics. An Earthship front entrance. Photo by author.

Some of the interns and employees more explicitly recognized that the aesthetics of an Earthship are also related to our bodily relationship to dwelling space. Mitch, who participated in the December internship, told me, "I recently had this talk with this eco-psychologist, for a play I'm currently writing. And they said, 'You know, the human species has been on the planet for like this long [with arms totally outstretched]. We've only been inside for that long [with just the thumb and index finger of one hand held up].' So we're— our DNA is built to be outside—we're built to function outside. So, it's almost as if the architecture has stumbled upon what the body's used to. To see curved walls, and no cubicles. . . . I think, I just find that fascinating." To this, a young and soft-spoken woman named Jen who had voluntarily participated in an Earthship build in British Columbia and then moved to Taos to work for the company, added, "There are no right angles in the human body. Why would you live in a square?"

The visual look of an Earthship is not an arbitrary aesthetic for whimsical emotive appeal: the design is based on the principles of using cheap and plentifully available materials (tires and dirt and glass bottles) and creating a comfortable dwelling space as efficiently as possible. The aesthetic of an Earthship interacts with the principled design to create a space that looks influenced by nature because it actually is. Further, the design principles of an Earthship connect the bodily actions of the dweller with the rhythms and resources of the natural world, reshaping the relationship between human dwellers and their natural environment. Earthships are "an architectural alternative to human alienation from the natural world which is embodied in the majority of present housing; an alienation which in itself is arguably the cause of much of the environmental damage being wreaked on the planet."[44]

Earthships require that the humans dwelling within them are attentive to the natural world. With the passive solar design, it's hard to ignore the daily rhythms of the sun. With plants inside for the gray water treatment system, nature is also literally inside your home. With a view of plants inside and sky outside on the south side of the home, the outside and inside blend together much more than in a conventional house.[45]

Further, Earthships require that a dweller live with the rhythms of nature. If it's been cloudy for a week, you may not be able to vacuum without sacrificing your ability to watch a movie later. If rain has been scarce, you may need to be extra attentive to your water usage. As an intern named Matthew told me, living in an Earthship "puts you back in touch with your surroundings and you integrate yourself to that ecosystem that's all around you, instead of just taking from it." For the Earthship homeowners I met, this connection to nature was not a sacrifice, but something to celebrate. As Jane told me, "The needs of humans don't exempt the needs of the earth, there's a balance, and here I can interact with my environment with minimal impact." And as Sandy said,

> When it's raining, everybody's excited, because we're getting water, you know. That's fun. And if it's been cloudy for three days and we get sun, it changes your feelings. And you're very connected to time and seasons and the light of the day, you know, in the summer we have a lot of light coming into our house and we know it's really long days, and in the winter we have less light. . . . By living here I've become way more aware of the climate. When it's cloudy for four days we're really aware because it affects our power, it affects our solar water heater. . . . Whereas, as a contrast, you could not even know what the temperature was outside or whether it was raining or snowing in a lot of places.

Expressing a similar idea, Margaret and Maggie told me that they liked living in an Earthship because it allowed them "to live as close to the earth as we can, and to support what we want the future to be, one where we're connected to and not constantly in conflict with the earth."

Nancy touched on several of these ways Earthships reshape the bodily relationship between dwelling and dweller when she said,

> I don't even really know how to describe it. Just walk in, and just the fact that there are plants, the planter and all the greenery and it's almost just like being outside. And then this sort of nonstop view of the sky, which is killer, which is almost like being outside. And then, I think the fact that you walk in and it just

kind of, it just kind of felt like you were being held, it's not like walking into a sterile house or an all done up and decorated house that's just a house. There's something for me that feels very, I don't even know what the word is. Plus the fact that you walk in and, knowing about Earthships, you know that you have your own power, you have your own water, you don't rely on anything outside the house. It's kind of like the house is gonna take care of you, and I think that was the sense that was like, this is it. . . . It's about living lightly. And living in tune instead of, you know the old paradigm of man against nature, you know that doesn't work. So this is man is living with nature, and how do you live with nature lightly . . . that's what I love about it.

The sociotechnical networks used in an Earthship reshape the daily actions of the dweller, and reshaping practice simultaneously reshapes thought. As Nancy told me, "I think that people who live in those normal systems don't think about them. So yeah, I think about them differently, because I think about them. I think about the interconnectedness. I'm conscious." An intern named Matthew said, "That's what Earthships forces you to do, is like, reexamine your priorities." Tasha, another intern, told me, "We haven't used a microwave for a month now, and you don't need it. Living differently, you see you don't need so many conveniences that we have. We don't need all that stuff." Changed practices and habits, changed thinking, and changed conceptions of normal action and thought simultaneously emerge through the use of different technologies: technologies shape bodily practices, which shape knowledge, knowing, and normal.

The Radicalism of Earthships: Changing Bodies, Changing Thoughts

Earthships are highly efficient living systems, an example of ephemeralization, meeting the needs and treating the waste of human dwellers in a largely closed-loop system. They are arguably significantly more radical than other cradle-to-cradle or environmentally conscious design schemes.[46] If the carpet in your home is made of recycled fibers and recycled again when you're ready to replace it, it does not change the way you walk across your carpeted floor. Yet when the water you use to drink, cook, and bathe comes from the sky, lands on your roof and is collected in a cistern, you not only need efficient systems to filter and reuse this water, but you also have to change your bodily behavior in a drought so that you do not run out of water. As Shiloh said, "When you're living in your own water and energy ecosystem, you may or may not run out of it, and I don't think it's difficult from there to make the larger leap to understand the bigger picture. And that's the problem, is that not enough people experience something that makes them think about doing something different." Earthship design requires that dwellers actively participate in the maintenance of their own comfort. This active participation reshapes both habit and thought in very profound ways, reconnecting humans to the rhythms and cycles of the natural world.

Nancy told me that she and her partner came to Taos to build "an alternative kind of home, and actually alternative kind of life, really." Riley, an Australian working on a PhD in architecture, told me, "I reckon it's kinda like, if you can live in an Earthship, then the Earthship has actually somehow redesigned your mind, the way your mind works and when you think about the limits that an Earthship imposes on you. . . . Doesn't it reconfigure how you think about, I mean think about getting in touch with nature, observing what's going on, you know I can't turn on the washing machine because the sun hasn't been out in three days. All that." When an intern named Matthew said, "When you start building your own home and building homes with garbage, it forces your mind to reexamine convention," an employee named Jess added, "but that opens up so many possibilities too."

Earthship homeowners said that to live comfortably in an Earthship, dwellers have to be willing to change both actions and minds. Stephanie told me, "There is a lifestyle change that I would think . . . some people would need to know that you can't just leave your water running, or your lights on, or two computers on and laptops running all the time. . . . I think for average people, it's something they may need to think about more often." Nancy said, "Why not everybody? I think that's habit, and I think that's also a cultural norm. We think we're so great in the U.S. that we can have everything instantly, you know, you don't have to save money just put it on a credit card. It's all a piece of this giant idea that you can have it, you can have it now, and you don't have to think about it. You know, I don't feel in any way that I'm deprived. . . . But, I think that's different than our culture."

Margaret and Maggie only have DC power in their home (they don't have an inverter to run conventional AC appliances), but they have a special DC TV and DVD player and both like to watch television. As Maggie told me, "Before you watch TV on a cloudy day, check the power levels. It's that simple. But some people want to do everything at the same time." Maggie also acknowledged that she, her mom, and her sister, who all live in the Greater World Community, were all willing and able to be "purposefully informed, while most people are too busy drowning their realities." Thus, Earthships require that people actively participate in rethinking their priorities and live, act, and be different in order to exist within the means of their own resources; people who live in Earthships talked about this reshaping of practice and thought as a benefit of living in this kind of alternative dwelling, not as a sacrifice. As Nancy told me,

I want to live like that. I don't want to just do it and not think about it. So I think about it, and I think that's what makes me feel *I'm alive*, is that I'm conscious of my choices, conscious of what I'm doing, and frankly I think too many people live subconsciously. I'm using the term kind of teasingly. It's just that they just don't have to be conscious. . . . Do you have to be conscious if you're living in a big house in the city? You don't. I suppose you're conscious when you're flippin bill comes, your utility bills, but hey . . . if they're making enough money they don't worry about that either. So, it's just too easy to be unconscious. And I don't choose to be unconscious. . . . Part of me thinks, you know, so what are you doing in your life that it's so important that you not spend any time thinking about what you're doing? I don't get that, I really don't. . . . Unconsciousness is too big a price to pay for comfort.

This captures precisely what the dominant technological systems that currently support residential dwelling cultivate in modern dwellers—isolation and dependence, alienation and unconsciousness—and the radicalism of the alternative technologies, designs, and practices embodied by Earthship dwelling.

Place and Policies Shaping Earthships as Alternative Dwelling Practice

The New Mexico license plate advertises the state as the "land of enchantment" and it is indeed enchanting.[47] A high desert climate with amazing mountain views, Taos is a unique place that attracts unique individuals. Artists and free spirits of all kinds flock here, and the longstanding culture of self-sufficiency and connection to the natural environment suits their independent souls. Nancy, a passionate and talkative older woman who came to Taos to build her own Earthship fifteen years ago, told me that the Pueblo in Taos is the longest continually occupied settlement on the continent; she felt that this long-term, living relationship between human beings and this harsh natural climate affected the spirit of the area. She also said,

Taos is very diverse, and very accepting. So some little mad man like Mike out here on his dirt bike and zooming around and crazy looking and building with garbage, he was accepted. . . . And I think there is a real connection to the environment here. Maybe because we have giant skies, you know what I mean. And northern New Mexico, it's very dramatic. High desert is a dramatic place. You know, it can be -12° at night and it can be then in the morning when it starts to warm up there can be a 40° degree difference. That's huge. So it's very dramatic, adjustments are huge, and I think when you learn to live in a place that's so dramatic like that it kind of colors everything.

Others say that Taos has its own unique hum.[48] The cultural hum is certainly one that values independence and connection to the natural world, likely shaped at least in part by the practical necessities of the harsh desert climate and rural remoteness as well as by the unique beauty of the landscape, the spiritual connections to indigenous traditions, and the free thinking artistic propensities of many of the people who live there. The unique landscape and culture in Taos has long attracted free thinkers.

Further, the local architecture of residential dwelling and other factors related to resource use also correspond to the principles of Earthship living. Many homes in the region are, and even more used to be, adobe structures. Adobe utilizes the same principles of thermal mass that Earthships rely on, providing a natural means of heating and cooling, as well as providing the same external aesthetic. Rainwater catchment is legal in New Mexico; some states do not allow homeowners to collect the rainwater that falls on their roof, which would clearly make relying upon rain as your sole source of water impossible. As Sandy told me,

> I think it's hard, just generally, for many people to understand the Southwest. Like you come to Taos, and regardless if you're looking at an Earthship, you see that people don't have yards, they don't have green grass, they don't have gardens. You know this traditional sense, even the architecture looks different and I think it's very hard for most people to take their concept of house to that next level. . . . But, I think you look at the Native Americans that have been here, and a lot of the principles in this building, it's pretty much what's been happening on the Pueblo for the past 900 years. So there's an acceptance of quote unquote alternative architecture, and New Mexico building codes are pretty progressive and they've adopted a lot of alternative architecture and building materials as accepted materials. I think that has to do with the history, with the Pueblo and the kind of frontier living. I think the extreme climate affects it, the climate is so harsh, it's so hot and so cold and so extreme you just can't get away with certain types of architecture here. And I think it's just a progressive, it's kind of a little bit of a free-for-all, end of the empire, kind of like unregulated area that has allowed a lot of experimentation.

All this is to say: spatial context matters.[49] There are factors unique to Taos, as a place and a culture, that arguably shape the culture, identity, and success of Earthship Biotecture. There is what can be called an elective affinity between the local geography and culture on the one hand and, on the other, the acceptance of this alternative form of dwelling and all that it entails.[50] As Nancy said, "It makes perfect sense that Mike would come to a place where there were people saying 'no, you know, we don't buy into that stuff.' And the community is still like that. . . . I think that was sort of a perfect place for Mike. In Cincinnati, maybe not so much."

Yet although the culture and climate of Taos is ideal for an experiment in dwelling like Earthships, there are certainly challenges in Taos and elsewhere. In Taos County, Mike was forced to transform the Greater World Community into a legal subdivision with plotted land parcels and increased red tape that makes it practically impossible to build your own home as was originally intended (with small home plots purchased inexpensively and the community collectively owning the vast majority of the land), drastically limiting the financial accessibility of owning an Earthship. In addition to restrictions on rainwater collection, building codes also often prevent grey water and black water treatment on site.[51] Coding and zoning issues that privilege standardized (and normalized) structures and systems present a challenge to Earthship construction in many parts of the world. Arguably, this is because both locally and globally standardized coding and permitting processes are systematically biased in favor of the entrenched, large-scale, money-driven systems that currently operate to meet the needs and comforts of modern dwellers. Mike Reynolds recognizes this as a challenge faced across the world, but to varying degrees, based on localized building codes; building code and zoning issues are often not federal or state issues but county and city issues, yet counties and cities all over the world have codes that privilege certain kinds of buildings at the expense of others.

The Earthship Ethic: Self-Sufficiency as Social Justice

Anthony, an Earthship employee, described why Earthships appeal to him in the following way: "For me, it's about the utilitarian existence. It's about the practical aspect of it. It is taking care of you, which is along my lines of thinking. I'm responsible for my own life, you know if everyone had their own water source, I mean, duh, then what am I doing? I'm not paying water, I'm not paying electricity, I'm not paying heat. And I think that's what's so great, of course it's good for the environment, but there are other aspects too, and it's great that those things aren't opposing forces." The kind of self-sufficiency Anthony is describing involves a profound sovereignty at the residential scale. By building your own home, you are empowered to do physical things you did not believe you were capable of. Detaching yourself from the economic systems of utility bills and mortgage payments provides unprecedented economic freedom. As Sandy told me, "Luckily, by living in Taos, we don't have the opportunity to consume the way people in the rest of the United States— we don't have shopping malls. . . . We don't have the things that I think affect the way people live in that way. . . . And I think living this way is definitely different that the way most people in the United States live. You know, we don't pay any utility bills. Same with a mortgage, we don't have a mortgage. That affords a certain freedom." Earthships are "a passport to freedom."[52] This sense of freedom was a primary motivator for those living in, working on, or inspired by Earthships. As an intern named Crysta aptly put it, "Nobody wants to pay an electric bill."

Living in an Earthship not only changes the way humans dwell; it also opens up the potential for additional changes to what else we do with our time, bodies and lives. Tasha put it this way: "If you have a home which is, I mean the whole cycle of life is based on, like, you're working for money and you're earning money to get what you want, keep yourself alive and get food in the house and so on, but if your house is taking care of all those things, then you can do so much more than just run behind money; you have so much more time and opportunity to pursue other things." An intern named Nate summarized it by saying, "You are no longer a slave to that centralization. Centralization creates a false slavery concept. And once that centralization is gone, there's a freedom. There's a freedom from, I don't have to work, no one's going to zap you or drain you of what you value, you're working for yourself and doing your thing, you're catching your water for yourself and making sure you have enough energy for yourself. It's a freedom in a vast sense." To this, Matthew added, "And that's the issue right there that will come from Earthships. . . . It would change the economic structure, if everyone had an Earthship."

Arguably, the orienting ethic underlying the design of Earthships and the actions of Mike's company, Earthship Biotecture, involves an ethical claim of what ought to be. Humans ought to live in environmentally responsible, self-sufficient dwellings that free them from technological and economic dependencies.[53] This is the ethic of self-sufficiency as social justice, a belief that humans have the right to construct their own dwellings that will provide for their needs and comforts without the intervention of building codes, electric utilities, mortgage payments, and monthly bills. Accordingly, each human being has the right to live with the water and energy resources they can harvest in their own dwelling space, without impinging on the rights of other humans to do the same or causing unnecessary environmental degradation.[54] In order to pursue this right, human beings must act responsibly, attentive to their own actions and providing for their own needs. Nancy described this ethic when talking about Mike and his business. She said,

> Mike at his best is in the Andaman Islands, or Haiti, or Japan, putting up some structure that's sustainable, that's gonna catch water so that people can catch water, and it's made out of all the plastic bottles that are just trash and tires that are thrown out. He makes a house, and these are people that desperately need a place to live, and they don't have a grid so they need water and they need their own electricity. And how are they gonna do it? Then Mike shows up and shows them how they can do it. Not only does he show them how they can do it, but they can do it, and then he can go away and they can do more of those.

That's Mike at his best. No permits, no red tape, no stupid county commission going 'Mmm, well you're doing this with no universal building code, mmm, you don't have an outlet every three inches.' That stuff kills him, it does, 'cause he's creative, but also because you know you tell him that these poor people need a house, and not only a structure, but it should give them a supply of water, and deal with waste and how to process their waste, and electricity because there's no grid to tie into. That's Mike at his best, I love it when Mike does that. . . . But I think that most of that drive of his, that when he sees an opportunity not only to provide—because that's his huge thing, I wonder if he would agree, that it's everyone's right to have a home and have those basic needs met. That's what I believe. I think there's that part of him. . . . I wonder if he would agree . . . that everyone has a right to shelter, and water and electricity, enough to sustain them. And food, arguably. I think he would.

In an Earthship, "running costs are negligible, which could, perhaps, provide a solution for the weak and vulnerable in society who are at risk from extremes of heat and cold and cannot afford the energy hungry means of ameliorating extreme temperatures through either air conditioning or heating."[55] Those involved in the Earthship community, including Mike Reynolds and the homeowners, interns, and employees I met, are inspired by a social justice ethic in which environmentally responsible self-sufficiency in dwelling is arguably a human right. Sandy, who still works with Mike as a contracted architect, told me,

> We're trying to work with Habitat for Humanity in the town of Taos, which has been really interesting. The town of Taos has a bunch of land, and Habitat has the means for getting low-income housing, and Mike has the alternative housing. So we've been working on that together. And they've really been fantastic, and that's been a really interesting shift, because a while ago they didn't want anything to do with it. Because Habitat, part of their mission is basically to get people mainstream housing, like a house that is normal, so that they can kind of participate in normal American life or whatever. . . . So taking the risk, from Habitat's side of things, is pretty interesting. Because it's not, I mean this is not normal. This is not an average way of living. It's different.

The example Sandy describes above is just one example of how Earthship Biotecture attempts to pursue the orienting ethic that motivates not only organizational practice[56] (such as the early attempts to have communally owned and thus cheaper land acquisition possibilities in Taos, the disaster relief projects in places like Haiti and Japan, and the educational opportunities provided by the business) but also broader participation of those who are building, owning, or learning about Earthships.[57]

This ethic of self-sufficiency as social justice, the idea that humans have the right to self-sufficient dwelling, meeting their own needs without jeopardizing the ability of every human and all other creatures, critters, and systems of the planet to do the same, orients the motivation to participate in the Earthship community as a business, as a homeowner, and as a volunteer. If you believe in peak oil, climate change, or any other cause of potential impending doom for the planet, living as self-sufficiently as possible is clearly good preparation.[58] Yet more profoundly, this orientation offers a conception of social justice. As Sandy told me, "Human impacts on the environment breed not only environmental damage but deep social inequities. Here, we can think about all the issues related to production, the use of hazardous materials, and the dealing with waste that fall disproportionately on the poor, both poor people within our own country, and poor countries globally, categorized as issues of environmental justice." Or, as Margaret put it, "We are destroying the planet as well as our stupid selves and we just don't have the right. I don't care about the people who are destroying it, but the people and the other parts of the planet who don't have a say. This means plants, animals, almost everyone outside the U.S. and a lot of people in it." Taking a more optimistic outlook on the same problem, an intern named Matthew told me, "There are conscientious decisions that need to be made when you're buying something made with slave labor or buying something that's harmful to the environment. But the fact that we're working in a direction away from all of the things that we feel are leading to the demise and the destruction and the chaos of the world, you know, I think that's the most important point."

Pursuing an orienting ethic of self-sufficiency as social justice requires radically rethinking the sociotechnical arrangements used to support residential dwelling as well as reshaping our bodily practices to be aligned rather than in conflict with the resources available to us as dwellers.[59] Sarah, who used to live in an Earthship that she designed and built in the Greater World Community, told me that she was driven to build her own Earthship because, "I didn't wanna pay rent, and I liked the fact that it had a really low carbon impact, and I liked the fact that you could, I could, I as a woman could do it, you know?" Not paying rent or a mortgage or monthly utility bills creates a "freedom in a vast sense." Being able to build and maintain your own home, as a man or a woman with no building experience, is empowering in a way that opens up the potential for further change in action and practice. Earthships demonstrate the ability to dwell with more freedom, more empowerment, and more opportunity—or conversely, less dependency, less isolation, and less resignation to prevailing sociotechnical systems we do not wish to support—doing ever more with ever less, opening up more time, money, resources, and potential.

7

Dwelling in Resistance

• •

Attention to the relationships among technology and society, and thought and action, through the internalization and enactment of social practices that are "socially *pre*formed and individually *per*formed,"[1] illustrates that classical sociological thinking has continued relevance in the present. Emile Durkheim is considered one of the founders of the field of sociology. His nephew, student and protégé, Marcel Mauss, studied Eskimos and how the material patterning of their groups, which changed seasonally, related to changes in their social interactions, norms, and practices. Mauss was interested, in his own words, in "how the material form of human groups—the very nature and composition of their substratum—affects different modes of collective activity."[2] Mauss, learning from Durkheim, was one of the first sociologists to study how the material structures of society shaped patterns of social practice.

In this study, I also seek to find connections between the material and cultural organization of society. While Mauss examined how material organization interacts with cultural formations and modes of bodily interaction in temporally reoccurring ritual settings, I seek to understand the ritualized techniques of the body in residential dwelling, specifically when it comes to alternative sociotechnical arrangements in residential dwelling and the corresponding alternative forms of practice that resist modern notions of "normal" dwelling experience.[3]

Yet the analysis presented in this book departs from the kind of sociological study Mauss or Durkheim would have likely written. First, unlike Mauss or Durkheim, I am interested in understanding how sociotechnical systems contribute to, reinforce, and have the potential to reshape relations of power.

Mauss lends some insight by recognizing the role material objects play in shaping internalized bodily practice.[4] More critical thinkers like Herbert Marcuse and Lewis Mumford highlight the changes in mental concepts and forms of social domination that come from modern technological systems.[5] Borrowing insight from Michel Foucault and Pierre Bourdieu, understanding power as capillary, cultural, and both productive of and transmitted through action, connects the dots between technological systems, internalized habits and bodily practice, and relations of power.[6]

The dominant technological systems that support residential dwelling produce isolated, dependent dwellers.[7] Residential dwellers have little control over the technological systems upon which their lives depend. The bodily rituals and practices related to the material systems of residential dwelling reinforce this isolation and dependence—we unthinkingly use the water, flush the toilet, dispose of the trash, and pay the monthly utility bills with very little connection to the natural and organizational systems that support and maintain these sociotechnical networks.[8]

The theories of power employed to explain how material systems produce particular patterns of social practice as well as the stability of dominant technological systems are consistent with the other theoretical lenses used here to interpret these case studies, namely a pragmatist theory of action and scholarship on theories of practice. From the early Chicago School of sociology to the more recent work of Hans Joas, many sociologists recognize the value of pragmatism for understanding action.[9] Chicago School sociologists criticized the assumption "that men react in the same way to the same influences regardless of their individual or social past, and therefore it is possible to provoke identical behavior in various individuals by identical means."[10] Joas utilizes the pragmatist theory of action to "provide the first steps toward developing an adequate conception of the constitutive creativity of action."[11]

Understanding Action as Oriented Practice

Yet, if power operates through habitual actions that perpetuate relations of power, how can we understand the creative actions of living with alternative sociotechnical arrangements that involve dwelling in resistance and changing the relationships of isolation and dependence? The concept of orientation relies on an understanding of action that moves beyond the dualism between thought and action as well as between means and ends to understand how some dwellers make sense of dwelling in resistance, emphasizing the pragmatic nature of action.[12] Accepting the premise that ends do not always rationally follow from means and that thought and action are part of the same culturally-influenced process can "reconceptualize human agency as a temporally embedded process of social engagement, informed by the past (in its habitual aspect), but also oriented toward the future (as a capacity to imagine alternative possibilities) and toward the present (as a capacity to contextualize past habits and future projects within the contingencies of the moment)."[13] It is only then that we may "account for variability and change in actors' capacities for imaginative and critical intervention in the diverse contexts within which they act."[14]

Scholars from fields like sociology, economics, and psychology have long sought to understand consumption patterns and choices.[15] There are several distinct perspectives that seek to explain human choice and action when it comes to consumption, and specifically environmentally relevant consumption.[16] There are, furthermore, clear tensions among these perspectives.[17] While some see action as guided by rational economic calculation,[18] others see action as predictably irrational[19] based on emotive motivations, while others see consumers as locked-in to technologically determined social systems that are beyond their control or their ability to meaningfully change.[20]

Yet instead of seeing behavioral commitment to dominant technological systems or the choice to adopt alternative technologies as following from rational calculation of economic advantage or conscious consideration of consistency with values, these case studies suggest that there is value in instead asking, "What role do cultural meanings play in people's behavior?"[21] These cultural meanings are often unconscious, and shape both thought and action simultaneously. The concept of orientations, underlying ethics related to environmental values but also related to broader sets of social values and relationships, is consistent with "a renewed acceptance of practice" as an analytically valuable framework for understanding both stability and change in sociotechnical networks.[22]

The concept of an orienting ethic involves interrogating how cultural meanings come to shape technology adoption, and it suggests that individuals and groups of individuals pursue practices consistent with the way they see the world (although perhaps not consciously), both in terms of relationships with the natural world and regarding relationships among humans. The concept "integrates justificatory and motivational approaches" by recognizing how thought and practice reinforce and work to reshape one another.[23] The alternative practices involved in living with alternative technologies can be both motivated and justified by these orientations, but the orientations themselves are silent, often unconscious. Importantly, these orientations shape both thought and action. Further, in the process of guiding action, these orientations also reshape future action, thought, and practice; "[a]s we transform our lifestyles, we transform ourselves."[24]

The orientation concept is intended to serve as a theoretical tool, invoking a visual metaphor; our orientation shapes what we see and how we see it, as well as what is out of sight.[25] It shapes perceptions of the world and choices within it, actions that relate to the natural world and environmental impacts but also simultaneously to social relationships and social patterns of practice. Connecting the two, in the case studies presented here, are the sociotechnical systems used to meet residential needs and comforts.

The orientation concept borrows from well-known conservationist Aldo Leopold, who espoused a vague "land ethic" that oriented right action toward the environment.[26] Leopold's land ethic is not specific enough to predict particular actions in every situation, but it does orient attentiveness to human practices and choices that affect the natural world. However, the orientations operating in the case studies presented here suggest that how we view the natural world and ideas about "right action" toward it are not isolated statements of "environmental values." Instead, the concept of orientations suggests a broader view, recognizing actions toward the environment as integrally related to how people understand relationships among people, including both small-scale social interactions and larger questions about social structures.

The concept of orientations also invokes Bourdieu's notion of habitus, underlying and internalized dispositions that reflect particular class and status positioning.[27] Yet in our highly differentiated modern world, class and status are only one possible point of departure for understanding social fields. Early on, Lewis Mumford recognized this when he wrote, "What is typical of the machine is the fact that these ideals, instead of being confined to a class, have been vulgarized and spread—at least as an ideal—in every section of society."[28] Internalized orientations are organized around a multiplicity of categories, and orientations arguably can and do span class positions.[29] The concept of orientations also differs from the idea of an "ecological habitus" because they encompass more than environmental considerations.[30] Isolating choice and action based on categories like "environmental responsibility" or "ecological awareness" or even "economic calculation" obscures the holism necessary to accurately understand choice, context, and change.

Further, the concept of orienting ethics draws upon the work of classical sociologist Max Weber, who discussed the Protestant ethic and its "elective affinity" with modern capitalism.[31] Weber discusses how worldviews about things like economics and religion come together in a "practical ethic."[32] Weber claimed that, "Behind [ethics] always lies a stand towards something in the actual world."[33] Weber highlights how orienting ethics are, above all else, about orientations to action. He writes, "The term 'economic ethic' points to the practical impulses for action. . . . An economic ethic is not a simple 'function' of a form of economic organization; and just as little does the reverse hold, namely, that economic ethics unambiguously stamp the form of the economic organization."[34] Weber was interested in "those features in the total picture . . . which have been decisive for the fashioning of the *practical* way of life."[35] Understanding practical ethics is important for understanding action in the world.[36]

The use of orientation as a conceptual tool to explain behavior draws on and contributes to theories of social practices.[37] This approach is "united by a split from rational choice theory (revealed preferences) and behavioral drive theory (attitudes and values cause behavior)," and recognizes that "a whole system of knowledge and social practice grows up around technologies making them deeply entrenched and difficult to replace."[38] One intellectual foundation for theories of social practices is the concept of lifestyle as put forth by Anthony Giddens.[39] Theories of social practice recognize that "single consumption acts are merely the visible parts of social practices evolving from and reproducing socially shared understandings of normal consumption behavior."[40] The analysis presented here extends theories of social practices by recognizing the possibility for changes in practices, shaped by orientations that are not necessarily fully conscious, and by highlighting how changes in practice contribute to changes in thoughts about the kinds of possible practice in residential life.

Yet as Weber himself recognized, "by approaching in this way, we do not claim to use the only possible approach nor do we claim that all empirical structures of domination must correspond to one of these 'pure' types."[41] There is no reason to think that we could discover and name every orientation operating to shape the patterns of practice among all human groups, or any reason to think the particular language of "orientation" is useful for all cultural accounts of empirical discovery. Instead, it is meant to capture the ideas that underlie this study. First, both power, and resistance, can be internalized and enacted through the bodily practices that correspond to particular kinds of sociotechnical systems. Further, consciousness is a many-layered onion that is not necessary to explain either power or resistance. Finally, changes in the sociotechnical networks that sustain residential life can be shaped by orientations that do not require strong environmental values or environmental concern and do not contradict with many traditionally held values regarding independence, freedom, and how community connections help to facilitate both. Instead, humans can, and are, employing sociotechnical systems to support the kinds of social organization and social practice collectively desired, by recognizing that alternative technologies and alternative forms of community can promote individual freedom.

Values and Value in Dwelling in Resistance

Alexis de Tocqueville claimed that human beings "cannot become absolutely equal unless they be entirely free, and consequently equality, pushed to its furthest extent, may be confounded with freedom."[42] In writing about his observations of America in his 1840 *Democracy in America*, Tocqueville claimed that equality and freedom should be examined as two distinct things, and that Americans loved equality above all else, including freedom. Yet he also recognized that equality of conditions, the unique characteristic of the new nation of which Americans were so proud, was built on a foundation of freedom.[43]

Fast forward to contemporary America. In this book, I argue that the sociotechnical systems that support everyday life in America contribute to a lack of freedom, a dependence on monthly bills and material systems over which residential dwellers have no control. These systems create consequences of dependence on outside experts and on monthly payments for the systems that support residential life, as residential dwellers rely on the institutional, political, managerial, technological, and economic systems that provide for their needs and comforts. In this dependence, dwellers are also isolated from the natural world and from one another, with very little understanding of what makes the lights turn on, where the clean water comes from, or where our water or waste goes when we're done with it, and how these processes impact environmental systems as well as other people.

Our isolation and dependence, arguably, also perpetuate inequalities. In the first sense, this is because modern residential dwellers need to make money to pay the monthly bills for their electricity, water, and the removal of waste. There is also inequality in the ability to control, shape, influence, or step away from the dominant technological systems that support residential life. Further, there is inequality in abilities of individuals and groups to weather the volatilities that come from changes to global climate, deteriorating infrastructures, or polarizing politics that all make these systems less stable and secure. These consequences of isolation and dependence are already apparent across America, in incidences of toxic exposure and lacking water quality, or even lacking water at all, impacting most severely the most economically vulnerable groups in society.

The Americans I've met who are dwelling in resistance by living with alternative technologies in alternative communities are not doing so based on alternative or countercultural values regarding environmental concern; they are not universally contradicting the values of the society in which they live. Arguably, the orientations that motivate the constellations of technology and practice adopted in each of the case studies actually draw upon values many Americans once shared, values first observed by Tocqueville in 1840. These Americans are looking to increase their freedom, and their equality, through the adoption of alternative sociotechnical systems, and by recognizing these two values as interrelated (see table 7.1 for a summary of the cases).

In the case of The Farm, community members value stewardship of land and the sacrament of life processes. They have created a social world[44] where no one lives with the burden of a mortgage, where access to land is shared, where the birth of babies is seen as spiritual practice, and where children are educated as whole human beings, not just test takers. Guided by principles of Right Livelihood and acknowledging, as Tocqueville claimed the Americans

Table 7.1
Case Study Summary

Case	The Farm	Twin Oaks	Dancing Rabbit	Earthships
Electricity provision	Some solar; tied to local utility grid	Some solar; tied to local utility grid	Some solar and wind; some off-grid; maintain community grid; tied to local utility grid	Off-grid
Water provision	Community well	Community well	Rainwater catchment; local water available from pumps in community	Rainwater catchment
Waste treatment	Septic tanks	Community biological waste treatment	Humanure nutrient capture	Biological waste treatment
Transportation	Individual vehicles	Community vehicles	Community vehicles	Individual vehicles
Food	Individual food production; largely vegetarian	Community produces up to 70% of food consumed; rest bulk purchased; individual dietary choices vary	Individual food production and animal husbandry; individual dietary choices vary	Some food production in internal greenhouses; individual dietary choices
Work	Estimated 50% in community; others commuting	100% in community (labor and income sharing)	Some in community; some online; very limited commuting	Individual work choices
Other alternative practices	Alternative birth, education, death; shared community spaces	Shared housing; alternative education; shared community spaces and resources (tools, clothing, recreation)	Shared kitchen, bathing, and community spaces	Building techniques that minimize raw material input (recycled tires, bottles, and cans, and earthen walls)
Shared spaces	Community building; land; access to community grocery store	Shared housing; community buildings; land	Shared land; community building	Intended to be shared land; majority private space
Orientation	Custodial ethic	Plenitude ethic	Symbiotic mutualism and individualism	Self-sufficiency as social justice

he observed understood, that "it is the duty as well as the interest" of human beings "to make themselves useful to their fellow-creatures,"[45] their shared custodial ethic is about recognizing that by tending to the earth and to one another without pretense of ownership or control, humans can live more freely and equally. Through a custodial orientation, members of this community see the value of stewardship for helping individuals and communities flourish.

At Twin Oaks, community members recognize that sharing does not have to be a form of sacrifice. Through a plenitude ethic, community members embrace sharing systems as a form of abundance. Here, people have access to a wealth of resources while sharing in a way that reduces the overall economic and environmental burdens of resource use. People here live in full economic equality and recognize that as a way to increase their personal freedom. Their ethic of abundance, a shared orientation to a politics of plenty, recognizes the inextricable link between equality and freedom and involves adoption of technological systems and forms of material organization that honor that relationship and promote both values simultaneously.[46]

By recognizing the inevitable impacts of human activity on the natural world and on other humans, members of Dancing Rabbit Ecovillage seek a relationship of symbiotic mutualism, honoring the impacts of their individual and collective behaviors while seeking to make those impacts mutually beneficial on a broad scale (beyond impacts on humans, to consider impacts on species, ecosystems, and climate). Members of this community also value individuality, and the community is organized based on requiring individuals to meet their own subsistence needs. Life at Dancing Rabbit is oriented to the dance between individualism and symbiotic mutualism.

The case of Earthship Biotecture demonstrates how residential dwelling systems that differ radically from the mainstream can actually correspond to more traditional values regarding freedom, particularly the freedom to meet individual needs, as a form of social justice. Tocqueville lamented that, "It is easy to foresee that the time is drawing near when man will be less and less able to produce, of himself alone, the commonest necessaries of life."[47] Earthships work against that tendency by allowing individual residential dwellers to provide for their electricity, water, and waste treatment needs on site. Also allowing for some food production and sometimes built with innovative systems for food preservation, these off-grid homes substantially diminish the negative environmental impacts of residential dwelling. Yet for those living in, working for, and volunteering with Earthships, these homes are not primarily about their minimized environmental footprint; they are about providing a way to meet individual needs and comforts through individual responsibility as a form of social justice. This case suggests that self-sufficiency, in terms of having water and electricity and warmth without having to pay monthly bills, is an integral part of social justice.

These case studies each demonstrate ways to reorganize sociotechnical networks to lessen isolation and dependence. These case studies each demonstrate potential ways that sociotechnical systems can promote the values of freedom and equality and above all recognize the positive relationship between the two. In each community, individuals are adopting an attitude that I believe Tocqueville would recognize as "self-interest rightly understood"—demonstrating how technologies that reshape the source, scale, and scope of resource usage can be viewed as simultaneously beneficial to individuals, communities, and the natural world.

As the orientation concept suggests, one of the commonalities among these case studies is that participants in each consider their choice entirely reasonable. The people I met thought their alternative lifestyles "just made sense." While their decisions may seem irrational or maybe just unusual, their adoption of alternative sociotechnical arrangements corresponds to their orientations so as to make their actions reasonable, appropriate even, for facilitating the freedom and connection that corresponds to their underlying ethics. While labeling these alternative practices as environmentally motivated or based on alternative environmental motivations may work to marginalize and diminish potential adoption of alternative sociotechnical arrangements in residential life,[48] recognizing their congruence with broader and more traditional values regarding freedom and equality encourages a deeper engagement with what may facilitate, or hinder, opportunities for sociotechnical change in residential dwelling.

Connections among Cases

This engagement may start with asking: If the orientations that motivate adoption of alternative sociotechnical networks in residential dwelling represent articulations of longstanding cultural values about equality and freedom, why are there so few people living these alternative potentials? The answers, I believe, brings us back to some of the foundational arguments made here, and highlight some of the commonalities among the case studies. First and foremost, the co-constitutive relationship between technological and social organization means that "specific social practices, especially of people's dwellings, … produce, reproduce and transform different natures and different values."[49] Taking seriously the relationship between technology and society, articulating the politics of technology and the role of technologies in perpetuating particular relations of power, is necessary to begin to comprehend possible changes in practice and power.

In doing so, we can start to make sense of the claims made by many of the people I met while conducting this research, claims that the sociotechnical relationships in modern society encourage values like consumerism, superficiality, and self-centered isolationism. They recognized, as Bill McKibben wrote, that modern Americans "are not just individualists; we are hyper-individualists such as the world has never known."[50] Addressing the degradation of both environmental resources and social connections means contending with what is arguably the foundational issue of a systemic crisis, "the idea that the American 'system' as a whole is in real trouble—that it is heading in a direction that spells the end of its historic values."[51] This crisis plays out in the daily enactments of social practice, reinforcing social relationships and social values that contradict the historic values of freedom, equality, and "self-interest rightly understood." Changing sociotechnical arrangements, and thus practices, may require recognizing the role of current sociotechnical networks in reinforcing practices, norms, and forms of social organization that are internalized as deeply engrained unconscious understandings of normal life, and recognizing that these internalized norms and practices contradict shared, foundational cultural values.

Understanding practice as an essential unit of analysis is also important for promoting possibilities for dwelling in resistance. Rather than seeing people as guided by a rational means-end calculation in which thought precedes action, a practice perspective suggests that rationality is learned through action and that opportunities to change practice are part of planting seeds for future motivations. Learning new practices is part of embracing new normative stances. This points to one commonality among the cases: the importance of physical, bodily engagement with the dwelling experience and opportunities to experience such engagement as part of creating social change. In each case, people learned new norms of residential practice by engaging in new experiences, and creating opportunities for new forms of bodily engagement may be fertile ground for planting seeds of change.

A related point is that these alternative forms of dwelling are best seen as experiments, not representing the single or best path forward but rather the value in embracing pluralism in the dwelling experience as part of furthering individual freedom and social connection while also minimizing the environmental impacts of residential life.[52] This suggests a need "to expand the variables of design practice itself" to "develop environmental futures that are not only technologically possible but also socially desirable."[53] This highlights the role of policy, operating at multiple scales; the regulatory frameworks that shape residential dwelling are key in either creating possibilities for bodily engagement and experimentation in dwelling in resistance, or in hindering opportunities to imagine and implement the multiple possibilities.

Bodily Doing, Bodily Knowing

The case studies presented here demonstrate how living with alternative sociotechnical systems involves new bodily practices that reshape internalized orientations. In each case study, people talked about "learning by doing" and the importance of active learning for living with alternative technologies. The Earthship concept and the business model of Earthship Biotecture both center on the importance of informal education: people who have never built a home can build their own Earthship and people interested in Earthships can come to participate in builds, seminars, internships, and an academy program at Earthship Biotecture. One of the satisfying aspects of living in an intentional community (an element that increases personal freedom) according to folks at The Farm, Twin Oaks, and Dancing Rabbit is that they have myriad opportunities to learn new things in informal ways through community businesses and other community practices such as natural building, gardening, animal husbandry, food preservation, and mechanical maintenance. This importance of voluntary, informal, and do-it-yourself education corresponds to Juliet Schor's observations on the plenitude economy: "Self-provisioning is a way to spread the skills and practices we will need. It's especially important because it represents a return to more widespread capacity among the population to feed, clothe, shelter, and provide for itself."[54]

These case studies suggest that people learn by doing, and that it is by learning new bodily practices that people come to reconsider their ideas regarding normal residential practice. From this, it seems that there are many potential opportunities to create new forms of doing and knowing, without having to move to an intentional community. Experiences that allow people to temporarily change their dwelling practices may trigger new kinds of rationalities and new perspectives on the possibilities for dwelling. Educational opportunities that give children and adults alike the chance to grow food, build shelters, learn about and learn to use technologies for solar-powered electricity, water heating, and cooking, and generally learn to meet their own dwelling needs through their own bodily engagement are arguably valuable because they offer opportunities to shift practice, which has the potential to shift ways of knowing through changed ways of doing.

Recognizing the value of bodily engagement for creating new forms of knowing also helps to explain why, in each of these case studies, communities are adopting simplified, what might even be called primitive, forms of appropriate technology. Certainly, it is possible to lessen the environmentally damaging consequences of residential life through sophisticated modern technologies like programmable building systems, smart meters, and even low flow water fixtures. However, these kinds of technological refinements can be implemented without any required changes in bodily practice, and it is arguably the changes in bodily practice that create new rationalities to shape human interest, empowerment, and connection. Simplified sociotechnical systems may be appropriate systems not only because they are low cost and thus more accessible, and are adaptable to local conditions and circumstances, but also because they create opportunities for change in doing, knowing, and being.[55]

Against Technological Determinism, for Technological Experimentation

It is hard to discuss the relationship between technology and society without falling into language that suggests a deterministic relationship between the two. Both Marcuse and Mumford stand accused of deterministic thinking. Mumford wrote ardently on the relationship between technology and capitalism, believing that "although capitalism and technics must be clearly distinguished at every stage, one conditioned the other and reacted upon it." He claimed, "It was because of certain traits in private capitalism that the machine—which was a neutral agent—has often seemed, and in fact has sometimes been, a malicious element in society, careless of human life, indifferent to human interests. The machine has suffered for the sins of capitalism; contrariwise, capitalism has often taken credit for the virtues of the machine."[56]

Marcel Mauss also slipped into deterministic language when he wrote this methodological rule: "[S]ocial life in all its forms—moral, religious, and legal—is dependent on its material substratum and that it varies with this substratum."[57] Yet James Fox, in introducing Mauss's study of the Eskimos, uses the contrasting language of "strict determinism" versus "complex dependency" to help clarify Mauss's ecological approach. Fox writes, "Although biological and technological factors are extremely important, 'they are insufficient to account for the total phenomenon.' . . . Mauss's concern with a total social phenomenon will not allow him to accept these factors as an adequate explanation of the characteristic cultural features of these different periods."[58]

More recent writers on the relationship between technology and society have also been accused of technological determinism as they attempt to elucidate the political implications of technological systems.[59] The approach offered here, focusing on the cultural and spatial contextuality of action as well as the internalized orientations that shape action, provides a potential means of avoiding such determinism. As Mary Douglas once wrote, understanding the relationship between the material and the social world "demands an ecological approach in which the structure of ideas and of society, the mode of gaining a livelihood and the domestic architecture are interpreted as a single interacting whole in which no one element can be said to determine the others."[60]

An ecological approach to understanding social phenomena is similar to the methodological ideas of the Chicago School of sociology, for which pragmatism provided a foundation for studying social action, emphasizing spatial and contextual contingencies that shape social experience to understand how individuals navigate the social waters. As Andrew Abbott writes, for the Chicago School of sociology, "contextuality was so important that one could no longer focus on a single process. Instead one must study a whole network of intertwined processes."[61] This intellectual tradition can shed light on how to study the relationship between the material and social world without falling into determinism, by recognizing the spatial and cultural embeddedness of both the material and the social and viewing both as aspects of the same cultural processes. The sociotechnical systems that humans utilize are not temporally or conceptually distinct elements of society that cause predictable or universal outcomes. Rather, they are simply one other piece of the social world. Further, pragmatism emphasizes creativity in action, suggesting that individuals reinterpret situations and change courses of action as they meet situations that conflict with internalized understandings of the world.[62]

In each of the case studies described in this book, people made sure to tell me that they did not view their particular example of dwelling in resistance as the only possible option for change. Although the orientations operating in each case suggest shared values regarding the relationships among sociotechnical networks, freedom and equality, dependence and isolation, and environmental and social systems, each one represents a different form of experimentation with alternative sociotechnical arrangements. Furthermore, none of these cases represent perfection when it comes to either self-sufficiency or environmental responsibility.

Given the possibilities for creativity in action and understanding practice as embedded in temporal and spatial context, this study indicates the real existence and the real value of experimentation in dwelling in resistance. Coupled with the argument about the role of bodily engagement for creating changed ways of knowing, this suggests that alternative forms of residential dwelling may be limited because people do not have opportunities to experiment with different forms of social practice in residential life. We simply do not know how many residential dwellers would prefer to live in a neighborhood where homes are constructed based on passive solar design, or are equipped with rainwater catchment and grey water treatment, or where homeowner or community associations share the responsibility of managing a community-scale biological waste treatment facility and encourage rather than prevent clotheslines and gardens—because these options are largely absent from the American residential landscape. This highlights the role of policy and regulation in shaping opportunities for alternative residential dwelling. Experimentation with forms of dwelling in resistance can, and arguably should, be supported by policies that allow for possibilities among the possible in alternative residential dwelling.

Policy for Promoting Alternatives in Residential Dwelling

Policies, from community norms and rules to local, state, and federal regulations dictating the materiality of residential practice, shape the possibilities for new practice, working to either help or hinder the options available for resisting normalized sociotechnical arrangements in residential life. In each one of the communities described in this book, local policies operate to allow for alternative forms of sociotechnical arrangements in residential life. Considering the role of policies, including state, county, and local zoning, and coding requirements as well as both formal and informal community policies, demonstrates the extent to which standardization that shapes the use of sociotechnical systems in residential life largely works to prevent alternatives.[63]

For example, while the success of Earthship Biotecture is arguably linked to its location in the unique community of Taos, New Mexico, and many people involved with the Earthship community identify with this artistic social scene and dramatic natural environment, the localized building codes and zoning in Taos County also contribute to Earthship's success there. Further, the case of Earthship Biotecture demonstrates the importance of localized building codes as well as the way standardized building expectations have become transnational. Adjacent counties might have very different building requirements and permitting processes, while people from opposite sides of the globe may face similar limitations from conservative architectural design standards.

These case studies all demonstrate the value of rethinking private land ownership for promoting more environmentally benign and economically accessible residential dwelling arrangements.[64] In each case, shared ownership of land is key to reshaping sociotechnical systems. The case of Earthships demonstrates how this can be challenged by local policy, as Earthship communities in Taos were required to transition to privately owned lots.

Local building codes also make the Twin Oaks community possible, since rural zoning there allows multiple, nonrelated individuals to cohabitate, a rule leftover from eras of slavery and then sharecropping. Local regulations in many places prevent multiple unrelated people from cohabitating. When questions of scale are put to the pragmatic test, the local locus of control is substantial in shaping the use of sociotechnical systems in everyday life. This seems even more significant considering that many of the alternative technologies considered here are also appropriately scaled for local communities; they are what can be called soft path technologies:[65] small scale, distributed, locally controlled and locally adaptable. The importance of local context suggests the value of promoting policies and practices geared toward alternative technology adoption at the community scale.

A related point is that sharing systems, such as car sharing or tool sharing, do come with transaction costs.[66] The Twin Oaks community has managed to give value to these transaction costs, honoring the time it takes to manage such systems through labor credits. In mainstream residential life, this could perhaps translate into such systems being organized and managed by homeowner associations, community organizations, or local businesses, with those responsible for their oversight rewarded for their time and efforts, although local policies including bank lending and insuring are often ill-equipped for promoting shared ownership of shared resources. These sharing systems have tremendous potential for reducing resource use and bringing people together in meaningful communities.

More generally, this research suggests that the regulation and standardization that shape residential dwelling often stand in the way of alternative technology adoption and radical change in residential dwelling technology. Earthship Biotecture confronts these obstacles when dealing with the legality of rainwater collection and the permitting of alternative sewage treatment methods, and found such policy obstacles insurmountable when trying to organize more economically accessible communities through shared land ownership. Building codes and local zoning rules can also limit how many unrelated people are allowed to dwell together. When remodeling a home, codes may restrict the replacement of old windows with high-efficiency ones on account of historical character. They can limit where solar panels or small-scale wind turbines are installed and whether you can have chickens, a garden, or even a clothesline. My point is not (necessarily) that all these rules need eradicated, but that we should look closely and carefully at existing policies and question how they might unnecessarily limit the use of sociotechnical systems that promote freedom and social justice as well as social connection and environmental stewardship.

The case studies presented here suggest that some technological systems are "technologies of more." They meet Buckminster Fuller's goal of ephemeralization, doing ever more with ever less. In other words, they serve multiple benefits. For example, consider the multiple benefits that can be provided by technologies like solar electricity and bicycles as a form of transportation. Solar electric technology is good for the planet, and it decreases the amount of money a family must have every month to comfortably survive, thus decreasing dependence on a volatile and fragile money economy and a deteriorating electricity grid infrastructure. Solar energy can also improve community resiliency through localized energy security, and apparently it can also help to insulate your roof (I've been told by PV adopters). Bike riding is good for the earth, good for your body, good for your wallet, and good for planning and managing dense urban spaces.[67]

In general, appropriate technologies operate as technologies of more. Community control of the sociotechnical systems that support residential dwelling can benefit the natural and the social world through material systems and forms of organization with multiple benefits from multiple perspectives. Policy that supports technologies of more, and not just "green technologies," could provide more comprehensive benefits for human beings, communities, and the natural world.

Possibilities for Dwelling Differently

Some people remember college as the best years of their lives; environmental writer and activist Bill McKibben suggests that this fond reminiscence is due to the sociotechnical structure of college life, when many people live in denser communities and with more shared resource use than at any other time in their lives. It is, as he writes, "when we live roughly as we're evolved to live."[68] One purpose for embarking on this study is, simply put, to highlight the importance of the material world in shaping our social experience. The technological systems that support residential dwelling are integral to understanding social practices and the consequences of modern social life. As Lewis Mumford wrote long ago, "In discussing the modern technics, we have advanced as far as seems possible in considering mechanical civilization as an isolated system: the next step toward reorienting our technics consists in bringing it more completely into harmony with the new cultural and regional and societal and personal patterns we have coordinately begun to develop. It would be a gross mistake to seek wholly within the field of technics for an answer to all the problems that have been raised by technics."[69] Social scientists seeking to understand both social problems and possibilities for social change must take the materiality of life, including sociotechnical networks and their particular arrangements, seriously, for "it is only when we take the materiality of the world seriously that we can appreciate and preserve the resources"[70] that support human life on earth.

Herbert Marcuse once wrote, "Social theory is supposed to analyze existing societies in the light of their own functions and capabilities and to identify demonstrable tendencies (if any) which might lead beyond the existing state of affairs."[71] I have argued that the dominant technological systems that support residential dwelling produce both isolated and dependent dwellers, conflicting with traditional values regarding freedom and equality, and revealing, as Marcuse posed, "the question is no longer: how can the individual satisfy his own needs without hurting others, but rather: how can he satisfy his needs without hurting himself, without reproducing, through his aspirations and satisfactions, his dependence on an exploitative apparatus which, in satisfying his needs, perpetuates his servitude?"[72]

Some modern dwellers are reacting to this dilemma by living with sociotechnical systems that correspond to their shared orientations, ways of viewing the world and organizing ethical priorities that tend to shape action. These case studies suggest, as Mumford did, that "it is just at the moment of cultural and social dissolution that the mind often works with a freedom and intensity that is not possible when the social pattern is stable and life as a whole is more satisfactory."[73] The individuals I met who are dwelling in resistance are in pragmatic fashion responding to conflicts between orientations and practice in mainstream residential life, and by changing practices are creating both changed action and changed thought.

These cases demonstrate the possibility for "a quiet revolution begun by ordinary people with the stuff of our everyday lives."[74] The kinds of sociotechnical systems being used in these alternative communities tend toward self-provisioning, moving away from dependence and isolation by allowing and sometimes requiring that inhabitants provide for themselves. The idea of self-provisioning highlights how localized sociotechnical systems that work with rather than against the resources provided by the natural world can move communities and individuals away from dependencies on experts and a money economy and away from their isolated positions as alienated producers and consumers. Instead, the technologies represented in the case studies here offer opportunities to move toward self-provisioning at either the individual or community scale.

The technologies that support dwelling in resistance are technologies of more; they involve tending to community needs through community-driven knowledge and practice in ways that produce multiple, overlapping benefits for the individual, the community, and the natural world. Examining the values and practices of those living in alternative communities demonstrates that alternatives in residential dwelling are motivated by shared orientations, operating to shape both thought and practice while acknowledging how thought and practice are themselves inseparable. By changing practices through engagement with alternative sociotechnical systems, we can transform both our societies and ourselves. In this transformation, there are more possibilities than we have thus far allowed ourselves to comprehend.

Appendix

Reflections on Method

●●●●●●●●●●●●●●●●●●●●●●

The case studies presented here are part of what has been a larger exploration of the lives of people whose residential life entails living with alternative
technologies—how they talk about their decisions to use alternative technologies and how they differ in their practices of residential life. I have personally been drawn to alternative forms of dwelling, such as natural building
techniques and materials, distributed renewable energy generation for home
use, and Rainbow Gatherings and intentional communities, for all my adult
life. Growing up in suburbia, the epitome of normalized dwelling in America,
sparked an early interest in searching for alternatives. My first professional
interests as a sociologist focused on understanding the social context of residential solar electricity (PV) adoption.[1] Some of these early efforts deepened
my sense that in order to understand how behavioral change is motivated, how
sociotechnical choices in residential life are made, it was important to hear the
fully contextual stories that people tell about their process of learning to dwell
differently. Statistical analyses can tell us that people with higher incomes are
more likely to adopted PV, but having the higher income does not explain
why they adopted. Economic, demographic, or spatial correlations also do not
shed light on the way users themselves articulate or understand their decisions,
or how they actively live with sociotechnical systems through engaged social
practice.

When awarded an EPA STAR fellowship to study motivations for residential PV adoption using qualitative methods, I argued for an approach that
would consider the role of more obvious (or at least obviously accepted by
existing scholarship) influences like environmental concern and the factors

shaping the early adoption of innovative technologies. Most importantly, I also wanted to focus on how power, articulated through conventional understandings of normalized residential dwelling, shaped how adopters talked about their decision to adopt; I aimed to focus attention on how these alternative residential dwellers expressed their own understandings of how sociotechnical systems relate to systems of power. I did not have the language to know it at the time, but I was really interested in studying practice, and how alternative technologies in residential life translated to confronting power through changed practice.

Mustafa Emirbayer, my PhD advisor, inspired me to think beyond my first academic love of residential PV adoption, to consider broader cases that would expand my thinking and my interests. He knew about my experience with Rainbow Gatherings, and encouraged me to consider studying them ethnographically, using a similar lens attentive to alternative sociotechnical networks and alternative social practices.[2] I already had an interest in Earthships, and Mustafa encouraged me to study them too. Twin Oaks and The Farm presented interesting contrasts to the other cases, because they are, to my knowledge, the oldest still existing secular intentional communities formed during the 1960s era in the United States, offering cases of alternative technology adoption at the community scale.

Each of these case studies represent an example of a more environmentally responsible way to live, lessening the negative environmental consequences of the sociotechnical arrangements that dominate modern residential life. At a very basic level, I wanted to know whether these people living more environmentally responsible lives were primarily motivated by environmental values. At a broader level, I wanted use these case studies to examine the relationship between motivation, thought, action, and practice. I discovered that a similar logic unfolded across this very diverse set of cases.

To explore these issues, I spent time in each community, meeting the people who live there and asking them firsthand about their motivations and experiences while watching them live their lives. In the three intentional communities (The Farm, Twin Oaks, and Dancing Rabbit), this involved staying in the community as an official visitor. Twin Oaks and Dancing Rabbit both have formal visitor programs, and I spent time in each of these communities during a summer visitor session; in both cases the community knew I was coming as a researcher wishing to write about the community. In each case, I submitted my application to participate in a visitor program with a clear statement of my research project and aims, and follow-up correspondence initiated a process of community approval for my research plans. As both a visitor and an identified researcher, I had the opportunity to learn about the community through organized visitor activities (informational sessions, community tours, and social events) as well as through both formal and informal scheduled and unscheduled one-on-one conversation. I also had the opportunity to observe daily life in each community.

The Farm does not have a formalized visitor session, but members can and do host visitors. During my stay there, two longtime members of the community hosted me in their home. My hosts knew I was there as a researcher, although there was no formal process for sharing information about my visit with the community or getting community approval, I started each conversation with someone new by announcing my identity as a researcher and my research aims. I spent my days talking with community members that I saw while out walking the community, sometimes in more formal and prearranged conversations but often just through casual conversion, and visiting community spaces, including the school and the midwifery center. I also attended a community wedding, two spiritually based ritual gatherings, and a workshop on vegan cooking.

My study of Earthships is based on two months spent in Taos. I spent July 2011 interviewing homeowners (I contacted and recruited homeowners by leaving short letters on the doors),[3] getting to know interns, and participating in a three-day educational seminar (hosted by Earthship Biotecture) that involved hours of presentation from Mike Reynolds. I spent December of the same year participating in an Earthship internship, meeting and talking to other interns as well as employees. As with the intentional communities, I was clear about my identity as a research and my plans for research, in my conversations with other seminar participants, in my application to participate in the internship program, and in all informal conversations (formal interviews started with a written introduction of my research aims and the signing of a written consent form to participate). I did formal interviews with homeowners and former homeowners still in the area, focus groups with two groups of interns,[4] and had many informal conversations with employees, interns, and seminar participants. I learned about Earthships through these interviews and conversations and by spending time at Earthship Biotecture as an intern and seminar attendee.

One unique thing about my ethnographic research is the focus on a single overarching set of questions and interests. In most cases, when ethnographers enter a new place for field research, they are interested in understanding the entire social context in which people live and the processes of contextualized social meaning–making that shape daily life. Many times, ethnographers enter a field site prepared for surprises in what kinds of meanings and practices are most salient for the community members of study. For my work, however, I entered each field site with my interests already in mind; given my focus on the relationship among sociotechnical systems, the language of motivation and meaning making, and active bodily engagement in social practices, my processes of data collection and analysis were more targeted, arguably more narrow, than for many ethnographic researchers. On the one hand, this meant that my relatively short stay in each community, never more than one month at a time, did not feel like a limit on my ability to capture the essence of my interests; by the end of my time in each case study site, I felt I had reached "saturation" in the sense that new data and information was not contradicting my developed model of explanation. As saturation is the goal in qualitative data collection, I believe that the explanations presented here are accurate representations of the orientations and experiences of those in each community.

On the other hand, this targeted interest in alternative technologies and alternative practices means that some interesting and potentially relevant, or at the very least revealing, elements of each case study site have been overlooked or downright ignored in my analysis. Further, since sociotechnical systems and motivations to adopt alternative technologies in residential life are perhaps most accurately viewed as just one part of the larger social whole, there are likely aspects of social practice and social relationships that have analytical importance and explanatory power for understanding the orientations operating to support dwelling in resistance in each case study, but that are underemphasized here. For example, each of the intentional communities discussed here have unique decision making and planning structures. Studying consensus-based decision making, or the use of a system of planners and managers in an egalitarian and trust-based community, may be highly relevant for a holistic approach to pursuing environmental and social sustainability, but it is not something discussed at length here. Those interested in decision-making processes could return to these places and learn a lot through their own firsthand inquiry, even after reading this book, because that has not been the focus of my own ethnographic explorations. The same could be said generally for understanding communication and conflict resolution, which was only relevant for me to the extent that it related to the alternative organization of sociotechnical networks in each case. I also pay little attention to variables that are typically of interest to sociologists, like race, class, and gender. In each case study, I learned intimate details about people's life, including their sexual and romantic lives, but I do not discuss those details here, although they may be of great interest to sexuality scholars. In each case, I also learned a lot about differing ways of raising children and about some of the ways life in these alternative communities has impacted the children raised there, but that was also outside the scope of my scholarly interests, although it deeply affected me on a personal and emotional level.

Arguably, all ethnographic research ends up focusing in on particular aspects of social life, while downplaying or ignoring others, at least partially based on the researchers' interests, ethics, and access. The difference in my work is that I have selected my focus before entering the field for research, rather than either consciously or unconsciously selecting a focus during the course of the research. One benefit of this approach is that it opens up opportunities for comparability among multiple case studies; the cases I've studied, including those presented here as well as residential solar electricity adoption and Rainbow Gatherings, are dissimilar in a lot of ways, but they are comparable in that they all represent examples of human beings living with alternative sociotechnical systems and alternative social practices in residential life. By selecting the social contexts and social meanings I am most interested in (those related to use of technologies), I am able to engage in comparable research across otherwise very diverse cases.

My interests in understanding the social meanings and social practices involved in using alternative sociotechnical systems also shape the way I approach ethnographic inquiry. In additional to both formal, recorded interviews and informal conversations, ethnographers use field notes as one of primary tool for data collection and subsequent analysis. In these detailed descriptions and recollections of daily life in these communities, I paid particular attention to action, actual bodily doing, in addition to noting what people said, where, when, and how. Attentiveness to practice arguably requires shifting the ethnographic lens to focus on bodily engagement with the material world, and to pay careful attention to how regularized, normalized techniques of the body vary among different social groups in different social worlds.

There are multiple ways to approach ethnographic research, which I understand as a toolkit, a suite of methodological options that all focus on understanding social meaning from the perspective of participants in whatever case you are studying. For me, this was limited to understanding how processes of meaning making interact with technological choices and social practices. Other ethnographers do not start with a targeted focus, but as I hope this book has made clear, I believe a focus on sociomaterial systems and the technologies we use in everyday life is highly relevant for both sociological study and for creating social change.

Acknowledgments

This book would not have been possible without the support of many, many folks, including mentors, colleagues, friends, family, and the acquaintances who let me into their lives to share the stories presented here. In attempt to thank them all, I feel like I must begin at the beginning. As a sociologist who aims to help introductory students understand the role of their own social positions in shaping the outcomes of their own lives (in an attempt to help them empathize with those who grew up in a different social context), I am acutely aware of how family has shaped my ability to engage in pursuits I find meaningful and rewarding. Without the love and support of my family, I would not be where I am today, in a career I enjoy, pursuing work I find fulfilling on so many levels.

With an upbringing that gave me the privilege to attend college, I met two women who changed the course of my life. Dr. Jean Blocker and Dr. Susan Chase at the University of Tulsa provided my first introduction to the field of sociology; it was their influence and inspiration that sparked my own desire to teach sociology at the college level. They also encouraged my bourgeoning substantive interests; as a college freshman I was already interested in alternative ways of living, and the support of these early mentors helped encourage me that these interests were worthy of pursuit. I continued to explore these interests at the University of Wisconsin-Madison, where I finished my undergraduate degree. I was learning about the specific case studies presented here even then; I did an early project on Earthships (see chapter 6) in an environmental studies class taught by Dr. Cal DeWitt, and he too encouraged my interest in understanding opportunities for changing the environmental and social consequences of residential life. I am so thankful for this early support. My interests in alternative technologies continued as I furthered my education at Colorado State University, where Dr. Peter Taylor, an incredible advisor, was instrumental in developing my commitment to studying real existing opportunities for social change. I am forever grateful for his insightful mentoring.

Much of the empirical research for this book (with the exception of chapter 5) was originally conducted as part of my PhD dissertation. Dr. Mustafa Emirbayer at the University of Wisconsin–Madison was central in making this work a reality. He encouraged me to explore the social worlds that I find so compelling and intriguing with an ethnographic attentiveness and theoretical depth I could not have developed without his advising. I am so grateful for his wisdom, imagination, and advice.

At Michigan Technological University, I've had the support of many colleagues, and am especially grateful to Sarah Fayen Scarlett and Laura Rouleau for time spent in writing group conversations discussing this project, and to LouAnn Wurst for sharing meals, drinks, ideas, and editorial insight. This manuscript has received much attention and insightful reflection from Peter Mickulas, editor at Rutgers University Press. I am so appreciative of his time, impeccable editorial advice, and support for this project. Emily Prehoda, a graduate student in Environmental and Energy Policy at Michigan Tech, and a fantastic human being, has been so helpful in working on the reference list and formatting of this manuscript.

The fieldwork required to complete the chapters based on my dissertation research was partially supported by the Robert F. and Jean E. Holtz Center for Science and Technology Studies at the University of Wisconsin–Madison and by a National Science Foundation Integrative Graduate Research and Education Traineeship program, NSF-CHANGE, at UW-Madison. The fieldwork completed for chapter 5 was conducted with support from a Rural Sociological Society Early Career Research Award and Michigan Technological University's Research Excellence Fund. From my time doing dissertation fieldwork through the transition to academic appointment, the birth of our daughter and the publication of this book, I've had my partner in life by my side, and I am forever grateful for his patience, commitment, and our shared love of Avery.

In each of the places I visited described in these chapters, I had the opportunity to meet incredible, inspiring people who let me into their lives during my visit. Through shared meals, shared work, and shared time, I got to know many people whose authenticity and openness allowed me to see the real possibilities for changing how we live. In search of understanding possibilities for alternatives to the material and technological organization of daily living, I met people who embodied the social possibilities for less alienated, isolated, and dependent lives. I am so grateful to them for opening up their worlds to me, to show me these possibilities so worthy of pursuit.

Notes

Chapter 1 What Does It Mean to Dwell in Resistance?

1 On habitual rituals of the body, see Marcel Mauss, "Techniques of the Body," [1934] *Economy & Society* 2 (1973): 70–88; Erving Goffman, *Interaction Ritual: Essays on Face-to-Face Behavior* (New York: Pantheon, 1967); Randall Collins, *Interaction Ritual Chains* (Princeton, NJ: Princeton University Press, 2004).

2 Mauss, "Techniques of the Body," 70–88. See also Marcel Mauss, *Seasonal Variations of the Eskimo: A Study in Social Morphology* (London: Routledge & Kegan Paul, 1979).

3 Herbert Marcuse, "Some Social Implications of Modern Technology," *Studies in Philosophy and Social Sciences* 9 (1941): 138–162.

4 Ibid.

5 Theodore Schatzki, *Social Practices: A Wittgensteinian Approach to Human Activity and the Social* (New York: Cambridge University Press, 1996); Andrea Reckwitz, "Toward a Theory of Social Practices: A Development in Culturalist Theorizing," *European Journal of Social Theory* 5 (2002): 243–263; Alan Warde, "Consumption and Theories of Practice," *Journal of Consumer Culture* 5 (2005): 131–153.

6 Elizabeth Shove, *Comfort, Cleanliness and Convenience: The Social Organization of Normality* (New York: Berg, 2003); Gert Spaargaren, "Sustainable Consumption: A Theoretical and Environmental Policy Perspective," *Society and Natural Resources* 16 (2003): 687–701; Kirsten Gram-Hanssen, "Understanding Change and Continuity in Residential Energy Consumption," *Journal of Consumer Culture* 11 (2011): 61–78.

7 C. Wright Mills, "Situated Actions and Vocabularies of Motive," *American Sociological Review* 5 (1940): 904–913; Mustafa Emirbayer and Anne Mische, "What Is Agency?" *American Journal of Sociology* 103 (1998): 962–1023; Stephen Vaisey, "Motivation and Justification," *American Journal of Sociology* 114 (2009): 1675–1715.

8 See http://www.ic.org/, accessed 10/20/2015.

9 Michel Foucault, "Truth and Power," Interview reprinted in *The Foucault Reader*, ed. Paul Rabinow (New York: Pantheon Books, 1984), 51–75.

10 Lewis Mumford, *Technics and Civilization* (New York: Harcourt, Brace, 1934).

11 Foucault, "Truth and Power," 51–75.

12 Ibid., 63.
13 In Foucault's terms, power is "capillary"—it flows throughout all the localized, narrow networks in modern society.
14 Samer Alatout, "Towards a Bio-Territorial Conception of Power: Territory, Population, and Environmental Narratives in Palestine and Israel," *Political Geography* 25 (2006): 601–621, 603.
15 Allan Pred, "Philadelphia," *American Historical Review* (1982): 268–270, 269. See also George H. Daniels, "The Big Questions in the History of American Technology," *Technology and Culture* 11 (1970): 1–21.
16 Graham Burchell, Colin Gordon, and Peter Miller, *The Foucault Effect: Studies in Governmentality* (Chicago: University of Chicago Press, 1991); Michel Foucault, *Discipline and Punish* (New York: Vintage Books, 1980); Michel Foucault, *The History of Sexuality Volume 1: An Introduction* (New York: Random House, 1990).
17 Pierre Bourdieu, "Rethinking the State: Genesis and Structure of the Bureaucratic Field," *Sociological Theory* 12 (1994): 1–18.
18 As I have suggested, technology use is one type of ritual practice. See Randall Collins, "Stratification, Emotional Energy, and Transient Emotions," in *Research Agendas in the Sociology of Emotions*, ed. Theodore D. Kemper (New York: New York University Press, 1990), 27–34. Reprinted in Mustafa Emirbayer, *Emile Durkheim: Sociologist of Modernity* (Boston: Blackwell, 2003), 129–133.
19 As Karl Marx commented in a well-known footnote, "Technology discloses man's mode of dealing with Nature, the process of production by which he sustains his life, and thereby also lays bare the mode of formation of his social relations, and of the mental conceptions that flow from them." So, just as I have attempted to argue here, the technologies we use "disclose" our way of seeing the natural world and our relationship to it, as well as production processes, social relations, and mental categories. See Karl Marx, *Capital: Volume I* (1887; rpt. New York: International Publishers, 1967).
20 Bourdieu, "Rethinking the State," 3–4.
21 See Theodore Schatzki, *Social Practices: A Wittgensteinian Approach to Human Activity and the Social* (New York: Cambridge University Press, 1996); Andreas Reckwitz, "Toward a Theory of Social Practices: A Development in Culturalist Theorizing," *European Journal of Social Theory* 5 (2002): 243–263; Alan Warde, "Consumption and Theories of Practice" *Journal of Consumer Culture* 5 (2005), 131–153.
22 Emily Kennedy, Harvey Krahn, and Naomi T. Krogman, "Taking Social Practice Theories on the Road: A Mixed-Methods Case Study of Sustainable Transportation," in *Innovations in Sustainable Consumption: New Economics, Sociotechnical Transitions and Social Practices*, ed. M. J. Cohen, H. S. Brown, and P. J. Vergragt (Cheltenham and Northampton, MA: Edward Elgar, 2013), 252–276.
23 Spaargaren, "Sustainable Consumption," 687–701; Gert Spaargaren, "Theories of Practices: Agency, Technology, and Culture: Exploring the Relevance of Practice Theories for the Governance of Sustainable Consumption Practices in the New World-Order," *Global Environmental Change* 21 (2011): 813–822; Elizabeth Shove, "Beyond the ABC: Climate Change Policy and Theories of Social Change," *Environment and Planning A* 42 (2010): 1273–1285.
24 Shove, *Comfort, Cleanliness and Convenience*; Martin Hand, Elizabeth Shove, and Dale Southerton, "Explaining Showering: A Discussion of the Material, Conventional, and Temporal Dimensions of Practice," *Sociological Research Online*

10 (2005), accessed July 7, 2015; Elizabeth Shove and Gordon Walker, "What Is Energy For? Social Practice and Energy Demand," *Theory, Culture, and Society* 31 (2014): 41–58.

25 This research has a long history. See, for example, T. S. Robertson, *Innovative Behavior and Communication* (New York: Holt, Rinehart, and Winston, 1971); G. Zaltman and N. Lin, "On the Nature of Innovations," *American Behavioral Scientist* 14 (1971): 651–673; L. E. Ostlund, "Perceived Innovation Attributes as Predictors of Innovativeness," *Journal of Consumer Research* 1 (1974): 23–29; D. F. Midgley and G. R. "Downing, Innovativeness: The Concept and Its Measurement," *Journal of Consumer Research* 4 (1978): 229–242; D. C. Labay and T. C. Kinnear, "Exploring the Consumer Decision Process in the Adoption of Solar Energy Systems," *Journal of Consumer Research* 8 (1981): 271–278.

26 Two popular theories for how values influence behavior include Ajzen and Fishbein's theory of planned behavior. See I. Ajzen and M. Fishbein, "Attitude-Behavior Relations: A Theoretical Analysis and Review of Empirical Research," *Psychological Bulletin* (1977): 888–918; I. Ajzen and M. Fishbein, *Understanding Attitudes and Predicting Social Behavior* (Englewood Cliffs, NJ: Prentice-Hall, 1980); I. Ajzen, "From Intentions to Actions: A Theory of Planned Behavior," in *Action-Control: From Cognition to Behavior*, ed. J. Kuhl and J. Beckman (New York: Springer, 1985), 11–39; I. Ajzen, "The Theory of Planned Behavior," *Organizational Behavior and Human Decision Processes* 50 (1991): 179–211. For Shalom Schwartz's norm activation model see S. H. Schwartz, "Normative Explanations of Helping Behavior: A Critique, Proposal, and Empirical Test," *Journal of Experimental Social Psychology* (1973): 349–364; S. H. Schwartz, "Normative Influences on Altruism," in *Advances in Experimental Social Psychology*, ed. L. Berkowitz (New York: Academic Press, 1977), 221–279. Both theories have been applied in attempts to understand and predict environmentally motivated behavior.

27 P. C. Stern, "When Do People Act to Maintain Common Resources? A Reformulated Psychological Question for Our Times," *International Journal of Psychology* (1978): 149–158; P. C. Stern, "What Psychology Knows about Energy Conservation," *American Psychologist* 47 (1992): 1224–1232; N. Fransson and T. Gärling, "Environmental Concern: Conceptual Definitions, Measurement Methods, and Research Findings," *Journal of Environmental Psychology* 19 (1999): 369–382; Charlie Wilson and Hadi Dowlatabadi, "Models of Decision Making and Residential Energy Use," *Annual Review of Environment and Resources* 32 (2007): 169–203; Tom Heberlein, *Navigating Environmental Attitudes* (New York: Oxford University Press, 2012).

28 See Emirbayer and Mische, "What Is Agency?"

29 Juliet Schor, *Plenitude: The New Economics of True Wealth* (New York: Penguin, 2010).

30 Mills, *Situated Actions and Vocabularies of Motive*, 907.

31 Ibid., 913.

32 Schor, *Plenitude*, 11.

33 Shove, "Beyond the ABC."

34 Emirbayer and Mische, "What Is Agency?," 963.

35 For exceptions, see Brown, Halina Szejnwald, Philip J. Vergragt, and Maurie J. Cohen, "Societal Innovation in a Constrained World: Theoretical and Empirical Perspectives," in *Innovations in Sustainable Consumption: New Economics, Socio-technical Transitions, and Social Practices*, ed. M. J. Cohen, H. S. Brown, and

P. J. Vergragt, (Cheltenham and Northampton, MA: Edward Elgar, 2013), 1–30; Elizabeth Shove, "Energy Transitions in Practice: The Case of Global Indoor Climate Change," in *Governing the Energy Transition: Reality, Illusion, or Necessity?* ed. G. Verbong and D. Loorbach (New York: Routledge, 2012), 51–74; Chelsea Schelly, "How Policy Frameworks Shape Environmental Practice: Three Cases of Alternative Dwelling," in *Putting Sustainability into Practice: Advances and Applications of Social Practice Theories*, ed. Emily Huddart Kennedy, Maurie J. Cohen, and Naomi Krogman (Cheltenham and Northampton, MA: Edward Elgar, 2016), 185–203.

36 Martin Bulmer, *The Chicago School of Sociology: Institutionalization, Diversity, and the Rise of Sociological Research* (Chicago: University of Chicago Press, 1986); Andrew Abbott, "Of Time and Space," *Social Forces* 75 (1997): 1149–82; Andrew Abbott, *Department and Discipline: Chicago Sociology at One Hundred* (Chicago: University of Chicago Press, 1999).

37 See John Dewey, *How We Think* (London: Heath, 1910); John Dewey, *Experience and Nature* (London: Open Court, 1925); John Dewey, *The Quest for Certainty: A Study of the Relation of Knowledge and Action* (New York: Minton, Balch, 1929).

38 Emirbayer and Mische, What Is Agency?, 969.

39 Ibid., 967.

40 Ibid., 968; citing Hans Joas, *Pragmatism and Social Theory* (Chicago: University of Chicago Press, 1993).

41 Emirbayer and Mische, "What Is Agency?," 969–970.

42 Abbott's next sentence reads: "The latter are a figment of our unsociological imagination." Abbott, *Department and Discipline*, chapter 2.

43 W. I. Thomas, *W. I. Thomas on Social Organization and Social Personality: Selected Papers* (Chicago: University of Chicago Press, 1966), chapter 2. This description of social life is very similar to Foucault's claims that power is concerned with "the conduct of conduct." See Gordon Burchell, and Miller, eds., *The Foucault Effect: Studies in Governmentality.*

44 Conservation biologist Aldo Leopold developed this statement as his land ethic: "A thing is right when it tends to preserve the integrity, stability, and beauty of the biotic community. It is wrong when it tends otherwise." Aldo Leopold, *A Sand Country Almanac* (1949; rpt. New York: Oxford University Press, 1966), 262.

45 Sociologist Pierre Bourdieu uses the word *habitus* to mean "a system of lasting dispositions which, integrating past experiences, functions at every moment as a matrix of perceptions, appreciations and actions." Each individual person occupies and acts within a particular habitus, which "could be considered as a subjective but not individual system of internalized structures, schemes of perception, conception, and action common to all members of the same group or class." See Pierre Bourdieu, *Outline of a Theory of Practice* (Cambridge: Cambridge University Press, 1977), 82–83, 86; See also Pierre Bourdieu, "Social Space and Symbolic Space" and "Structures, Habitus, Practices," in *Contemporary Sociological Theory*, ed. Craig Calhoun et al. (CITY?, United Kingdom: Wiley-Blackwell, 2007), 259–289.

46 Max Weber, "The Social Psychology of the World Religions" [1946], in *From Max Weber: Essays in Sociology*, ed. H. H. Gerth and C. Wright Mills (New York: Oxford University Press, 1958), 267–301.

47 See Reckwitz, "Toward a Theory of Social Practices"; Shove, "Beyond the ABC."

48 See Noah Goldstein, Robert Cialdini, and Vladas Griskevicius, "A Room with a Viewpoint: Using Social Norms to Motivate Environmental Consent," *Journal of Consumer Research* 35 (2008); Dena Gromet, Howard Kunreuther, and Richard

P. Larrick, "Political Ideology Affects Energy-Efficiency Attitudes and Choices," *Proceedings of the Academy of Science* (2013): 9314–9319.

49   Tom Hargreaves, Michael Nye, and Jacquelin Burgess, "Making Energy Visible: A Qualitative Field Study of How Householders Interact with Feedback from Smart Energy Monitors," *Energy Policy* (2010): 6111–6119; Janet Lorenzen, "Going Green: The Process of Lifestyle Change," *Sociological Forum* 27 (2012); Chelsea Schelly, "Residential Solar Electricity Adoption: What Motivates, and What Matters? A Case Study of Early Adopters," *Energy Research and Social Science* 2 (2014): 183–191; Chelsea Schelly, "Transitioning to Renewable Sources of Electricity: Motivations, Policy, and Potential," in *Controversies in Science and Technology*, vol. 4, ed. Daniel Lee Kleinman, Karen Cloud-Hansen, and Jo Handelsman (New York: Oxford University Press, 2014), 62–72.

50   S. Bamberg and G. Möser, "Twenty Years after Hines, Hungerford, and Tomera: A New Meta-Analysis of Psycho-Social Determinants of Pro-Environmental Behavior," *Journal of Environmental Psychology* 27 (2007): 14–25; Heberlein, *Navigating Environmental Attitudes*, 2012.

51   Mauss, "Techniques of the Body," 1973; Shove, *Comfort, Cleanliness and Convenience*.

52   T. Hargreaves, N. Longhurst, and G. Seyfang, "Up, Down, Round and Round: Connecting Regimes and Practices in Innovation for Sustainability," *Environmental and Planning A* 45 (2013): 402–20; Shove, *Comfort, Cleanliness, and Convenience*; Spaargaren, "Sustainable Consumption"; G. Spaargaren, S. Martens, and T. Beckers, "Sustainable Technologies and Everyday Life," in *User Behaviour and Technology Development: Shaping Sustainable Relations Between Consumers and Technologies*, ed. P. P. Verbeek and A. Slob (Dordrecht, the Netherlands: Springer, 2006), 107–118.

53   Bellah et al., *Habits of the Heart: Individualism and Commitment in American Life* (1985; rpt. Berkeley: University of California Press, 1985), xvi; see Alexis de Tocqueville, *Democracy in America* [1835], trans. Arthur Goldhammer (New York: Penguin, 2004). See also Chelsea Schelly, "Everyday Household Practice in Alternative Residential Dwellings: The Non-Environmental Motivations for Environmental Behavior," in *The Greening of Everyday Life: Challenging Practices, Imagining Possibilities*, ed. John Meyer and Jens Kersten (New York: Oxford University Press, 2016), 265–280.

54   Bellah et al., *Habits of the Heart: Individualism and Commitment in American Life*, xiv.

55   Ibid., 23.

56   Ibid., 43.

57   See Karen T. Liftin's, *Eco-Villages: Lessons for Sustainable Community* (Malden, MA: Polity Press, 2013).

58   Chelsea Schelly, "Everyday Household Practice," 269.

59   Ibid.

Chapter 2   What "Normal" Dwelling Looks Like: The History of Home Technologies

1   Mischa Hewitt and Kevin Telfer, "Earthships as Public Pedagogy and Agents of Change," in *Handbook of Public Pedagogy: Education and Learning Beyond Schooling*, ed. Jennifer A Sandlin, Brian D. Schultz, and Jake Burdick (New York: Routledge, 2010), 171–178.

2   As Lewis Mumford notes, the "realities" of modern technological systems "were power, prices, capital, shares: the environment itself, like most of human existence, was treated as an abstraction." Lewis Mumford, *Technics and Civilization* (New York: Harcourt, Brace, 1934), 168.

3   Economic technologies shape how one makes a living; the term *labor* technologies applies to the conditions of actual work. Economic technologies include all the sociotechnical relationships required to get to work and successfully earn a paycheck. In the United States, this often means driving and parking a car, sitting in an office that is spatially and socially distinct from home, and so forth. Labor technologies include the conditions of actual work, such as working for a boss, performing one task for approximately forty hours a week.

4   See Amory Lovins, "Energy Strategy: The Road Not Taken?" *Foreign Affairs* 55 (1976): 65–96; Amory Lovins, *Soft Energy Paths: Toward a Durable Peace* (New York: Harper & Row, 1977).

5   Alienation here is meant in the Marxist sense. For Karl Marx, alienation from production (meaning that we no longer have ownership of or a sense of connection to that which we produce) creates an objective condition of alienation, in which we are alienated from production, but also from our own "species being" as well as from the natural world and one another. See, for example, Karl Marx, *Economic and Philosophic Manuscripts of 1844*, ed. D. J. Struik (London: Lawrence and Wishart, 1970); Bertell Ollman, *Alienation: Marx's Conception of Man in a Capitalist Society* (Cambridge: Cambridge University Press, 1971).

6   An examination of consequences has been the domain of sociology since its very early days, with Emile Durkheim claiming sociology's territory as the examination of functions, or objective consequences. See Emile Durkheim, *Rules of Sociological Method* [1895], ed. Steven Lukes (New York: Free Press, 1982). The emphasis on functions or consequences dominated American sociology during the mid-twentieth century reign of Talcott Parsons; Robert Merton further emphasized that the most important contribution of sociology is the study of latent or unintended consequences. See Robert K. Merton, *On Theoretical Sociology* (New York: The Free Press, 1967). While my approach is certainly not aligned with the tenants of structural-functionalism or any contemporary variant, I do take seriously the need to examine unintended consequences through a sociological perspective. My perspective is one that also takes seriously relations of power and the importance of discursive formations in the production of bodily action, social organization, technological inertia, and the unintended consequences impacting the material and social world.

7   Mumford, *Technics and Civilization*, 161–163.

8   See Jim Johnson, "Mixing Humans and Non-Humans Together: The Sociology of a Door Closer," *Social Problems* 35 (1988): 298–310. As Bruno Latour (under the pseudonym Jim Johnson) writes, "The most liberal sociologist often discriminates against nonhumans. Ready to study the most bizarre, exotic, or convoluted social behavior, he or she balks at studying nuclear plants, robots, or pills" (298). Latour has spent much of his career explicating why sociologists can and should pay attention to what he calls the 'non-human,' or technologies.

9   William Freudenburg, Scott Frickel, and Robert Gramling, "Beyond the Nature/Society Divide: Learning to Think about a Mountain," *Sociological Forum* 3 (1995): 361–392.

10 Paul Edwards, "Infrastructure and Modernity: Force, Time, and Social Organization in the History of Sociotechnical Systems," in *Modernity and*

Technology, ed. Thomas J. Misa, Philip Brey, and Andrew Feenberg (Cambridge, MA: MIT Press, 2003), 185–225.

11 Eric Schatzberg, "Technik Comes to America: Changing Meanings of Technology Before 1930," *Technology and Culture* 47 (2006): 486–512; Leo Marx, "Technology: The Emergence of a Hazardous Concept," *Technology and Culture* (2010): 561–577.

12 Martin Heidegger, *The Questions Concerning Technology and Other Essays* (New York: Harper & Row, 1977).

13 President's Commission on Critical Infrastructure Protection, *Critical Foundations: Protecting America's Infrastructures* (Washington, DC: US Government Printing Office, 1997).

14 Infrastructures rely on other infrastructures. For more on the idea of "systems of systems" or second-order technological systems, see Ingo Braun and Bernward Joerges, "How to Recombine Large Technical Systems: The Case of European Organ Transplant," in *Changing Large Technical Systems*, ed. Jane Summerton (Boulder, CO: Westview Press, 1994), 25–52. See also Thomas J. Misa, "The Compelling Tangle of Modernity and Technology," in *Modernity and Technology*, ed. T. J. Misa, P. Brey, and A. Feenberg (Cambridge, MA: MIT Press, 2003), 1–30.

15 Edwards, *Infrastructure and Modernity*, 186.

16 Ibid. See also Bruno Latour, *We Have Never Been Modern* (Cambridge, MA: Harvard University Press, 1993); Bruno Latour, *Pandora's Hope: Essays on the Reality of Science Studies* (Cambridge, MA: Harvard University Press, 1999).

17 Gert Spaargaren, "Sustainable Consumption: A Theoretical and Environmental Policy Perspective," *Society and Natural Resources* 16 (2003): 687–701.

18 Edwards, *Infrastructure and Modernity*, 185.

19 Ruth Cowan, *A Social History of American Technology* (New York: Oxford University Press, 1997).

20 Ibid., 150.

21 See, for example, Sheila Jasanoff, ed., *States of Knowledge: The Co-Production of Science and Social Order* (London: Routledge, 2004).

22 Karl Marx, *The Poverty of Philosophy* (London, Martin Lawrence Limited, 1847).

23 Langdon Winner, "Do Artifacts Have Politics?" *Daedalus* 109 (1980): 121–136.

24 Lovins, "Energy Strategy: The Road Not Taken?"

25 Elizabeth Shove, "Efficiency and Consumption: Technology and Practice," *Energy & Environment* 15 (2004): 1053–1065.

26 T. J. Misa, "The Compelling Tangle of Modernity and Technology," in *Modernity and Technology*, ed. T. J. Misa, P. Brey, and A. Feenberg (Cambridge, MA: MIT Press, 2003), 1–30.

27 Philip Brey, "Theorizing Modernity and Technology," *Modernity and Technology* (2003), 33–71.

28 Wiebe E. Bijker, Thomas P. Hughes, and Trevor Pinch, eds., *The Social Construction of Technological Systems* (Cambridge, MA: MIT Press, 1989); Wiebe Bijker and John Law, eds., *Shaping Technology / Building Society: Studies in Sociotechnical Change* (Cambridge, MA: MIT Press, 1992); Michel Callon, "The Sociology of an Actor Network: The Case of the Electric Vehicle," in *Mapping the Dynamics of Science and Technology*, ed. Michel Callon, John Law, and Arie Rip (London: Macmillan Press, 1986), 19–34; Michel Callon, "Some Elements of a Sociology of Translation: Domestication of the Scallops and Fishermen of St. Brieuc Bay," in *Power, Action, and Belief: A New Sociology of Knowledge*, ed. John Law (London: Routledge, 1986) 196–233; Bruno Latour, "On Recalling ANT," in *Actor Network*

Theory and After, ed. John Law and John Hassard (Oxford: Blackwell Publishers, 1999), 15–25.

29   Brey, "Theorizing Modernity and Technology," 47. See also Bijker and Law, *Shaping Technology / Building Society: Studies in Sociotechnical Change*; Joel A. Tarr and Gabriel Dupuy, eds., *Technology and the Rise of the Networked City in Europe and America* (Philadelphia: Temple University Press, 1988). Tarr and Dupuy use the language of 'sinews' to describe urban technological networks, which "guide and facilitate urban functioning and urban life in a multitude of ways, some positive, others negative, some visible and others invisible" (xiii).

30   Brey, "Theorizing Modernity and Technology."

31   Heidegger, *Questions Concerning Technology and Other Essays*, 19.

32   Jacques Ellul, *The Technological Society*. (New York: Vintage Books, 1964). Cited in Joel A. Tarr, "Sewerage and the Development of the Networked City in the United States, 1850–1930," in *Technology and the Rise of the Networked City in Europe and America*, ed. Joel A. Tarr and Gabriel Dupuy (Philadelphia: Temple University Press, 1988), 159–185.

33   Herbert Marcuse, *One-Dimensional Man: Studies in the Ideology of Advanced Industrial Society* (Boston: Beacon Press, 1964); see also Jürgen Habermas, "Technology and Science as 'Ideology,'" in *Toward a Rational Society: Student Protest, Science, and Politics*, ed. Jürgen Habermas (Boston: Beacon Press, 1970), 90–122.

34   Marcuse, *One-Dimensional Man*.

35   Critical theorists argue that faith in modern technologies contributes to an ideological belief in reason and rationality, which is used in modern societies as a form of domination and oppression. See Max Horkheimer and Theodor W. Adorno, *The Dialectic of Enlightenment* (New York: Herder and Herder, 1972). See also Marcuse, *One-Dimensional Man*. Their argument is similar to Michel Foucault's claim that, "The Enlightenment's promise of attaining freedom through the exercise of reason has been turned upside down, resulting in a domination by reason itself, which increasingly usurps the place of freedom." Michel Foucault, *The Essential Works of Michel Foucault, 1954–1984*, vol. 3: *Power*, ed. Paul Rabinow (London: Allen Lane, 2000), 273.

36   See, in particular, Michel Foucault, *Discipline and Punish* (New York: Vintage Books, 1980).

37   Alan Sheridan, *Foucault: The Will to Truth* (London: Routledge, 1990).

38   See Andreas Reckwitz, "Toward a Theory of Social Practices: A Development in Culturalist Theorizing," *European Journal of Social Theory* 5 (2002): 243–263; Gert Spaargaren, "Sustainable Consumption: A Theoretical and Environmental Policy Perspective," *Society & Natural Resources* 16 (2003): 687–701; Gert Spaargaren, "Theories of Practices: Agency, Technology, and Culture: Exploring the Relevance of Practice Theories for the Governance of Sustainable Consumption Practices in the New World-Order," *Global Environmental Change* 21 (2011), 813–822; A. Warde, "Consumption and Theories of Practice," *Journal of Consumer Culture* 5 (2005): 131–153.

39   T. Shatzki, *Social Practices: A Wittgensteinian Approach to Human Activity and the Social* (New York: Cambridge University Press, 1996); Warde, "Consumption and Theories of Practice," 131–153.

40   Marcel Mauss, "Techniques of the Body" [1934], *Economy & Society* 2 (1973): 70–88.

41   Randall Collins, "Stratification, Emotional Energy, and Transient Emotions," in *Research Agendas in the Sociology of Emotions*, ed. Theodore D. Kemper (Albany:

State University of New York Press, 1990), 27–34. Reprinted in Mustafa Emirbayer, *Emile Durkheim: Sociologist of Modernity* (Malden, MA: Blackwell, 2003).

42 Horace Miner, "Body Ritual among the Nacirema" *American Anthropologist* 58 (1956): 503–507.

43 See, for example, Ruth S. Cowan, *A Social History of American Technology* (New York: Oxford University Press, 1997); Claude S. Fischer, *America Calling: A Social History of the Telephone to 1940* (Berkeley: University of California Press, 1992); Kenneth T. Jackson, *Crabgrass Frontier: The Suburbanization of the United States* (New York: Oxford University Press, 1985); Thomas Hughes, *Networks of Power: Electrification in Western Society, 1880–1930* (Baltimore: Johns Hopkins University Press, 1993); David E. Nye, *Electrifying America: Social Meanings of a New Technology, 1880–1940* (Cambridge, MA: MIT Press, 1997); David E. Nye, *Consuming Power: A Social History of American Energies* (Cambridge, MA: MIT Press, 1998); Eric Schatzberg, *Wings of Wood, Wings of Metal: Culture and Technical Choice in American Airplane Materials, 1914–1945* (Princeton, NJ: Princeton University Press, 1999).

44 Berger and Luckmann's claim seems fitting here: "It is impossible to understand an institution adequately without an understanding of the historical process in which it was produced. Institutions also, by the very fact of their existence, control human conduct by setting up predefined patterns of conduct, which channel it in one direction as against the many other directions that would theoretically be possible." Peter L. Berger and Thomas Luckmann, *The Social Construction of Reality: A Treatise in the Sociology of Knowledge* (New York: Irvington, 1966). As Mary Douglas also claims, "Any institution that is going to keep its shape needs to gain legitimacy by distinctive grounding in nature and reason: then it affords to its members a set of analogies with which to explore the world and with which to justify the naturalness and reasonableness of the instituted rules, and it can keep its identifiable continuing form." Mary Douglas, *How Institutions Think* (Syracuse, NY: Syracuse University Press, 1986). Thanks to Peter Mickulas for the suggestion of how to phrase this point more clearly.

45 Jackson, *Crabgrass Frontier*, 3.

46 Richard Rudolph and Scott Ridley, *Power Struggle: The Hundred-Year War Over Electricity* (New York: Harper & Row, 1986); S. W. Usselman, "From Novelty to Utility: George Westinghouse and the Business of Innovation During the Age of Edison," *Business History Review* 66 (1992): 251–304.

47 Thomas P. Hughes, "The Electrification of America: The System Builders," *Technology and Culture* 20 (1979): 124–161.

48 Thomas Edison to Edward Hibbard Johnson, 1886. From the Thomas Edison papers at Rutgers University. Special Collections Series, Vail Papers (1885–1888, 1900): [ME004; TAEM 148:5].

49 Harold P. Brown in the *New York Tribune*, December 1888. From the Thomas Edison Papers at Rutgers University, Document File Series D8828—1888, [D8828AEA; TAEM 122:990]. See also letter from Alfred Tate to Harold Brown, November 11, 1889. From the Thomas Edison Papers at Rutgers University, International Business Operations, General Letterbook Series, [LB-034022, TAEM 139:910].

50 See Rudolph and Ridley, *Power Struggle*; On Edison providing AC electricity to customers, see Thomas Edison's Testimony in Zipernowsky v. Edison, 20 May 20, 1890, National Archives, Washington, DC, Records of the Patent and Trademark Office (Record Group 241). From the Edison Archives at Rutgers University. On the danger of AC technology, see Harold P. Brown's published letter in the *New*

York Tribune, December 1888. From the Edison Archives, Document File Series D8828—1888, [D8828AEA; TAEM 122:990].

51 Usselman, "From Novelty to Utility," 303.

52 Hughes, "The Electrification of America: The System Builders."

53 Samer Alatout and Chelsea Schelly, "Rural Electrification as a 'Bioterritorial' Technology: Redefining Space, Citizenship, and Power during the New Deal," *Radical History Review* 107 (2010): 127–138.

54 Ibid.

55 See Ruth S. Cowan, *More Work for Mother: The Ironies of Household Technology from the Open Hearth to the Microwave* (New York: Basic Books, 1983).

56 As Amory Lovins writes, "In an electrical world, your lifeline comes not from an understandable neighborhood technology run by people you know who are at your own social level, but rather from an alien, remote, and perhaps humiliatingly uncontrollable technology run by a faraway, bureaucratized, technical elite who have probably never heard of you. Decisions about who shall have how much energy at what price also become centralized—a politically dangerous trend because it divides those who use energy from those who supply and regulate it." Lovins, "Energy Strategy: The Road Not Taken?," 54.

57 Tarr, "Sewerage and the Development of the Networked City in the United States," 160.

58 Ibid., 161.

59 James Salzman, "Thirst: A Short History of Drinking Water," *Yale Journal of Law & the Humanities* 18 (2006): 94–121, http://digitalcommons.law.yale.edu/yjlh/vol18/iss3/6.

60 Tarr, "Sewerage and the Development of the Networked City in the United States," 165, 172. See also John Duffy, *The Sanitarians: A History of American Public Health* (Urbana: University of Illinois Press, 1990).

61 This reality is quickly changing for many residential dwellers, even in America. See Julia Lurie, "California's Drought Is So Bad That Thousands Are Living without Running Water," *Mother Jones*, July 31 2015, http://www.motherjones.com/environment/2015/07/drought-5000-californians-dont-have-running-water, accessed September 22, 2015.

62 Tarr, "Sewerage and the Development of the Networked City in the United States," 161.

63 Ibid., 166.

64 Ibid., 170.

65 Adam Rome, *The Bulldozer in the Countryside: Suburban Sprawl and the Rise of American Environmentalism* (Cambridge: Cambridge University Press, 2001), 87.

66 Ibid., 87–118.

67 Ibid., 93; Robert M Brown, "Urban Planning for Environmental Health" *Public Health Reports* (1964): 202.

68 Kathleen E. Halvorsen and Hugh S. Gorman, "Onsite Sewage System Regulation Along the Great Lakes and the US EPA 'Homeowner Awareness' Model," *Environmental Management* 37 (2006): 395–409.

69 Jackson, *Crabgrass Frontier*, 161n10.

70 Ibid., 162.

71 Ibid., 168.

72 Ibid., 164.

73 Ibid., 168.

74 Ibid., 249.

75 Ibid., 172.

76 Lewis Mumford, *The Highway and the City* (New York: New American Library, 1964).

77 James J. Flink, *America Adopts the Automobile, 1895–1910* (Cambridge, MA: MIT Press, 1970); James J. Flink, *The Car Culture* (Cambridge, MA: MIT Press, 1975); James J. Flink, *The Automobile Age* (Cambridge, MA: MIT Press, 1990); Ronald Kline and Trevor Pinch, "Users as Agents of Technological Change: The Social Construction of the Automobile in the Rural United States," *Technology and Culture* 37 (1996): 763–795; Ralph Nader, *Unsafe at Any Speed: The Designed-In Dangers of the American Automobile* (New York: Grossman, 1965).

78 Cowan, *A Social History of American Technology*; Richard Wightman Fox and T. J. Jackson Lears, eds., *The Culture of Consumption: Critical Essays in American History, 1880–1980* (New York: Pantheon Books, 1983); Jackson, *Crabgrass Frontier.*

79 See Cowan, *A Social History of American Technology*, 224–248.

80 John Meyer, *Engaging the Everyday: Environmental Social Criticism and the Resonance Dilemma* (Cambridge, MA: MIT Press, 2015).

81 Ibid., chapter 6.

82 Cowan, *A Social History of American Technology*. See also Bruno Latour, *Pasteurization of France* (Cambridge, MA: Harvard University Press, 1988); Melanie E. DuPuis, *Nature's Perfect Food: How Milk Became America's Drink* (New York: New York University Press, 2002).

83 Michael Pollan, *Omnivore's Dilemma: A Natural History of Four Meals* (New York: Penguin Press, 2006); Michael Pollan, *In Defense of Food: An Eater's Manifesto* (New York: Penguin Press, 2008).

84 Cowan, *A Social History of American Technology*, 150.

85 Ibid., 151.

86 Pollan, *In Defense of Food*, 137–201.

87 Bruno Latour's classic *We Have Never Been Modern* explains this fallacy in detail; while many human societies claim that modernity has diminished dependence on nature (a claim that continues to be supported by so-called "eco-modernists" who call it "decoupling"; see The Breakthrough Institute, "The Ecomodernist Manifesto," http://www.ecomodernism.org/manifesto-english/, accessed May 12, 2016), humans are still completely dependent on natural systems for subsistence and comfort, even if that dependence is often largely invisible.

88 Mark H. Rose, "Urban Gas and Electric Systems and Social Change, 1900–1940," in *Technology and the Rise of the Networked City in Europe and America*, ed. Joel A. Tarr and Gabriel Dupuy (Philadelphia: Temple University Press, 1988), 229–245, 231–232.

89 Lewis Mumford, *The City in History: Its Origin, Its Transformations, and Its Prospects* (New York: Harcourt, Brace & World, 1961), 486.

90 See Joseph L. Arnold, *New Deal in the Suburbs: A History of the Greenbelt Town Program, 1935–1954* (Columbus: Ohio State University Press, 1971); Robert Fishman, *Bourgeois Utopias: The Rise and Fall of Suburbia* (New York: Basic Books, 1987); Jackson, *Crabgrass Frontier*; Rome, *The Bulldozer in the Countryside.*

91 Jackson, *Crabgrass Frontier.*

92 "The average single-family dwelling built in 1980 was 1,740 square feet. Twenty years later, it had expanded 45 percent, to 2,521. Ninety-five percent of these homes have two or more bathrooms, 90 percent have air-conditioning, and 19 percent have

three-car or larger garages." Juliet Schor, *Plenitude: The New Economics of True Wealth* (New York: Penguin, 2010), 45.

93 William Cronon, "Conservation and Patriotism: A Conservation with William Cronon." Interview by William Poole, *Land and People* (Trust for Public Land, 2006), 46–50.

94 Jackson, *Crabgrass Frontier*; Douglas S Massey, "American Apartheid: Segregation and the Making of the Underclass," *American Journal of Sociology* (1990): 329–357; Arnold R. Hirsch, "Less Than *Plessy*: The Inner City, Suburbs, and State-Sanctioned Residential Segregation in the Age of Brown," in *The New Suburban History*, ed. Thomas J. Sugrue and Kevin M. Kruse (Chicago: University of Chicago Press, 2006), 33–56; Andrew Wiese, "The House I Live In: Race, Class, and African American Suburban Dreams in the Postwar United States," in *The New Suburban History*, ed. Thomas J. Sugrue and Kevin M. Kruse (Chicago: University of Chicago Press, 2006), 99–119.

95 Mumford, *Technics and Civilization*, 313.

96 Mumford, *The City in History*, 494.

97 Tarr, "Sewerage and the Development of the Networked City in the United States," 159.

98 In *Technics and Civilization*, Mumford satirically writes that people living in the suburbs "were stable, 'normal,' 'adjusted' people. . . . In relation to the entire environment in which they worked and thought and lived, they merely behaved *as if* they were in a state of neurotic collapse, *as if* there were a deep conflict between their inner drive and the mechanical environment they had helped to create, *as if* they had been unable to resolve their divided activities into a single consistent pattern" (313).

99 Foucault, *Power/Knowledge: Selected Interviews and Other Writings*, ed. Colin Gordon (New York: Pantheon Books, 1980).

100 Lovins, "Energy Strategy: The Road Not Taken?"; Lovins, "Soft Energy Paths"; Winner, "Do Artifacts Have Politics?"; Langdon Winner, *The Whale and the Reactor: A Search for Limits in the Age of High Technology* (Chicago: University of Chicago Press, 1986).

101 Mumford, *Technics and Civilization*, 150.

102 E. F. Schumacher, *Small Is Beautiful: Economics as if People Mattered* (New York: Perennial, 1973).

103 "Truman's Point Four Program, June 24, 1949," in *Documents of American History*, 7th ed., ed. Henry Steele Commager (New York: Appleton, 1963), 558–559.

104 Carol Pursell, "The Rise and Fall of the Appropriate Technology Movement in the United States, 1965–1985," *Technology and Culture* 34 (1993): 629–637, 361.

105 Jordon B. Kleiman, "The Appropriate Technology Movement in American Political Culture" (PhD diss., University of Rochester, 2001).

106 Kleiman, "The Appropriate Technology Movement in American Political Culture."

107 Andre G. Kirk, *Counterculture Green: The Whole Earth Catalog and American Environmentalism* (Lawrence: University of Kansas Press, 2001), 29. See also David Dickson, *Alternative Technology and the Politics of Technical Change* (Glasgow: Fontana Collins, 1974), 148–173.

108 Allan Schnaiberg, "Redistributive versus Distributive Politics: Social Equity Limits in Environmental and Appropriate Technology Movements," *Social Inquiry* 53 (1983): 200–215.

109 The Farallons Institute (Point Reyes, CA, 1974), 3; quoted in Pursell, "The Rise and Fall of the Appropriate Technology Movement in the United States," 632.

110 Ibid.

111 Ibid.

112 Lovins, "Energy Strategy: The Road Not Taken?," 78.

113 Ivan Illich, *Tools for Conviviality* (New York: Harper & Row, 1973). See also Ivan Illich, *Toward a History of Needs* (New York: Pantheon Books, 1978).

114 Kleiman, "The Appropriate Technology Movement in American Political Culture."

115 Kirk, *Counterculture Green*, 28.

116 Malcolm Hollick, "The Appropriate Technology Movement and its Literature: A Retrospective," *Technology and Society* 4 (1982): 213–229.

117 The Farallons Institute, in "The Rise and Fall of the Appropriate Technology Movement in the United States," 632.

118 Hollick, "The Appropriate Technology Movement and Its Literature."

119 Pursell, "The Rise and Fall of the Appropriate Technology Movement in the United States," 630.

120 Mumford, *Technics and Civilization*, 269.

Chapter 3   Custodians of the Earth, Witnesses to Transition: The Story of The Farm

1   Stephen Gaskin, *Monday Night Class* (1970; rpt. Summertown, TN: Book Publishing Company, 2005).

2   Visit www.thefarmcommunity.com to learn more about The Farm today. Accessed March 1, 2016.

3   Stephen Gaskin in the new introduction (2005) to *Monday Night Class*, originally published in 1970, by The Farm's book publishing company, 10.

4   Ina May Gaskin, *Spiritual Midwifery* (Summertown, TN: Book Publishing Company, 1977).

5   Ibid.

6   Outstanding critic of technology Lewis Mumford recognized modern reproductive technologies as a material system deserving of criticism long ago; he wrote, "the mechanical interference of the obstetrician, eager to resume his rounds, has apparently been largely responsible for the current discreditable record of American physicians, utilizing the most sanitary hospital equipment, in comparison with midwives who do not attempt brusquely to hasten the processes of nature." Mumford, *Technics and Civilization* (New York: Harcourt, Brace, 1934).

7   See Albert Bates, "Technological Innovation in a Rural Intentional Community, 1971–1987," *Bulletin of Science, Technology, and Society* 8 (1988): 183–199.

8   *The Holy Bible: Containing the Old and New Testaments* (Philadelphia: A. J. Holman, 1942), Acts 2: 44–45, p. 1121.

9   Including the community land and through land trusts associated with The Farm.

10   In her examination of fourteen eco-villages across the globe, Karen Litfin writes that the "shift from 'all for one and one for all' to private ownership" is "common among ecovillages that trace their roots to the 1960s and 1970s." Karen Litfin, *Ecovillages: Lessons for Sustainable Community* (Malden, MA: Polity Press, 2014), 98.

11   See http://www.thefarm.org/etc/inn.html, accessed April 27, 2016.

12   There is also a one-time new membership fee for those who join The Farm, which was $3,000 at the time of my visit.

13   Written by individuals serving on the Farm's membership committee and based on statements found in "This Season's People" and other Farm documents. Printed in a Farm brochure, "A spiritual community based on the principles of nonviolence and respect for the earth."

14 During my time there, one golf cart passed me on the road blaring Amy Winehouse from its speakers. I saw another golf cart loudly playing children's music while a little boy followed on his bike.

15 Swan Conservation Trust, http://swantrust.org/, accessed April 6, 2016.

16 Gaskin, *Monday Night Class*, Introduction.

17 Ibid.

18 Stephen Gaskin, *The Caravan* (1972; rpt. Summertown, TN: Book Publishing Company, 2007).

19 Gaskin, *Monday Night Class*, Introduction.

20 Bates, "Technological Innovation in a Rural Intentional Community, 1971–1987."

21 Charismatic leadership, in the Weberian sense, is leadership based on emotive and personal appeal. See Max Weber, "The Social Psychology of the World Religions" [1946], in *From Max Weber: Essays in Sociology*, ed. H. H. Gerth and C. Wright Mills (New York: Oxford University Press, 1958), 267–301, 295. Describing the work of Philip Slater, Anne Swidler claims, "the group creates a rudimentary though inadequate solidarity through its shared infatuation with the leader. Later, when the leader has disappointed their fantasies of love and salvation, the group members rebel, breaking the spell of enchantment with the leader, coming to evaluate their own capacities more realistically." Swidler, *Organizations without Authority* (Cambridge, MA: Harvard University Press, 1979), 75. See Philip Slater, *Microcosm: Structural, Psychological and Religious Evolution in Groups* (New York: Wiley, 1966).

22 Rupert Fike, ed., *Voices from the Farm: Adventures in Community Living* (Summertown, TN: Book Publishing Company, 1998).

23 In the early years, The Farm relied upon and advocated for appropriate technologies as a means of addressing both global inequities and sustainable development. On the potential of appropriate technologies for equitable global development, see Rowan A. Wakefield and Patricia Stafford, "Appropriate Technology: What It Is and Where It's Going" *The Futurist* 11 (1977): 72–77; Billie R. DeWalt, "Appropriate Technology in Rural Mexico: Antecedents and Consequences of an Indigenous Peasant Innovation," *Technology and Culture* 19 (1978): 32–52; Heather M. Murphy, Edward A. McBean, and Khosrow Farahbakhsh, "Appropriate Technology—A Comprehensive Approach for Water and Sanitation in the Developing World," *Technology in Society* 31 (2009): 158–167; Joshua M. Pearce, "The Case for Open Source Appropriate Technology," *Environment, Development, and Sustainability* 14 (2012): 425–431.

24 Kozeny, Geoph. *Visions of Utopia: Experiments in Sustainable Culture* [motion picture], Volume 2 (Rutledge, MO: Community Catalyst Project, Fellowship for Intentional Community, 2009).

25 Litfin, *Ecovillages*, 89.

26 In *The Class Struggles in France, 1848–1850* (1850), Marx develops the idea of permanent revolution, arguing that because economic markets are connected, socialism succeeding in only one country is impossible, and it must be a worldwide phenomenon. See Karl Marx, *The Class Struggles in France, 1848–1850* (New York: International Publishers: New York, 1964). See also Leon Trotsky, *The Permanent Revolution*, trans. Max Schachtman (New York: Pioneer Publishers, 1931). While this certainly helps explain the failure of The Farm's early model of cooperative living, the next community that I visited (Twin Oaks, see chapter 4) presents a nuanced challenge.

27 A longtime Farm member told me that of the approximately 200 members of The Farm, maybe 30 people will show up for these workdays. Sometimes, community members make and serve lunch after the workday as a draw to attend.

28 The company's operating procedures are also principled: employees log their time but don't have set hours.

29 See S. E. International, http://www.seintl.com/, accessed April 6, 2016.

30 Preston moved to The Farm before becoming school principal. He told me that he is still considered a new member, demonstrating the lack of new memberships on The Farm.

31 See Karl Marx, *Capital: Volume I* [1887] (New York: International Publishers, 1967); Karl Marx, *Economic and Philosophic Manuscripts of 1844*, ed. D. J. Struik (London: Lawrence and Wishart, 1970).

32 Herman Daly, ed., *Toward a Steady-State Economy* (San Francisco: Freeman, 1973); Herman Daly, *Steady-State Economics: The Economics of Biophysical Equilibrium and Moral Growth* (San Francisco: Freeman, 1977); Robert Goodland, Herman E. Daly, and Salah El Serafy, eds., *Population, Technology, and Lifestyle: The Transition to Sustainability* (Washington, DC: Island Press, 1992).

33 See Jim Merkel, *Radical Simplicity: Small Footprints on a Finite Earth* (Gabriola Island, BC: New Society Publishers, 2003); David Wann, *Simple Prosperity: Finding Real Wealth in a Sustainable Lifestyle* (New York: St. Martin's Press, 2007).

34 See Leslene Della Madre, *Midwifing Death: Returning to the Arms of the Ancient Mother* (Austin, TX: Plain View Press, 2005).

35 *Visions of Utopia, Volume 2.*

36 Fike, ed., *Voices from the Farm.*

37 *Visions of Utopia, Volume 2.*

38 Ibid.

39 Only some referred to it by name, but they were all referencing the Henslow's Sparrow, a grassland bird species in decline that has since been sighted in very few other locations in Tennessee.

40 The Farm community has done this through a nonprofit organization that has purchased land surrounding The Farm's 1,750 acres and placed it in a conservation trust. See Swan Conservation Trust, http://swantrust.org/, accessed April 6, 2016.

41 *Visions of Utopia, Volume 2.*

42 Fike, ed., *Voices from the Farm*, Introduction.

43 *Visions of Utopia, Volume 2.*

44 Ibid.

45 Peter C. Brown and Geoffrey Garver, *Right Relationship: Building a Whole Earth Economy* (San Francisco: Berret-Kohler Publishers, 2008), 1.

46 This story involves some context specific to this particular community, and I am grateful to a longtime member of The Farm for sharing his perspective on this issue. Quoting this personal correspondence directly: "The Farm began experiments with composting toilets while still on The Caravan. Clearly, some negative associations were formed then, but there has likely never been a single year in The Farm's history when at least one compost toilet has not been in operation. For a number of years in the 70s and 80s, there were extensive experiments with producing biogas from home brew digesters. Many experiments with radical designs of all kinds, like tilt trucks and rotating barrels, 'failed' in the sense they were smelly or hard to manage, but experiments are only a failure if you fail to learn from them. There is a general fault line that is pointed to by several families who came to The Farm . . . designed

a new neighborhood, and then held a public meeting to 'ask permission' to build compost toilets and use them to fertilize neighborhood food gardens. Big mistake. Putting such a controversial suggestion up for popular approval virtually guaranteed they would be rejected by the elderly minority with bad associations of smelly buses crossing the desert. . . . Even as those newcomers stomped away mad . . . elsewhere on The Farm any number [of] residents quietly composted their humanure and did not raise a stink." It should be noted here, as this member pointed out to me, that The Farm community continues to practice informal consensus-based decision making, so the minority who are opposed to composting human waste were able to block adoption for this proposed neighborhood. It should be noted that the practice of composting human waste is not legal in Tennessee, but composting of human waste does continue to take place on The Farm, at the Ecovillage Training Center, using a system they refer to as "earth closets."

47 The ETC has used alternative materials for blown insulation from materials derived from used dollar bills, sold as insulation material by the US Mint in Fort Knox, Kentucky. Personal communication, April 21, 2016. For more on the Ecovillage Training Center, see http://www.thefarm.org/lifestyle/albertbates/akb.html, and http://www.thefarm.org/etc/, accessed April 28, 2016.

48 *Visions of Utopia, Volume 2.*

49 Andrew Kirk, *Counterculture Green: The Whole Earth Catalog and American Environmentalism* (Lawrence: University of Kansas Press, 2001).

50 Buckminster Fuller, *Operating Manual for Spaceship Earth* (Carbondale: Southern Illinois University Press, 1969).

51 *Visions of Utopia, Volume 2.*

52 Ibid.

53 Daly, ed., *Toward a Steady-State Economy*; Daly, *Steady-State Economics.*

54 Max Weber, *The Protestant Ethic and the Spirit of Capitalism* [1940] (New York: Scribner, 1930).

Chapter 4 The Abundance of the Commons: Twin Oaks and the Plenitude Ethic

1 Jessica Ravitz, "Utopia: It's Complicated," CNN, http://www.cnn.com/interactive/2015/09/us/communes-american-story/, accessed September 16, 2015.

2 You can read more about Twin Oaks at www.twinoaks.org, accessed April 30, 2016.

3 See https://funologist.org/2012/08/14/collective-bikes/, accessed April 29, 2016.

4 See Thomas Princen, *The Logic of Sufficiency* (Cambridge, MA: MIT Press, 2005); Michael Maniates and John M. Meyer, eds., *The Environmental Politics of Sacrifice* (Cambridge, MA: MIT Press: 2010).

5 "How Sustainable Is Twin Oaks?" https://funologist.org/2014/11/08/how-sustainable-is-twin-oaks/, accessed April 29, 2016.

6 Juliet Schor, *Plenitude: The New Economics of True Wealth* (New York: Penguin, 2010).

7 Ibid., 7.

8 See Thomas Princen, *The Logic of Sufficiency* (Cambridge, MA: MIT Press, 2005). In this book, Princen rightly suggests that people living based on the logic of sufficiency do not see it as a sacrifice, but rather as thriving (3). However, Princen contracts sufficiency with efficiency, claiming that these are "two fundamentally opposed worldviews" (342) and suggesting that the logic of sufficiency (unlike the logic of efficiency) does require constraint. However, Twin Oaks demonstrates that

people who live based on a logic of sufficiency do not need to consciously articulate their lives as organized by or involving consideration of constraint. That is why, in this chapter, I choose to employ Juliet Schor's language of "plenitude."

9   Schor, *Plenitude*, 99.

10   Due to Garrett Hardin's infamous 1968 article, the language of tragedy has long dominated conversations about common, public, or shared resources. Garrett Hardin, "The Tragedy of the Commons," *Science* 162 (1968): 1243–1248.

11   B. F. Skinner, *Walden Two* (New York: Macmillan, 1948).

12   Ibid., 14.

13   Ibid., 263. This seems to directly correspond to Thomas Princen's claim that the logic of sufficiency becomes a principle of management and the logic of sufficiency can become an organizing principle for social groups "when rules and procedures regularize collective behavior." See Princen, *The Logic of Sufficiency*, 7. Certainly, there are many policies, rules, and procedures that help organize sharing systems at Twin Oaks so that they do not impose on individual freedoms and abilities. These range from the informal social norm that eye contact is not a required politeness when people pass on the trails (allowing people to be alone with their thoughts even though they are rarely physically alone) to formal policies like providing labor credit hours for managing and organizing the car-sharing system. This all suggests that sociotechnical networks that involve high degrees of sharing can be coordinated without a sense of individual sacrifice, but they must be proactively coordinated and shaped by rules and procedures that allow for individual freedom within the shared structures.

14   Kat Kinkade in Geoph Kozeny, *Visions of Utopia: Experiments in Sustainable Culture* [motion picture], Volume 1 (Rutledge, MO: Community Catalyst Project, Fellowship for Intentional Community, 2004).

15   At least in part due to the bureaucratic structure of governance at Twin Oaks, things take awhile to happen. Decisions happen slowly. For example, I was told during my visit that "out loud music" was "a new experiment" in the hammock shop, meaning only recently had the community approved of listening to music aloud while working in the shop.

16   Kathleen Kinkade, *A Walden Two Experiment; The First Five Years of Twin Oaks Community* (Scranton, PA: William Morrow, 1974); Kathleen Kinkade, *Is It Utopia Yet? An Insider's View of the Twin Oaks Community in its Twenty-Sixth Year* (Louisa, VA: Twin Oaks Publishing, 1994).

17   Skinner, *Walden Two*, 237.

18   The community was taken to court in the 1980s for their legal standing. The case (*Twin Oaks v. IRS*) took six years, but the community won.

19   Now, most children at Twin Oaks are born under the care of a midwife and most children are homeschooled.

20   See Acorn Community. www.acorncommunity.org, accessed January 2, 2013.

21   See Living Energy Farm. www.livingenergyfarm.org, accessed January 2, 2013.

22   Twin Oaks Community, "Life in a Feminist Ecovillage" (Community Pamphlet, 2004).

23   As one member put it, "Guests are like parasites, they need a host"—meaning a member of the community must know and vouch for them. Guests can stay for up to three months.

24   During my visit, I was told that acceptance rates used to be approximately 75 percent accept, 12–15 percent "visit again" (meaning potential members are told

to attend a second visitor period, sometimes with a time interval specified, such as "visit again in three months" or "visit again in a year"), and 10–12 percent reject. Now, with a growing waiting list and relatively high population stability, rates have become 50 percent accept, 30–35 percent visit again, and 15–20 percent reject. More people are interested in the lifestyle of Twin Oaks than the community has room for; this is good news! This is also why communities such as Acorn (www. acorncommunity.org) and Living Energy Farms (www.livingenergyfarm.org) emerged very near Twin Oaks, and why other income-sharing communities such as East Wind (see www.eastwind.org) and Sandhill Farm (www.sandhillfarm.org) are longstanding demonstrations of the possibilities of communal living.

25 See "Is Twin Oaks Ageist." https://funologist.org/2013/09/24/is-twin-oaks-ageist/, accessed April 29, 2016.

26 Members wishing to have children must submit an application to the community. In order to be considered, they must be members for two years and have at least one year of experience working with children in the community. Couples can apply together, and single women and even single men are allowed to apply (although only one man ever has, and his application was accepted). One measure of success for intentional communities is whether they are able to create multigenerational membership, with children raised in the community staying as adults. See Benjamin Zablocki, *The Joyful Community* (Baltimore: Penguin, 1971); Rosabeth Moss Kanter, *Commitment and Community: Communes and Utopias in Sociological Perspective* (Cambridge, MA: Harvard University Press, 1972); Clifford F. Thies, "The Success of American Communes," *Southern Economic Journal* 67 (2010): 186–199. During my visit in 2011, only one child who had been raised at Twin Oaks has made the transition to adult member, and only very recently. As of April 2016, there are three community members who grew up at Twin Oaks and have transitioned from child member to full adult member.

27 You also may not use any individual assets such as savings accounts, inheritances, child support checks, or personal vehicles while living in the community.

28 See Chelsea Schelly, "How Policy Frameworks Shape Environmental Practice: Three Cases of Alternative Dwelling," in *Putting Sustainability into Practice: Advances and Applications of Social Practice Theories*, ed. Emily Huddart Kennedy, Maurie J. Cohen, and Naomi Krogman (Cheltanham, UK: Edward Elgar, 2016), 185–203.

29 See Maniates and Meyer, eds., *The Environmental Politics of Sacrifice*. See also John M. Meyer, *Engaging the Everyday: Environmental Social Criticism and the Resonance Dilemma* (Cambridge, MA: MIT Press, 2015). As Meyer writes, "All too often, public arguments for change are countered by the claim that people in postindustrial societies are unwilling to sacrifice anything to address environmental challenges. . . . [T]he premise that seems to underlie this sort of retort is that people in postindustrial societies live in something like the best of all possible worlds now, so there is little motivation for most people in these societies to alter 'contemporary lifestyles'" (*Engaging the Everyday*, 16–17).

30 Schor, *Plenitude*, 6, 21; see also Teresa Gowan and Rachel Slocum, "Artisanal Production, Communal Provisioning, and Anticapitalist Politics in the Aude, France," in *Sustainable Lifestyles and the Quest for Plenitude: Case Studies of the New Economy*, ed. Juliet Schor and Craig J. Thompson (New Haven, CT: Yale University Press 2014), 27–62.

31 Schor, *Plenitude*, 116.

32 Finn, who was doing much of the mechanical work when I visited Twin Oaks, told me that the community puts approximately 350,000 miles in a year on their vehicles.

33 *Visions of Utopia Volume 1*, 2004.

34 Ibid.

35 Schor, *Plenitude*, 3.

36 See Etsy. https://www.etsy.com/, accessed May 12, 2016.

37 Richard Sennet, *The Craftsman* (New Haven, CT: Yale University Press, 2008).

38 Ibid., 101–102.

39 *Visions of Utopia Volume 1*, 2004.

40 See also Meyer, *Engaging the Everyday*, chapter 7.

41 Schelly, "How Policy Frameworks Shape Environmental Practice."

42 Ibid.

43 Also, personal laptop computers can only be used in personal rooms, although the community has a communal computer room where people can access communally owned computer technology.

44 *Visions of Utopia Volume 1*, 2004.

45 Schor, *Plenitude*, 138.

46 Ibid., 138, 137.

47 See "Collective Bikes." https://funologist.org/2012/08/14/collective-bikes/, accessed April 29, 2016.

48 Schor, *Plenitude*, 136.

49 See Maniates and Meyer, eds., *The Environmental Politics of Sacrifice*.

50 A significantly larger proportion of Twin Oaks members identify as polyamorous compared to adults in mainstream society. I was told during my visit that approximately twenty of the adults living at Twin Oaks at the time considered themselves to be "poly"—meaning either in or open to transparently nonmonogamous relationships—and that the number of people in the community identifying at polyamorous was much higher twenty-five years ago. One explanation for the prevalence of this relationship style at Twin Oaks, as explained by one member, is that the community "embraces diversity" and "avoids making dictates regarding the benefits and privileges given to certain relationships" so many members feel comfortable "experimenting with different relationship styles" while at Twin Oaks. Arrow's explanation that "poly is part of the radical sharing perspective" seems more applicable to the topics explored here. As Rosabeth Moss Kanter writes, based on her extensive study of nineteenth-century communes and intentional communities, "exclusive two-person bonds within a larger group, particularly sexual attachments, represent competition for members' emotional energy and loyalty. The cement of solidarity must extend throughout the group." Kanter, *Commitment and Community*, 86.

51 Schor, *Plenitude*, 2.

52 Ibid. See also Debbie Kasper, "Contextualizing Social Practices: Insights into Social Change," in *Putting Sustainability into Practice: Advances and Applications of Social Practice Theories*, ed. Emily Huddart Kennedy, Maurie J. Cohen, and Naomi Krogman (Cheltanham, UK: Edward Elgar, 2016), 25–46.

53 As John Urry wrote, "to slow down, let alone reverse, increasing carbon emissions and temperatures requires the total reorganization of social life, nothing more and nothing less." John Urry, "Consuming the Planet to Excess," *Theory, Culture & Society* 27 (2010): 191–213.

54 Gar Alperovitz, *America Beyond Capitalism: Reclaiming Our Wealth, Our Liberty, and Our Democracy* (Hoboken, NJ: Wiley, 2005).

55 Ibid., 6.

56 See Maniates and Meyer, eds., *The Environmental Politics of Sacrifice.*

57 *Visions of Utopia Volume 1*, 2004.

58 Schor, *Plenitude*, 70.

59 R. Buckminster Fuller, *Operating Manual for Spaceship Earth* (Carbondale: Southern Illinois University Press, 1969).

60 Lewis Mumford, *Technics and Civilization* (New York: Harcourt, Brace, 1934), 399.

61 Beth Greenfield "Welcome to the Commune Where 100 Adults Raise 17 kids," Yahoo News. https://www.yahoo.com/parenting/welcome-to-the-commune-where-100-adults-raise-17–120635816292.html, accessed September 8, 2015; Byron Pitts, Jackie Pou, and Lauren Effron, "Inside the Off-the-Grid Virginia Commune Where Everything from Housing to Childcare Is Shared," ABC News via *Nightline.* http://abcnews.go.com/Lifestyle/inside-off-grid-virginia-commune-housing-child-care/story?id=33190577, accessed September 8, 2015; Tara Dodrill, "Twin Oaks Commune in Virginia Blends Off Grid Harmony and Business Savvy," *Inquisitor.* http://www.inquisitr.com/2354394/twin-oaks-commune-in-virginia-blends-off-grid-harmony-and-business-savvy/?fb_action_ids=10153156374266795&fb_action_types=og.comments, accessed September 8, 2015; Sophie Jane Evans. "Inside Off-the-Grid Virginia Commune Where EVERYTHING—from Housing to Childcare—Is Shared (but Residents Must Get Group PERMISSION before Having a Baby)," *Daily Mail.* http://www.dailymail.co.uk/news/article-3205601/Inside-grid-Virginia-commune-housing-pay-childcare-shared-residents-PERMISSION-having-baby.html, accessed September 8, 2015.

62 Pitts, Pou, and Effron, "Inside the Off-the-Grid Virginia Commune."

63 Ibid.

64 "Utopia: It's Not Happening." http://funologist.org/2015/09/15/utopia-its-not-happening/, accessed September 16, 2015.

65 "Re-post Island." http://funologist.org/2011/12/10/re-post-island/, accessed September 16, 2015.

Chapter 5 Individualism and Symbiosis: The Dance at Dancing Rabbit

1 See http://www.dancingrabbit.org/about-dancing-rabbit-ecovillage/, accessed May 12, 2016.

2 This corresponds to the ideal population size of human settlements argued for in *Human Scale* by Kirkpatrick Sale; members say it is an ideal population for economies of scale in sharing sociotechnical systems, a thriving local economy for exchange, and a socially sustainable system in which people know their neighbors and fellow community members, at least as familiar faces. See Kirkpatrick Sale, *Human Scale* (CITY?: India: New Catalyst Books, 2007). This population goal and plans for an organized system of integrated villages as sub-communities also corresponds to Gar Alperovitz's claim that "there is no way to achieve democracy . . . without radical decentralization, ultimately in all probability to some sort of regional units." Gar Alperovitz, *America Beyond Capitalism: Reclaiming Our Wealth, Our Liberty, and Our Democracy* (Hoboken, NJ: Wiley, 2005).

3 John Meyer, *Engaging the Everyday: Environmental Social Criticism and the Resonance Dilemma* (Cambridge, MA: MIT Press, 2015), 47.

4 This case study demonstrates (as one author recently wrote in specific reference to DR) that "when the ecovillage is viewed as a site for the production and integration (or "bundling" in the terminology of E. Shove, M. Pantzar, and M. Watson, *The Dynamics of Social Practice: Everyday Life and How It Changes* [Thousand Oaks, CA: Sage], 2012), of social practices . . . two things become clear. First, choice-based models of environmental change employed implicitly or explicitly by local governments miss opportunities for transitioning to more sustainable consumption. Second, social competencies of interpersonal communication and conflict resolution are critical to sustainable consumption." R. Boyer, "Achieving One-Planet Living Through Transitions in Social Practice: A Case Study of Dancing Rabbit Ecovillage," *Sustainability: Science, Practice, & Policy* (2016). Published online April. Accessed May 5, 2016. http://www.google.com/archives/vol12iss1/1506–007.boyer. html. Citing Shove, M. Pantzar, and Watson, *The Dynamics of Social Practice.*

5 Keith went on to say: "We're shitting in our own bed, and we're gonna end up swimming in shit. History is the length of an arm, and humankind is the tip of a fingernail. I don't think we're going to make it for the long haul, but if we are, we have to stop shitting in our own bed."

6 This has long been a theme in environmental literature; Rachel Carson's *Silent Spring* identified environmental problems as impacting both natural systems and human health. See Rachel Carson, *Silent Spring* (New York: Houghton Mifflin Harcourt, 1962).

7 John Meyer, *Engaging the Everyday*, 160, citing Dolores Hayden, *Redesigning the American Dream: The Future of Housing, Work, and Family Life* (New York: Norton, 1984).

8 In an internal economic survey conducted in the spring of 2015, the average annual income among respondents was $13,000 (median $12,000). Among the twenty-three respondents, 35 percent said they wanted more income generating work, 30 percent said they wanted less, and 35 percent said their amount of paid work was just right.

9 There were many shared memories, motivations, and other commonalities among the people I met, but most specific examples were not mentioned by more than two members each. For example: two members told me about an old Elm tree on the property that had been cut down, two talked about how they had wanted to start their own community before moving to DR (itself an indication of valuing individuality), and two mentioned finding DR through an Internet search for "vegan community" (interesting since DR is not and has never been a vegan community and does not dictate the diets of members).

10 The analysis in this chapter builds on the most recent book by John Meyer, who writes, "If we are to talk about a politics of material practices, we must first unmask this invisibility. . . . The challenge, then, is to respect—and at times even to expand upon—the sense of privacy and freedom that is so often associated with everyday material practices while simultaneously opening them up to more explicit political contestation. To do so requires that we recognize the diverse forms of exclusion from these opportunities for privacy and freedom that are the everyday experience of many and recognize that these exclusions are no accident but instead reflect the exercise of power. We must also highlight the often unacknowledged constraints that even those of us who are privileged enough to benefit from these practices (driving, living in comfortable homes, owning property, etc.) regularly experience." Meyer, *Engaging the Everyday*, 76.

11 See Dancing Rabbit Ecovilliage, http://www.dancingrabbit.org/about-dancing-rabbit-ecovillage/dancing-rabbit-inc-nonprofit/, accessed April 28, 2016.

12 In *Engaging the Everyday* (see particularly chapter 6), John Meyer points specifically to the case of private automobiles to demonstrate that things that seemingly support individual freedom, when looked at from another angle, also work to undermine it—by requiring, in the automobile example, insurance payments, state surveillance through registration, time spent in traffic jams, and so forth.

13 DRVC includes a fleet of diesel vehicles and one all-electric car. The community used to make its own biodiesel, but currently does not. The community currently purchases soy diesel, but does run conventional petro-diesel in its vehicles during the winter, because vegetable diesel gels in cold temperatures. Thus, the community is currently out of compliance with this covenant, something they are very aware of and actively attempting to change. Community member Ethan told me, "We're not proud of this. But we're building a full-feature village. We're just not full feature yet."

14 Note that hand tools are conspicuously absent from this list. During my visit, I saw gas-powered lawn mowers and Weedwackers in use, and was told that some members use chainsaws for building and wood fuel harvesting.

15 See http://www.dancingrabbit.org/about-dancing-rabbit-ecovillage/vision/ecological-covenants/, accessed May 12, 2016.

16 One well-known option is Forest Stewardship Council certification, https://us.fsc.org/en-us, accessed May 12, 2016.

17 The Milkweed Mercantile, referred to simply as "The Mercantile" by members, where outside visitors can stay and where members can come to have a drink during happy hour, by some small grocery items, or eat a pizza at the popular once a week pizza night. See http://www.milkweedmercantile.com/, accessed May 12, 2016.

18 Kayla Brooke Jones, "Toward Sustainable Community: Assessing Progress at Dancing Rabbit Ecovillage." (Master's thesis, University of North Texas, August 2014). http://digital.library.unt.edu/ark:/67531/metadc700019/m1/3/, accessed November 15, 2016.

19 Jones, "Toward Sustainable Community"; updated data provided in a presentation by Brooke Jones at Dancing Rabbit Ecovillage in July 2015.

20 Importantly, members adopt this value in their development of land as well, using organic and permaculture gardening and natural building, to allow even the land that is developed to continue to do its thing. The sub-community called "Critters" at Dancing Rabbit is aptly named for its focus on animal husbandry. In addition to working to increase the productivity of their animal production, the group also considers the value of having self-sustaining animal production (selecting some for broodiness, meaning that the females are likely to naturally want to reproduce and raise their own young). This "makes room for nature to do its thing" while also being economically advantageous in the long run, as the animals reproduce themselves without continuously purchasing new stock.

21 Individuals who choose to access extra land outside the densely populated village living lease lands from the Land Trust based on reduced garden or agricultural lease fees. (The garden rate for leased land is 1/10th of one penny, and the agriculture rate is 1/10th of that.) The leases on this land are protected such that they cannot be evoked without, as one member put it, "a lot of process." Even short-term leases are protected for 12 years, so that efforts to improve soil or land are worth the investment.

22 John Meyer argues that land can never be fully commodified: "A 'commodity' has been defined as a thing that is produced for the sale or exchange and so perfectly alienable—consistent with the absolutist concept. In this way, the distinction between the scope and the character of private property is elided; the observation that an object is private property is regarded as synonymous with a claim that it has become property in the absolutist sense.... I will argue that the formation of an alternative to the absolutist concept can begin from the recognition that property is rarely—and land is never—fully commodified." Meyer, *Engaging the Everyday*, 104.

23 There is a minimum payment of $5/month for small warrens.

24 Moose also discussed the misinformed view that smaller houses are always cheaper; he had recently helped someone completed a tiny dwelling structure that ended up costing $125 per square foot, while he had preciously completed a larger structure with very similar materials for $65 per square foot. Moose also said that the community was slowly recognizing the value of building efficiency, and focusing less on "natural" building and more on "green" building, which emphasizes resource usage across the life of a building and not just the materials that go into building it. He hoped that the community's mistakes, in terms of having inexperienced builders construct inefficient but natural buildings, could help evolve the community's own building practices and educate others about the different ways to achieve the goals of environmental responsibility in building practice.

25 Christopher Alexander, Sara Ishikiawa, and Murray Silverstein, *A Pattern Language: Towns—Buildings—Construction* (New York: Oxford University Press, 1977).

26 Ibid., xiii.

27 Jones, "Toward Sustainable Community."

28 Ibid.

29 The variety of cooperatively organized sharing systems means that members having access to Internet, laundry, a playroom for kids, a library, and a huge living room for gathering with friends in the common house, a large shared space for dancing/yoga/meditation/workshops, and even the bounty provided by goats, among many other things; this section focuses on transportation, electricity, and food.

30 A concept originally introduced by Ruth Schwartz Cowan in *A Social History of American Technology* (New York: Oxford University Press, 1997).

31 Meyer, *Engaging the Everyday*, 122.

32 The rates for DRVC at the time of my visit included a $100 DRVC refundable deposit, then usage charges of 65 cents/mile, 80 cents/mile in the truck, and the possibility of long distance rental for $50–75/day with a maximum distance of 300 miles/day.

33 Meyer, *Engaging the Everyday*, 138. Cites Anthony Westen, *Mobilizing the Green Revolution: An Exuberant Manifesto* (Gabriola Island, BC: New Society Publishers, 2012), 49–50.

34 From Dancing Rabbit Eco-Audit: The community consumed 592 kwh per person in 2012, 768 kwh per person in 2013, and 871 kwh per person in 2014, compared to the U.S. average of 4,168 per person. The consumption data from Dancing Rabbit is only based on BEDR usage and does not include the off-grid energy systems. See Jones, "Toward Sustainable Community."

35 Data provided in a presentation by Kayla Brooke Jones at Dancing Rabbit Ecovillage in July 2015.

36 Another way to put this is that, at Dancing Rabbit, labor is not de-commodified. Members are paid to make pizza at the café and paid to make food for the visitors

who come in the Visitor Program, members must pay for the own food and housing. I heard estimates of housing construction cost from $15,000 to $35,000, for homes that were largely built by their occupants; hiring builders would obviously increase costs substantially. In an internal economic survey conducted in the spring of 2015, members reported an average cost of home building of $41,386, with a range from $300 to $140,000, and reported an average number of unpaid hours that went into building their homes of 3,655. When talking to me about the DR economy, some members talked about having more time to do things they really loved or to do things for the community, but these members were also paid for at least some of the work that they did for DR. Some scholars suggest that decommodification of labor is necessary for a truly sustainable sharing economy (like Juliet Schor in *Plenitude: The New Economics of True Wealth* [New York: Penguin, 2010]), but the longevity and success of Dancing Rabbit suggest that alternative forms of residential dwelling can be organized around diverse forms of economic organization.

37 See Dancing Rabbit Ecovillage. http://www.dancingrabbit.org/about-dancing-rabbit-ecovillage/dancing-rabbit-inc-nonprofit/; http://www.dancingrabbit.org/introducing-the-center-for-sustainable-and-cooperative-culture-formerly-known-as-dr-inc/, accessed April 15, 2016.

38 See http://www.ic.org/, accessed May 12, 2016.

39 In an internal economic survey conducted by the community in the spring of 2015, completed by twenty-three members (ten women and thirteen men; age breakdown of respondents: eighteen to twenty-nine years old: 9 percent; thirty to thirty-nine: 48 percent; forty to forty-nine: 30 percent; fifty to fifty-nine: 5 percent; sixty-plus: 9 percent) who had lived in the community for varying amounts of time (less than two years: 22 percent; two to five: 26 percent; five to ten: 22 percent; ten-plus: 30 percent), respondents indicated that 32 percent work online/on phone, serve outside community; 32 percent work in person for business or individuals outside community:; 77 percent work for DR; 59 percent work for business at DR; 18 percent rely on income from investments; 36 percent rely on family support/inheritance; 27 percent rely on government or nonprofit assistance; and 41 percent reported other income such as loans or IRA payments. Among respondents, 17 percent said their relationship to money was a lot more stressful at DR than when living elsewhere, 17 percent said a little more stressful, 44 percent said they experienced the same amount of stress, 9 percent said a little less stress, and 13 percent said a lot less. The community is very conscious of the economic barriers to living at DR (due to the community vehicle requirement, it is very difficult for members to commute to work daily, the dominant model of income generation in America) and is actively working to build a sustainable village economy.

40 See http://www.milkweedmercantile.com/.

41 There are several other intentional and alternative communities very close to DR, including Sandhill Farm (see http://www.sandhillfarm.org/) and Red Earth Farm (http://www.redearthfarm.org/), both within walking distance.

42 Dancing Rabbit Ecovillage. http://www.dancingrabbit.org/about-dancing-rabbit-ecovillage/social-change/economy/local-currency/; see also Matador Network, accessed April 29, 2016. http://matadornetwork.com/life/8-best-areas-united-states-alternative-lifestyles/.

43 Like, for example, co-counseling. See http://cci-usa.org/. Dancing Rabbit member Sasha, who helped facilitate the session on inner sustainability, told me that community members are not dogmatic in the application of cocounseling although

many do use the core principles to resolve both inner and interpersonal conflict and to promote personal well-being.

44 Claire recognizes that, as political theorist John Meyer writes, "Home is not a wholly private realm disconnected from the public. Both materially and conceptually, it is embedded in a web of connections and flows with the outside." Meyer, *Engaging the Everyday*, 158.

45 See Schor, *Plenitude*.

46 The community also has a lovely pond, at the very back of the community's developed land, where nudity is socially normative and acceptable, which offers a space of reprieve for members who want to avoid the public gaze of touring visitors or others who see only the "front stage." See Erving Goffman, *Presentation of Self in Everyday Life* (New York: Random House, 1959).

47 Like The Farm, members of Dancing Rabbit cannot get bank loans for construction or mortgages, because the land is collectively rather than individually owned.

48 Prospective members must be residents of DR for six months prior to becoming a member, so this captures people who were new to the community as well.

49 For a review of practice theory and its importance in explaining consumption behavior and behavioral change, see Emily Huddart Kennedy, Maurie J. Cohen, and Naomi T. Krogman, eds., *Putting Sustainability into Practice: Applications and Advances in Research on Sustainable Consumption* (Cheltenham, UK: Edward Elgar, 2015). See, in particular, Chelsea Schelly, "How Policy Frameworks Shape Environmental Practice: Three Cases of Alternative Dwelling," 195–203.

50 See Debbie Kasper, "Contextualizing Social Practices: Insights into Social Change," in *Putting Sustainability into Practice: Applications and Advances in Research on Sustainable Consumption*, ed. Emily Huddart Kennedy, Maurie J. Cohen, and Naomi T. Krogman (Cheltenham UK: Edward Elgar, 2015), 25–46.

51 She also said, "Actually, someone who was showering and wearing clean clothes everyday would probably experience some reverse peer pressure; although no one would directly tell them not to, it's just not what we do here, and we feel like it's beneficial to be more open to diversity in how people look and dress and to not feel pressure to meet certain standards."

52 Meyer, *Engaging the Everyday*, 110.

53 As Meyer writes, "by living amid technologies that remain outside the mainstream experience of many, they also become acutely aware that these are necessarily entwined with decisions and priorities about how to live well—a broad and apt definition of politics itself." Meyer, *Engaging the Everyday*, 152.

54 Ibid., 145.

55 Boyer, "Achieving One-Planet Living Through Transitions in Social Practice."

56 As Gar Alperovitz writes, "there is no way to achieve meaningful individual liberty in the modern era without individual economic security and greater amounts of free time." Alperovitz, *America Beyond Capitalism*, 233.

Chapter 6 Self-Sufficiency as Social Justice: The Case of Earthship Biotecture

1 See R. Buckminster Fuller, *Operating Manual for Spaceship Earth* (Carbondale: Southern Illinois University Press, 1969); Donnella H. Meadows, Jorgen Randers, and Dennis L. Meadows, *The Limits to Growth: A Report for the Club of Rome's Project on the Predicament of Mankind* (New York: New American Library, 1974).

2   Mischa Hewitt and Kevin Telfer, "Earthships as Public Pedagogy and Agents of Change," in *Handbook of Public Pedagogy: Education and Learning Beyond Schooling*, ed. Jennifer A Sandlin, Brian D. Schultz, and Jake Burdick (New York: Routledge, 2010), 171–178.

3   Ibid., 173; See Fuller, *Operating Manual for Spaceship Earth*.

4   This move toward ephemeralization in design is not necessary unique to Earthships; others have long made arguments for "cradle-to-cradle" designs and "soft" energy paths. On cradle to cradle, see William McDonough, *The Hannover Principles: Design for Sustainability* (Charlottesville, VA: William McDonough & Partners, 1992); William McDonough and Michael Braungart, *Cradle to Cradle: Remaking the Way Things Work* (New York: North Point Press, 2002); on "soft" energy technology, see Amory Lovins, "Energy Strategy: The Road Not Taken?" *Foreign Affairs* 55 (1976): 65–96; Amory Lovins, *Soft Energy Paths: Toward a Durable Peace* (New York: Harper & Row, 1977); Amory Lovins, "Soft Energy Technologies," *Annual Review of Energy* 3 (1978): 477–517; Paul Hawken, Amory Lovins, and L. Hunter Lovins, *Natural Capitalism: Creating the Next Industrial Revolution* (Boston: Little, Brown, 1999); See also Timothy W. Luke, "Ephemeralization as Environmentalism: Rereading R. Buckminster Fuller's Operating Manual for Spaceship Earth," *Organization & Environment* 23 (2010): 354–362.

5   See Kevin Telfer, "Earth Mover (A profile of Mike Reynolds)" *Architects' Journal* (2003): 18–19.

6   See Ann Swidler, *Organization with Authority: Dilemmas of Social Control in Free Schools* (Cambridge, MA: Harvard University Press, 1979).

7   Earthship Biotecture runs three-week internship programs as well as a three-month academy program in Taos, New Mexico, where their headquarters are located. Individuals can also voluntarily participate in Earthship "builds"—where individuals are building Earthship homes (either with or without the professional assistance of Earthship Biotecture) or in instances when Earthship Biotecture is constructing a humanitarian project (examples from the past include living structures in disaster-stricken Haiti and schools in Africa, among many others)—across the world. See www.earthship.com, accessed March 16, 2013.

8   Michel Foucault, *Discipline and Punish* (New York: Vintage Books, 1980); Michel Foucault, "Governmentality," in *The Foucault Effect: Studies in Governmentality*, ed. G. Burchell, C. Gordon, and P. Miller (Chicago: University of Chicago Press, 1991), 87–104; Pierre Bourdieu, "Rethinking the State: Genesis and Structure of the Bureaucratic Field," *Sociological Theory* 12 (1994): 1–18.

9   Lewis Mumford, *Technics and Civilization* (New York: Harcourt, Brace, 1934).

10 Earthship Biotecture's website (www.earthship.com) and/or any and all of the books published by Michael Reynolds (see below) provide a wealth of information about building design and the history, activities, and culture of the company.

11 Mike Reynolds has a number of books, both technical and fanciful, about Earthships. See Michael Reynolds, *A Coming of Wizards: A Manual of Human Potential* (Taos, NM: High Mesa Foundation, 1989); Michael Reynolds, *Earthship Volume I: How to Build your Own* (Taos, NM: Solar Survival Press, 1990); Michael Reynolds, *Earthship Volume II: Systems and Components* (Taos, NM: Solar Survival Press 1990); Michael Reynolds, *Earthship Volume III: Evolution Beyond Economics* (Taos, NM: Solar Survival Press, 1993); Michael Reynolds, *Comfort in Any Climate* (Taos, NM: Solar Survival Press, 2001); Michael Reynolds, *Journey: Part One* (Taos, NM: Earthship Biotecture, 2008).

12   In the Northern hemisphere.

13   Hewitt and Telfer, "Earthships as Public Pedagogy and Agents of Change," 173, 174.

14   Foucault, "Governmentality." However, an isolationist or escapist worldview does not necessarily motivate the self-sufficiency promoted by the principles of design and construction in Earthships. Rather, architect Michael Reynolds and the employees, volunteers, and homeowners I met share a conception of justice that sees the right to dwell comfortably without the intervention of money-based technological systems and infrastructures and without unnecessary imposition on other humans, species, or the natural world at large as the right of every human being.

15   Sociologist Max Weber, one of three classical sociological theorists that every undergraduate learns about in their first social theory course, wrote about charismatic leadership as based, among other things, on emotive appeal. Weber wrote that charismatic leadership is based on "a certain quality of an individual personality by virtue of which he is set apart from ordinary men and treated as endowed with supernatural, superhuman, or at least specifically exceptional qualities." Max Weber, *The Theory of Social and Economic Organization*, trans. A. M. Henderson and Talcott Parsons (1925; rpt. New York: Free Press, 1964). See also Max Weber, *Economy and Society: An Outline of Interpretive Sociology*, trans. G. Roth and C. Wittich (1935; New York: Bedminster, 1968).

16   See Reynolds, *A Coming of Wizards*.

17   Ann Swidler in *Organization with Authority* writes that charismatic leaders "draw on collective symbols and emotions to sustain their personal prestige, and thus their influence" (80).

18   It is important to preemptively note here how this structure for land acquisition and home construction fits with the justice orientation of Earthship Biotecture: cutting down on the cost of land purchase through smaller plots and communally owned land increases access to the potential for building an Earthship without a mortgage and without the lifelong schedule of bills that plaque most homeowners. The STAR community (an acronym for Social Transformation Alternative Republic, which is located far down a dirt road in an extremely isolated part of Taos County) and REACH community (an acronym for Rural Earthship Alternative Community Habitat) were conceived to address the justice concerns regarding accessibility to not only homes, but homes that provide for the needs of dwellers.

19   During my time at the Greater World Community, two of the interns I met in July were still around in December, three other interns from the previous internship program were still living in The Hive during my December internship, and I met two former academy participants who were still lingering.

20   See Meadows, Randers, and Meadows, *The Limits to Growth*; Fuller, *Operating Manual for Spaceship Earth*.

21   See Ivan Illich, *Tools for Conviviality* (New York: Harper & Row, 1973); Malcom Hollick, "The Appropriate Technology Movement and its Literature: A Retrospective," *Technology and Society* 4 (1982): 213–229; Andrew Kirk, *Counterculture Green: The Whole Earth Catalog and American Environmentalism* (Lawrence: University of Kansas Press, 2001).

22   K. Darrow, and R. Pam, *Appropriate Technology Sourcebook* (Stanford, CA: Volunteers in Asia, Inc., 1976).

23   Allan Schnaiberg, "Redistributive Goals versus Distributive Politics: Social Equity Limits in Environmental and Appropriate Technology Movements," *Social Inquiry* 53 (1983): 200–215.

24 Kirk, *Counterculture Green*.

25 Lovins, "Energy Strategy: The Road Not Taken?"; Lovins, *Soft Energy Paths*; Lovins, "Soft Energy Technologies."

26 Hollick, "The Appropriate Technology Movement and its Literature," 215.

27 Luke, "Ephemeralization as Environmentalism," 358; C. Mitcham, "The Concept of Sustainable Development: Its Origins and Ambivalence," *Technology in Society* 17 (1995): 311–326.

28 Langdon Winner, *The Whale and the Reactor: A Search for Limits in the Age of High Technology* (Chicago: University of Chicago Press, 1986); See also Langdon Winner, "Do Artifacts Have Politics?" *Daedalus* 109 (1980): 121–136.

29 Lovins, "Energy Strategy: The Road Not Taken?"; Lovins, *Soft Energy Paths*; Lovins, "Soft Energy Technologies."

30 See, for example, Lester Brown, *Plan B: Rescuing a Planet under Stress and a Civilization in Trouble* (New York: Earth Policy Institute, 2003).

31 While a few of the employees and interns I met were expressively not motivated by environmental concern, all the homeowners were, and some of the homeowners could be described as eco-centric in their worldview, which is very different than mainstream culture. Nancy told me that she was concerned about the environment because "it's not ours to destroy." Margaret, similarly, told me to watch for snakes on the road in the Greater World Community, and to make sure to move them out of the way if I saw one. Stephanie is deeply concerned about the suffering of animals, plants, and the planet; she believes that a river can suffer, and this eco-centric worldview is one of her motivations for building and living in an Earthship. However, others were less concerned, and at least one employee admitted frankly that he was not motivated by care for the environment. Yet, as one intern named Shiloh told me, "The thing is, even if you don't do it for environmental reasons, it's the pragmatics of the benefits that you just can't avoid. I mean, if you buy the environmental stuff it helps, but the environmental benefits are there regardless. It's built into it, that's why it's such an ideal system."

32 See Foucault, *Discipline and Punish*.

33 This is a reference to environmental writer Henry David Thoreau. See Henry David Thoreau, *Walden, or Life in the Woods: And, On the Duty of Civil Disobedience* (New York: New American Library, 1999).

34 Not just, as an intern named Chris who was participating in the internship program along with her father, put it, "quote-unquote green" buildings.

35 Hewitt and Telfer, "Earthships as Public Pedagogy and Agents of Change," 173.

36 See Chris Shilling, "Embodiment, Experience, and Theory: In Defence of the Sociological Tradition," *Sociological Review* 49 (2001): 327–344; Chris Shilling, "Sociology and the Body: Classical Traditions and New Agendas," *The Sociological Review* 55 (2007): 1–18; Chris Shilling and Philip A. Mellor, "Cultures of Embodied Experience: Technology, Religion and Body Pedagogics," *Sociological Review* 55 (2007): 531–549.

37 Foucault, *Discipline and Punish*. See also Michel Foucault, *The History of Sexuality* (New York: Vintage, 1990).

38 See Marcel Mauss, "Techniques of the Body" [1934], *Economy & Society* 2 (1973): 70–88; Bourdieu, "Rethinking the State."

39 Mauss, "Techniques of the Body."

40 See, for example, Emily Huddart Kennedy, Maurie J. Cohen, and Naomi T. Krogman, eds., *Putting Sustainability into Practice: Applications and Advances in Research on Sustainable Consumption* (Cheltenham, UK: Edward Elgar, 2015).

41 Here, the word *how* should not be taken to imply causality. I am not arguing for a directly causal relationship, but simply noting that these things are integrally related.

42 See Reynolds, *Comfort in Any Climate*.

43 Matthew B. Crawford, *Shop Class as Soulcraft: An Inquiry into the Value of Work* (New York: Penguin, 2009).

44 Hewitt and Telfer, "Earthships as Public Pedagogy and Agents of Change," 174.

45 This is not unique to Earthships, but can be true in any passive solar building. Susan and Tom, solar homeowners who built a LEED certified passive solar home in Wisconsin, expressed this by saying: "The way you feel in a house is so critical, not just how it performs, but how you feel in it. And to be able to sit in it and get that nice, I mean we look out and just see blue sky, and when we get sunlight in here that really makes the difference."

46 Such as McDonough and Braungart, *Cradle to Cradle*. See also Luke, "Ephemeralization as Environmentalism."

47 This section explicitly considers what some sociologists increasingly acknowledge, that geography and physical space matter when it comes to the development of social ideas and social groups. See Andrew Abbott, "Of Time and Space," *Social Forces* (1997): 1149–1182.

48 I heard two explanations of the hum during my time in Taos: a unique frequency, and the sound of such expansive quiet.

49 Andrew Abbott, "Of Time and Space"; Mustafa Emirbayer, "Manifesto for a Relational Sociology," *American Journal of Sociology* 103 (1997): 281–317.

50 Max Weber, *The Protestant Ethic and the Spirit of Capitalism* (1904; rpt. New York: Scribner, 1930).

51 The gray water and black water treatment systems utilized in an Earthship have not been problematic for homeowners at all; the food grown in them is actually, according to one Earthship employee, cleaner and safer than food bought in a grocery store (based on their testing). Yet the systems that support residential dwelling in America are incredibly biased toward a model of concentration and, for lack of a better word, *away-ness*—we expect water and septic waste to go away from our homes and be treated along with the waste from 100 or 1,000 or 10,000 other homes. This is, arguably, in no way a better, safer, or cleaner system. Yet it is the system that arguably offers the most money-making potential, as well as the most reliance and dependency on the part of the dweller.

52 Hewitt and Telfer, "Earthships as Public Pedagogy and Agents of Change," 171.

53 In this way, Earthship culture is similar to Ann Swidler's "organization without authority"—because instead of establishing social control through status, Earthship Biotecture establishes cultural cohesion by setting itself up against the mainstream. See Ann Swidler, *Organization without Authority*, especially Introduction, 114, and chapter 6.

54 As an intern named Chris told me, "Here, you're responsible for your own resources."

55 Hewitt and Telfer, "Earthships as Public Pedagogy and Agents of Change," 175.

56 The unique organization of Earthship Biotecture—which is both for-profit and nonprofit, with both employed workers and many unpaid voluntary participants—is significant and arguably deserves more attention that it is given here; as Swidler writes, "Organizational innovation and cultural change are continually intertwined, since it is culture that creates the new images of human nature and new symbols with which people can move one another. Organizations, in turn, are the

contexts within which cultural meanings are used, tested, and made real." Swidler, *Organization without Authority*, viii.

57 In an "organization without authority" such as Earthship Biotecture, "What replaces formal structure is a collective capacity for continual, conscious attention to purpose. Indeed, it becomes inappropriate to speak of an organization and more appropriate to talk about capacities to undertake and implement purposes." Swidler, *Organization without Authority*, 178.

58 There is certainly a dash of this survivalist mentality within the culture of Earthships. One intern named Shiloh told me, "We have with our generation what nuclear war was during the Cold War, but in the mindset of environmental degradation or just societal downfall from economic factors." Jane told me that Mike says, "This is the house you want when the shit hits the fan." Yet listening to Mike's lectures during the summer seminar I attended, I believe this survivalist mentality is not the same as an isolationist, every man for himself, or all against all pretension. Mike told seminar participants that if every individual is able to provide for themselves, meeting many of their own needs in terms of comfort, electricity, and even food, then if and when scarcity hits, there will actually be less conflict, because each person has a little bit of what they need for themselves. So, if it is a survivalist mentality, perhaps it is best described as collective survivalism for the sake of one another, not just ourselves.

59 Every other species on the planet arguably already lives like this. For an interesting take on the "rift" in human-nature relationships, see John Bellamy Foster, "Marx's Theory of Metabolic Rift: Classical Foundations for Environmental Sociology," *American Journal of Sociology* 105 (1999): 366–405.

Chapter 7 Dwelling in Resistance

1 Melanie Jaefer-Erben and Ursula Offenberger, "A Practice Theory Approach to Sustainable Consumption," *GAIA* 23 (2014): 166–174; 168, italics in original.

2 Marcel Mauss, *Seasonal Variations of the Eskimo: A Study in Social Morphology* (London: Routledge & Kegan Paul, 1979). Mauss also claimed that sociological studies "are intended not just to describe but also to elucidate the material substratum of societies," 19–20.

3 Durkheim claimed that thorough sociological study "includes the mass of individuals who compose the society, the way in which they occupy the land, and the nature and configuration of objects of every sort which affect collective relations. . . . The constitution of this substratum affects, directly or indirectly, all social phenomena." Emile Durkheim in *L'Annee Sociologique* 2 (1899): 520; cited in James Fox's translator's introduction to *Seasonal Variations of the Eskimo*, 3.

4 Marcel Mauss, "Techniques of the Body" [1934], *Economy & Society* 2 (1973): 70–88.

5 See Lewis Mumford, *Technics and Civilization* (New York: Harcourt, Brace, 1934); Herbert Marcuse, "Some Social Implications of Modern Technology," *Studies in Philosophy and Social Sciences* 9 (1941): 138–162; Herbert Marcuse, *One-Dimensional Man: Studies in the Ideology of Advanced Industrial Society* (Boston: Beacon Press, 1964).

6 See Pierre Bourdieu, "Rethinking the State: Genesis and Structure of the Bureaucratic Field," *Sociological Theory* 12 (1994): 1–18; G. Burchell, Colin Gordon, and Peter Miller, eds., *The Foucault Effect: Studies in Governmentality* (Chicago:

University of Chicago Press, 1991). See also Jürgen Habermas, "Technology and Science as 'Ideology,'" in *Toward a Rational Society: Student Protest, Science, and Politics* (Boston: Beacon Press, 1970), 90–122.

7 This is not meant to imply a conscious intention.

8 Our awareness of this isolation and dependence, theoretically speaking, is irrelevant, because we experience it nonetheless.

9 See W. I. Thomas and Florian Znaniecki, *The Polish Peasant in Europe and America*, ed. Eli Zaretsky (Champagne: University of Illinois Press, 1926). Hans Joas claims pragmatism expanded the theory of pragmatism. See Hans Joas, *Pragmatism and Social Theory* (Chicago: University of Chicago Press, 1993). See also Robert Park and Ernest Burgess, *Introduction to the Science of Sociology* (Chicago: University of Chicago Press, 1921); Robert Park, Ernest Burgess, and Roderick McKenzie, *The City* (Chicago: University of Chicago Press, 1925); Andrew Abbott, "Of Time and Space," *Social Forces* 75 (1997): 1149–1182; Andrew Abbott, *Department and Discipline: Chicago Sociology at One Hundred* (Chicago: University of Chicago Press, 1999). From Hans Joas, see Joas, *Pragmatism and Social Theory*; Hans Joas, The *Creativity of Action* (Chicago: University of Chicago Press, 1996); Hans Joas, "The Creativity of Action: Pragmatism and the Critique of the Rational Actor Model" (Manuscript, Free University, Berlin, John F. Kennedy Institute for North American Study, n.d).

10 Thomas and Znaniecki, *The Polish Peasant in Europe and America*, 12.

11 Mustafa Emirbayer and Ann Mische, "What Is Agency?" *American Journal of Sociology* 103 (1998): 969; see also Joas, "The Creativity of Action."

12 See Andrew Abbott, "Transcending General Linear Reality," *Sociological Theory* 6 (1988): 169–186.

13 Emirbayer and Mische, "What Is Agency?," 963.

14 Ibid., 970.

15 See Paul C. Stern, *Environmentally Significant Consumption: Research Directions* (Washington, DC: National Academy Press, 1997).

16 See Rachael Shwom and Janet A. Lorenzen, "Changing Household Consumption to Address Climate Change: Social Scientific Insights and Challenges" *WIREs Climate Change* 3 (2012): 379–395.

17 See Elizabeth Shove, "Beyond the ABC: Climate Change Policy and Theories of Social Change," *Environment and Planning A* 42 (2010): 1273–1285; Lorraine Whitmarch, Saffron O'Neill, and Irene Lorenzoni, "Climate Change or Social Change? Debate Within, Amongst, and Beyond Disciplines," *Environment and Planning A* 43 (2010): 258–261; Elizabeth Shove, "On the Difference Between Chalk and Cheese: A Response to Whitmarch et al.'s comments on 'Beyond the ABC: Climate Change Policy and Theories of Social Change,'" *Environment and Planning A* 43 (2011): 262–264; Charlie Wilson and Tim Chatterton, "Multiple Models to Inform Climate Change Policy: A Pragmatic Response to the 'Beyond the ABC's' Debate," *Environment and Planning A* 43 (2011): 2781–2787.

18 See Carlo Jaeger, *Risk, Uncertainty, and Rational Action* (London: Earthscan, 2001).

19 See Steven D. Levitt and John A. List, "*Homo Economicus* Evolves," *Science* 319 (2008): 909–910.

20 See Allan Schnaiberg, *The Environment: From Surplus to Scarcity* (New York: Oxford University Press; 1980); Allan Schnaiberg and Kenneth A. Gould, *Environment and Society: The Enduring Conflict* (New York: St. Martin's Press,

1994); Kenneth A. Gould, David N. Pellow, and Allan Schnaiberg, "Interrogating the Treadmill of Production: Everything You Wanted to Know about the Treadmill but Were Afraid to Ask," *Organization and Environment* 17 (2004): 296–316. See also Gregory C. Unruh, "Understanding Carbon Lock-in," *Energy Policy* 12 (2000): 817–830; Christer Sanne, "Willing Consumers—Or Locked-in? Policies for a Sustainable Consumption," *Ecological Economics* 42 (2002): 273–287; Gregory C. Unruh, "Escaping Carbon Lock-in," *Energy Policy* 30 (2002): 317–325; Oksana Mont, "Institutionalisation of Sustainable Consumption Patterns Based on Shared Use," *Ecological Economics* 50 (2004): 135–153; Tim Jackson and Eleni Papathanasopoulou, "Luxury or 'Lock-in'? An Exploration of Unsustainable Consumption in the UK: 1968 to 2000," *Ecological Economics* 68 (2008): 80–95.

21 Stephen Vaisey, "Motivation and Justification," *American Journal of Sociology* 114 (2009): 1675–1715. See also William Sewell Jr., "A Theory of Structure: Duality, Agency, and Transformation," *American Journal of Sociology* 98 (1992): 1–29.

22 Stephen Toulmin, *Cosmopolis: The Hidden Agenda of Modernity* (Chicago: University of Chicago Press, 1992), 192, italics in original.

23 Vaisey, "Motivation and Justification," 1675.

24 Juliet Schor, *Plenitude: The New Economics of True Wealth* (New York: Penguin, 2010). Schor also provides a critique of the dominant ways of thinking about environmentally responsible behavior, and writes that, in moving toward a plenitude economy (which is her focus), "the question of well-being will begin to solve itself. In addition to, and perhaps more important than, the question of whether we are better or worse off in a quantitative sense—the issue to which the literature is addressed—we will discover that we are different" (179–180).

25 This conceptually borrows from Martin Heidegger, who argued that technology is "enframing"—also invokes a visual metaphor to suggest that technology frames our perception of reality, in some ways limiting the view. See Martin Heidegger, *Questions Concerning Technology and Other Essays* (New York: Harper & Row, 1977).

26 Aldo Leopold, *A Sand Country Almanac* (New York: Oxford University Press, 1949).

27 See Pierre Bourdieu, *Outline of a Theory of Practice* (Cambridge: Cambridge University Press, 1977); See also Debbie Kasper, "Contextualizing Social Practices: Insights into Social Change," in *Putting Sustainability into Practice: Applications and Advances in Research on Sustainable Consumption*, ed. Emily Huddart Kennedy, Maurie J. Cohen, and Naomi T. Krogman (Cheltenham, UK: Edward Elgar, 2015), 25–46.

28 Mumford, *Technics and Civilization*, 273.

29 Further, it is not my intention to attempt to precisely predict what does impact the formation of these orientations.

30 See Debbie V. S. Kasper, "Redefining Community in the Ecovillage," *Human Ecology Review* 15 (2008): 12–24; Debbie V. S Kasper, "Ecological Habitus: Toward a Better Understanding of Socioecological Relations," *Organization and Environment* 22 (2009): 311–326.

31 Max Weber, *The Protestant Ethic and the Spirit of Capitalism* (1904; rpt. New York: Scribner, 1930).

32 Max Weber, "The Social Psychology of the World Religions," in *From Max Weber: Essays in Sociology*, ed. H. H. Gerth and C. Wright Mills (1946; rpt. New York: Oxford University Press, 1958), 267–301. Weber argues that practical ethics are not "a simple 'function' of the social situation" and that it would be a misunderstanding of the concept to assume "it represents the stratum's 'ideology,' or that it is a

'reflection' of a stratum's material or ideal interest-situation." Weber, "The Social Psychology of the World Religions," 270.

33 Ibid., 281.

34 Ibid., 267–268. Weber also recognized the complexity of understanding the relationship between orienting ethics and action. Also on page 268, he writes, "We should lose ourselves in these discussions if we tried to demonstrate these dependencies in all their singularities. Here we can only attempt to peel off the directive elements in the life-conduct. . . . These elements have stamped the most characteristic features upon practical ethics, the features that distinguish one ethic from others."

35 Ibid., 294.

36 Ibid., 290–291.

37 See Andreas Reckwitz, "Toward a Theory of Social Practice: A Development in Culturalist Thinking," *European Journal of Social Theory* 5 (2002): 243–263; Gert Spaargaren, "Sustainable Consumption: A Theoretical and Environmental Policy Perspective," *Society & Natural Resources* 16 (2003): 687–701; Gert Spaargaren, "Theories of Practices: Agency, Technology, and Culture: Exploring the Relevance of Practice Theories for the Governance of Sustainable Consumption Practices in the New World-Order," *Global Environmental Change* 21 (2011): 813–822.

38 Shwom and Lorenzen, "Changing Household Consumption to Address Climate Change," 387 and 386. See also Emily Huddart Kennedy, Maurie J. Cohen, and Naomi T. Krogman, eds., *Putting Sustainability into Practice: Applications and Advances in Research on Sustainable Consumption* (Cheltenham, UK: Edward Elgar, 2015).

39 Anthony Giddens defines lifestyle as a "more or less integrated set of practices which an individual embraces, not only because such practices fulfill utilitarian needs but because they give material form to a particular narrative of self-identity." See Giddens, *Modernity and Self-Identity: Self and Society in the Late Modern Age* (Stanford, CA: Stanford University Press, 1991), 81. See also Anthony Giddens, *The Constitution of Society* (Berkeley: University of California Press, 1984). See also Janet A. Lorenzen, "Going Green: The Process of Lifestyle Change," *Sociological Forum* 27 (2012): 94–116. Arguably, the concept of lifestyle still puts too much emphasis on conscious choice to fully understand how practices emerge and evolve.

40 Jaefer-Erben and Offenberger, "A Practice Theory Approach to Sustainable Consumption," 167. See also Halina Szejnwald Brown, Halina Szejnwald, Philip J. Vergragt, and Maurie J. Cohen, "Societal Innovation in a Constrained World: Theoretical and Empirical Perspectives," in *Innovations in Sustainable Consumption: New Economics, Socio-Technical Transitions and Social Practices*, ed. M. J. Cohen, H. S. Brown, and P. J. Vergragt (Cheltenham and Northampton, MA: Edward Elgar, 2013), 1–30.

41 Weber, *The Social Psychology of the World Religions*, 299.

42 Alexis De Tocqueville, *Democracy in America* [1835], trans. Arthur Goldhammer (New York: Penguin, 2004).

43 Ibid. Gar Alperovitz claims that the great historic values of the United States are "above all equality, liberty, and meaningful democracy" (ix) and that these values are "the most important values that have given meaning to American history from the time of the Declaration of Independence" (1). Gar Alperovitz, *America Beyond Capitalism: Reclaiming Our Wealth, Our Liberty, and Our Democracy* (Hoboken, NJ: Wiley, 2005). This argument is not meant to suggest that the values of equality and freedom and uniquely American, but simply to highlight that they are

longstanding values in America, and that these American case studies of alternative residential dwelling, rather than being motivated by alternative values, seem to correspond to rather than contradict more widely held, traditional values.

44 The concept of social world is intended to highlight how different socially and spatially organized groups may develop into unique social worlds, with distinctive vocabulary, material systems, means of organizing them, cultural customs, and practices. The unique elements of social worlds shape the kinds of action and interaction that happen within them. See Nels Anderson, *The Hobo* (1923; rpt. Chicago: University of Chicago Press, 1961); Paul G. Cressey, *The Taxi-Dance Hall: A Sociological Study in Commercialized Recreation and City Life* (Chicago: University of Chicago Press, 1932); Paul S. P. Siu, *The Chinese Laundryman: A Study of Social Isolation* (New York: New York University Press, 1987); Anselm Strauss, *A Social World Perspective*, in *Creating Sociological Awareness*, ed. A. Strauss (1978; rpt. New Brunswick, NJ: Transaction Press, 1990), 233–244.

45 Tocqueville, *Democracy in America*.

46 See both Juliet Schor's *Plenitude* and Thomas Princen, *The Logic of Sufficiency* (Cambridge, MA: MIT Press, 2005).

47 Tocqueville, *Democracy in America*, 244.

48 Dena M. Gromet, Howard Kunreuther, and Richard P. Larrick, "Political Ideology Affects Energy-Efficiency Attitudes and Choices," *Proceedings of the National Academy of Science* (2013): 9314–9319.

49 P. Macnaughton and J. Urry, *Contested Natures* (Cambridge: Polity Press, 1998).

50 Bill McKibben, *Deep Economy: The Wealth of Communities and the Durable Future* (New York: Holt, 2007).

51 Alperovitz, *America Beyond Capitalism*, 2.

52 Simon Guy and Steven A. Moore, "Sustainable Architecture and the Pluralist Imagination," *Journal of Architectural Education* (2007): 15–23; David Schlosberg, "The Pluralist Imagination," in *Oxford Handbook of Political Theory*, ed. John S. Dryzek, Bonnie Honig, and Anne Phillips (2008). DOI: 10.1093/oxfor dhb/9780199548439.003.0007.

53 Guy and Moore, "Sustainable Architecture and the Pluralist Imagination." See also Peter Kroes, Andrew Light, Steven A. Moore, and Pieter Vermass, *Philosophy of Design: From Engineering to Architecture* (Berlin: Springer, 2007), 1–10.

54 Schor, *Plenitude*, 126. Schor also notes the importance of informal educational opportunities for furthering our ability to adopt new material systems and practices. She claims, "An informal education network has developed to foster permaculture, agroforestry, and biodynamic farming; cob, earthen, straw-bale, and other alternative construction; and solar and wind energy, biofuels, and other new ways of creating livelihood and meeting basic needs. Much of the skill transmission happens in short courses and workshops, under the auspices of a growing number of institutes, hands-on classes, and collaborative learning communities." Schor, *Plenitude*, 162.

55 Furthermore, simplified sociotechnical systems allow human beings to connect with, recognizing and living with the limits of, the natural world. Despite any claims for decoupling human subsistence with environmental resources, I would argue that nature is and will always be the ultimate source of sustenance for human beings, and thus there is value in connection with it.

56 Mumford, *Technics and Civilization*, 26–27.

57 Mauss, *Seasonal Variations of the Eskimo*, 80.

58 Fox, "Translator's Introduction," 9.

59 See Langdon Winner, "Do Artifacts have Politics?" *Daedalus* 109 (1980): 121–136; Bernward Joerges, "Do Politics Have Artefacts?" *Social Studies of Science* 29 (1994): 411–431; See also Langdon Winner, *The Whale and the Reactor: A Search for Limits in the Age of High Technology* (Chicago: University of Chicago Press, 1986).

60 Mary Douglas, "Symbolic Orders in the Use of Domestic Space," in *Man, Settlement, and Urbanism*, ed. P. J. Ucko, R. Tringham and G. W. Dimbleby (London: Duckworth, 1972), 513–514. Utilizing an ecological approach, as philosopher Stephen Toulmin claims, requires attentiveness to "particular, local, and timely circumstances. The Newtonian view encouraged hierarchy and rigidity, standardization and uniformity: an ecological perspective emphasizes, rather, differentiation and diversity, equality and adaptability." See Toulmin, *Cosmopolis*, 194.

61 Abbot, *Department and Discipline*.

62 Joas, *Pragmatism and Social Theory*; Joas, *The Creativity of Action*.

63 See Steven A. Moore and Barbara B. Wilson, *Questioning Architectural Judgment: The Problem of Codes in the United States* (London: Routledge 2014).

64 Meyer, *Engaging the Everyday*, chapter 5.

65 Lovins, "Energy Strategy: The Road Not Taken?"; Lovins, *Soft Energy Paths*.

66 Schor, *Plenitude*, 138.

67 Maggie L. Grabow et al., "Air Quality and Exercise-Related Health Benefits from Reduced Car Travel in the Midwestern United States," *Environmental Health Perspectives* 120 (2012): 66–76.

68 McKibben, *Deep Economy*, 109.

69 Mumford, *Technics and Civilization*, 434.

70 Schor, *Plenitude*, 6.

71 Herbert Marcuse, *An Essay on Liberation* (Boston: Beacon Press, 1969), 2.

72 Ibid., 4.

73 Mumford, *Technics and Civilization*, 45.

74 McKibben, *Deep Economy*, 3.

Appendix: Reflections on Method

1 See Chelsea Schelly, "Testing Residential Solar Thermal Adoption," *Environment and Behavior* 42 (2010): 151–170.

2 Which I did, as part of my dissertation research; see Chelsea Schelly, *Crafting Collectivity: American Rainbow Gatherings and Alternative Forms of Community* (Boulder, CO: Paradigm Publishers, 2014); reprinted in 2016 by Routledge.

3 All three Earthship communities are private. I only recruited homeowners living in the Greater World Community, the newest and most publicly visible Earthship community, which is located on the same land as the Visitor Center and internship program. The other two communities are older and much less accessible.

4 Focus groups are basically structured conversations involving several participants, and a total of twenty-one interns participated in these two focus groups. Earthship interns are, in my experience, a diverse group of people. Although most of the interns I met were younger than middle age, both groups included at least one person over forty (three interns total). In terms of geographical location and educational background, there was quite a range—from as far away as India, Sweden, New Zealand, and Australia, and with educational experiences ranging from no college education to a PhD.

Bibliography

Abbott, Andrew. *Department and Discipline: Chicago Sociology at One Hundred.* Chicago: University of Chicago Press, 1999.
———. "Of Time and Space." *Social Forces* 75 (1997): 1149–1182.
———. "Transcending General Linear Reality." *Sociological Theory* 6 (1988): 169–186.
Acorn Community website. www.acorncommunity.org. Accessed January 2, 2013.
Ajzen, I. "From Intentions to Actions: A Theory of Planned Behavior." In *Action-Control: From Cognition to Behavior*, edited by J. Kuhl and J. Beckman, 11–39. New York: Springer, 1985.
———. "The Theory of Planned Behavior." *Organizational Behavior and Human Decision Processes* 50 (1991): 179–211.
Ajzen, I., and M. Fishbein. "Attitude-Behavior Relations: A Theoretical Analysis and Review of Empirical Research." *Psychological Bulletin* 84 (1977): 888–918.
———. *Understanding Attitudes and Predicting Social Behavior.* Englewood Cliffs, NJ: Prentice-Hall, 1980.
Alatout, Samer. "Towards a Bio-Territorial Conception of Power: Territory, Population, and Environmental Narratives in Palestine and Israel." *Political Geography* 25 (2006): 601–621.
Alatout, Samer, and Chelsea Schelly, "Rural Electrification as a 'Bioterritorial' Technology: Redefining Space, Citizenship, and Power during the New Deal." *Radical History Review* 107 (2010): 127–138.
Alexander, Christopher, Sara Ishikawa, and Murray Silverstein. *A Pattern Language: Towns—Buildings—Construction.* New York: Oxford University Press, 1977.
Alperovitz, Gar. *America Beyond Capitalism: Reclaiming Our Wealth, Our Liberty, and Our Democracy.* Hoboken, NJ: Wiley, 2005.
Anderson, Nels. *The Hobo.* 1923; rpt. Chicago: University of Chicago Press, 1961.
Arnold, J. L. *New Deal in the Suburbs: A History of the Greenbelt Town Program, 1935–1954.* Columbus: Ohio State University Press, 1971.
Bamberg, S., and G. Möser. "Twenty Years after Hines, Hungerford, and Tomera: A New Meta-Analysis of Psycho-Social Determinants of Pro-Environmental Behavior." *Journal of Environmental Psychology* 27 (2007): 14–25.

Bates, Albert. "Technological Innovation in a Rural Intentional Community, 1971–1987." *Bulletin of Science, Technology, and Society* 8 (1988): 183–199.

Bellah, Robert N., Richard Madsen, William M. Sullivan, Ann Swidler, and Steven M. Tipton. *Habits of the Heart: Individualism and Commitment in American Life.* 1985; repr. Berkeley: University of California Press, 2007.

Berger, Peter L., and Thomas Luckmann, *The Social Construction of Reality: A Treatise in the Sociology of Knowledge.* New York: Irvington Publishers, 1966.

Bijker, Wiebe E., Thomas P. Hughes, and Trevor Pinch, eds. *The Social Construction of Technological Systems.* Cambridge, MA: MIT Press, 1989.

Bijker, Wiebe E., and John Law, eds. *Shaping Technology/Building Society: Studies in Sociotechnical Change.* Cambridge, MA: MIT Press, 1992.

Bourdieu, Pierre. *Outline of a Theory of Practice.* Cambridge: Cambridge University Press, 1977.

———. "Rethinking the State: Genesis and Structure of the Bureaucratic Field." *Sociological Theory* 12 (1994): 1–18.

———. "Social Space and Symbolic Space." In *Contemporary Sociological Theory* (2nd edition), edited by Craig Calhoun, Joseph Gerteis, James Moody, Steven Pfaff, and Indermohan Virk. West Sussex, UK: Wiley-Blackwell, 2007.

———. "Structures, Habitus, and Practice." In *Contemporary Sociological Theory* (2nd edition), edited by Craig Calhoun, Joseph Gerteis, James Moody, Steven Pfaff, and Indermohan Virk. West Sussex, UK: Wiley-Blackwell, 2007.

Boyer, Robert. "Achieving One-Planet Living Through Transitions in Social Practice: A Case Study of Dancing Rabbit Ecovillage." *Sustainability: Science, Practice, & Policy* 12 (1): 2016. http://www.google.com/archives/vol12iss1/1506–007.boyer.html. Accessed May 5, 2016.

Braun, Ingo, and Bernward Joerges, "How to Recombine Large Technical Systems: The Case of European Organ Transplant." In *Changing Large Technical Systems*, edited by Jane Summerton, 25–52. Boulder, CO: Westview Press, 1994.

Brey, Philip. "Theorizing Modernity and Technology." In *Modernity and Technology*, edited by Thomas J. Misa, Philip Brey, and Andrew Feenberg, 33–71. Cambridge: MIT Press, 2003.

Brown, Halina Szejnwald, Philip J. Vergragt, and Maurie J. Cohen, "Societal Innovation in a Constrained World: Theoretical and Empirical Perspectives." In *Innovations in Sustainable Consumption: New Economics, Socio-technical Transitions and Social Practices*, edited by M. J. Cohen, H. S. Brown, and P. J. Vergragt, 1–30. Cheltenham and Northampton, MA: Edward Elgar, 2013.

Brown, Peter C., and Geoffrey Garver. *Right Relationship: Building a Whole Earth Economy.* San Francisco: Berret-Kohler Publishers, 2008.

Bulmer, Martin. *The Chicago School of Sociology: Institutionalization, Diversity, and the Rise of Sociological Research.* Chicago: University of Chicago Press, 1986.

Burchell, Graham, Colin Gordon, and Peter Miller, eds. *The Foucault Effect: Studies in Governmentality.* Chicago: University of Chicago Press, 1991.

Callon, Michel. "The Sociology of an Actor Network: The Case of the Electric Vehicle." In *Mapping the Dynamics of Science and Technology*, edited by Michel Callon, John Law, and Arie Rip, 19–34. London: Macmillan Press, 1986.

———. "Some Elements of a Sociology of Translation: Domestication of the Scallops and Fishermen of St. Brieuc Bay." In *Power, Action, and Belief: A New Sociology of Knowledge*, edited by John Law, 196–233. London: Routledge, 1986.

Carson, Rachel. *Silent Spring.* New York: Houghton Mifflin Harcourt, 1962.

Collins, Randall. *Interaction Ritual Chains.* Princeton, NJ: Princeton University Press, 2004.

———. "Stratification, Emotional Energy, and Transient Emotions." In *Research Agendas in the Sociology of Emotions*, edited by Theodore D. Kemper, 27–34. Albany: State University of New York Press, 1990. Reprinted in Emirbayer, Mustafa. *Emile Durkheim: Sociologist of Modernity*. Malden, MA: Blackwell Publishing, 2003, 129–133.

Cowan, Ruth Schwartz. *More Work for Mother: The Ironies of Household Technology from the Open Hearth to the Microwave*. New York: Basic Books, 1983.

———. *A Social History of American Technology*. New York: Oxford University Press, 1997.

Crawford, Matthew B. *Shop Class as Soulcraft: An Inquiry into the Value of Work*. New York: Penguin, 2009.

Cressey, Paul G. *The Taxi-Dance Hall: A Sociological Study in Commercialized Recreation and City Life*. Chicago: University of Chicago Press, 1932.

Cronon, William. "Conservation and Patriotism: A Conservation with William Cronon." Interview by William Poole. In *Land and People*, 46–50. The Trust for Public Land (Wisconsin): December 2006.

Daly, Herman. *Steady-State Economics: The Economics of Biophysical Equilibrium and Moral Growth*. San Francisco: W. H. Freeman, 1977.

———, ed. *Toward a Steady-State Economy*. San Francisco: W. H. Freeman, 1973.

Dancing Rabbit Website. http://www.dancingrabbit.org/about-dancing-rabbit-ecovillage/dancing-rabbit-inc-nonprofit/. Accessed May 12, 2016.

Daniels, G. H. "The Big Questions in the History of American Technology." *Technology and Culture* 11 (1970): 1–21.

Darrow, K., and R. Pam. *Appropriate Technology Sourcebook*. Stanford, CA: Volunteers in Asia, 1976.

DeWalt, B. R. "Appropriate Technology in Rural Mexico: Antecedents and Consequences of an Indigenous Peasant Innovation." *Technology and Culture* 19 (1978): 32–52.

Dewey, John. *Experience and Nature*. London: Open Court, 1925.

———. *How We Think*. London: Heath & Co, 1910.

———. *The Quest for Certainty: A Study of the Relation of Knowledge and Action*. New York: Minton, Balch & Company, 1929.

Douglas, Mary. *How Institutions Think*. Syracuse NY: Syracuse University Press, 1986.

———. "Symbolic Orders in the Use of Domestic Space." In *Man, Settlement, and Urbanism*, edited by P. J. Ucko, R. Tringham, and G. W. Dimbleby, 513–514. London: Duckworth, 1972.

Duffy, J. *The Sanitarians: A History of American Public Health*. Urbana: University of Illinois Press, 1990.

DuPuis, Melanie E. *Nature's Perfect Food: How Milk Became America's Drink*. New York: New York University Press, 2002.

Durkheim, Emile. *Elementary Forms of Religious Life*. 1912; rpt. New York: The Free Press, 1995.

Earthship Biotecture website. www.earthship.com. Accessed March 16, 2013.

East Wind Community website. www.eastwind.org. Accessed January 4, 2013.

Edwards, P. "Infrastructure and Modernity: Force, Time, and Social Organization in the History of Sociotechnical Systems." In *Modernity and Technology*, edited by T. J. Misa, P. Brey, and A. Feenberg, 185–225. Cambridge, MA: MIT Press, 2003.

Ellul, Jacques. *The Technological Society*. New York: Vintage Books, 1964.

Emirbayer, Mustafa. "Manifesto for a Relational Sociology." *American Journal of Sociology* 103 (1997): 281–317.

Emirbayer, Mustafa, and Ann Mische. "What Is Agency?" *American Journal of Sociology* 103 (1998): 962–1023.

The Farm Community Website. www.thefarmcommunity.com. Accessed March 16, 2013.

Fike, Rupert, ed. *Voices from the Farm: Adventures in Community Living.* Summertown, TN: Book Publishing Company, 1998.

Fischer, Claude S. *America Calling: A Social History of the Telephone to 1940.* Berkeley: University of California Press, 1992.

Fishman, Robert. *Bourgeois Utopias: The Rise and Fall of Suburbia.* New York: Basic Books, 1987.

Flink, James J. *America Adopts the Automobile, 1895–1910.* Cambridge, MA: MIT Press, 1970.

———. *The Automobile Age.* Cambridge, MA: MIT Press, 1990.

———. *The Car Culture.* Cambridge, MA: MIT Press, 1975.

Foster, John Bellamy. "Marx's Theory of Metabolic Rift: Classical Foundations for Environmental Sociology." *American Journal of Sociology* 105 (1999): 366–405.

Foucault, Michel. *Discipline and Punish.* New York: Vintage Books, 1980.

———. *The Essential Works of Michel Foucault, 1954–1984,* vol. 3: *Power,* edited by Paul Rabinow. London: Allen Lane, 2000.

———. "Governmentality." In *The Foucault Effect: Studies in Governmentality,* edited by G. Burchell, C. Gordon, and P. Miller, 87–104. Chicago: University of Chicago Press, 1991.

———. *The History of Sexuality,* vol. 1: *An Introduction.* New York: Random House, 1990.

———. *Power/Knowledge: Selected Interviews and Other Writings, 1972–1977,* edited by Colin Gordon. New York: Pantheon Books, 1980.

———. "Truth and Power." Interview reprinted in *The Foucault Reader,* edited by Paul Rabinow, 51–75. New York: Pantheon Books, 1984.

Fox, James. "Translator's Introduction." In *Seasonal Variations of the Eskimo: A Study in Social Morphology* by Marcel Mauss. London: Routledge & Kegan Paul, 1979.

Fransson, N., and T. Gärling. "Environmental Concern: Conceptual Definitions, Measurement Methods, and Research Findings." *Journal of Environmental Psychology* 19 (1999): 369–382.

Freudenburg, William, Scott Frickel, and Robert Gramling. "Beyond the Nature/ Society Divide: Learning to Think about a Mountain." In *Sociological Forum,* vol. 10, no. 3, 361–392. New York: Plenum Publishers, 1995. http://www.jstor.org/ stable/684781?seq=1#page_scan_tab_contents. Accessed November 16, 2016.

Fuller, R. Buckminster. *Operating Manual for Spaceship Earth.* Carbondale: Southern Illinois University Press, 1969.

Gaskin, Ina May. *Spiritual Midwifery.* Summertown, TN: Book Publishing Company, 1977.

Gaskin, Stephen. *The Caravan.* 1972; rpt. Summertown, TN: Book Publishing Company, 2007.

———. *Monday Night Class.* 1970; rpt. Summertown, TN: Book Publishing Company, 2005.

Giddens, Anthony. *The Constitution of Society.* Berkeley: University of California Press, 1984.

———. *Modernity and Self-Identity: Self and Society in the Late Modern Age.* Stanford, CA: Stanford University Press, 1991.

Goffman, Erving. *Interaction Ritual: Essays on Face-to-Face Behavior.* New York: Pantheon, 1967.

———. *Presentation of Self in Everyday Life.* New York: Random House, 1959.

Goldstein, Noah J., Robert B. Cialdini, and Vladas Griskevicius. "A Room with a Viewpoint: Using Social Norms to Motivate Environmental Conservation in Hotels." *Journal of Consumer Research* 35 (2008): 472–482.

Goodland, R., Herman E. Daly, and Salah El Serafy, eds. *Population, Technology, and Lifestyle: The Transition to Sustainability.* Washington, DC: Island Press, 1992.

Gould, Kenneth A., David N. Pellow, and Allan Schnaiberg. "Interrogating the Treadmill of

Production: Everything You Wanted to Know about the Treadmill but Were Afraid to Ask." *Organization and Environment* 17 (2004): 296–316.

Gowan, Teresa, and Rachel Slocum, "Artisanal Production, Communal Provisioning, and Anticapitalist Politics in the Aude, France." In *Sustainable Lifestyles and the Quest for Plenitude: Case Studies of the New Economy*, edited by Juliet Schor and Craig J. Thompson, 27–62. New Haven, CT: Yale University Press, 2014.

Grabow, Maggie L., Scott N. Spak, Tracey Holloway, Brian Stone Jr., Adam C. Mednick, and Jonathan A. Patz. "Air Quality and Exercise-Related Health Benefits from Reduced Car Travel in the Midwestern United States." *Environmental Health Perspectives* 120 (2012): 66–76.

Gram-Hanssen, Kirsten. "Understanding Change and Continuity in Residential Energy Consumption." *Journal of Consumer Culture* 11 (2011): 61–78.

Gromet, Dena M., Howard Kunreuther, and Richard P. Larrick. "Political Ideology Affects Energy-Efficiency Attitudes and Choices," *Proceedings of the National Academy of Science* 110 (2013): 9314–9319.

Guy, Simon, and Steven A. Moore. "Sustainable Architecture and the Pluralist Imagination." *Journal of Architectural Education* (2007): 15–23.

Habermas, Jürgen. "Technology and Science as 'Ideology.'" In *Toward a Rational Society: Student Protest, Science, and Politics*, 90–122. Boston: Beacon Press, 1970.

Halvorsen, Kathy E., and Hugh S. Gorman. "Onsite Sewage System Regulation Along the Great Lakes and the US EPA 'Homeowner Awareness' Model." *Environmental Management* 37 (2006): 395–409.

Hand, Martin, Elizabeth Shove, and Dale Southerton. "Explaining Showering: A Discussion of the Material, Conventional, and Temporal Dimensions of Practice." *Sociological Research Online* 10 (2005). http://www.socresonline.org.uk/10/2/hand/hand.pdf. Accessed July 7, 2015.

Hardin, Garrett. "The Tragedy of the Commons." *Science* 162 (1968): 1243–1248.

Hargreaves, T., N. Longhurst, and G. Seyfang. 2013. "Up, Down, Round and Round: Connecting Regimes and Practices in Innovation for Sustainability." *Environmental and Planning A* 45: 402–20.

Hargreaves, T., M. Nye, and J. Burgess. "Making Energy Visible: A Qualitative Field Study of How Householders Interact with Feedback from Smart Energy Monitors." *Energy Policy* 38 (2010): 6111–6119.

Hawken, Paul, Amory Lovins, and L. Hunter Lovins. *Natural Capitalism: Creating the Next Industrial Revolution*. Boston: Little, Brown, 1999.

Heberlein, Thomas A. *Navigating Environmental Attitudes*. New York: Oxford University Press, 2012.

Heidegger, Martin. *Questions Concerning Technology and Other Essays*. New York: Harper & Row, 1977.

Hewitt, Mischa, and Kevin Telfer. "Earthships as Public Pedagogy and Agents of Change." In *Handbook of Public Pedagogy: Education and Learning Beyond Schooling*, edited by Jennifer A Sandlin, Brian D. Schultz, and Jake Burdick, 171–178. New York: Routledge, 2010.

Hirsch, Arnold R. "Less Than *Plessy*: The Inner City, Suburbs, and State-Sanctioned Residential Segregation in the Age of *Brown*." In *The New Suburban History*, edited by Thomas J. Sugrue and Kevin M. Kruse, 33–56. Chicago: University of Chicago Press, 2006.

Hirsch, Richard F. *Power Loss: The Origins of Deregulation and Restructuring in the American Electric Utility System*. Cambridge, MA: MIT Press, 1999.

Hollick, Malcom. "The Appropriate Technology Movement and its Literature: A Retrospective." *Technology and Society* 4 (1982): 213–229.

Horkheimer, Max, and Theodor W. Adorno. *The Dialectic of Enlightenment*. New York: Herder and Herder, 1972.

Hughes, Thomas P. "The Electrification of America: The System Builders." *Technology and Culture* 20 (1979): 124–161.

———. *Networks of Power: Electrification in Western Society, 1880–1930*. Baltimore: John Hopkins University Press, 1983.

Illich, Ivan. *Tools for Conviviality*. New York: Harper & Row, 1973.

———. *Toward a History of Needs*. New York: Pantheon Books, 1978.

Jackson, Kenneth T. *Crabgrass Frontier: The Suburbanization of the United States*. New York: Oxford University Press, 1985.

Jackson, Tim, and Eleni Papathanasopoulou. "Luxury or 'Lock-in'? An Exploration of Unsustainable Consumption in the UK: 1968 to 2000." *Ecological Economics* 68 (2008): 80–95.

Jaefer-Erben, Melanie, and Ursula Offenberger. "A Practice Theory Approach to Sustainable Consumption." *GAIA* 23 (2014): 166–174.

Jaeger, Carlo. *Risk, Uncertainty, and Rational Action*. London: Earthscan, 2001.

Jasanoff, Sheila, ed. *States of Knowledge: The Co-Production of Science and Social Order*. London: Routledge, 2004.

Joas, Hans. *The Creativity of Action*. Chicago: University of Chicago Press, 1996.

———. "The Creativity of Action: Pragmatism and the Critique of the Rational Actor Model." Manuscript. Free University, Berlin. John F. Kennedy Institute for North American Study, n.d.

———. *Pragmatism and Social Theory*, Chicago: University of Chicago Press, 1993.

Joerges, Bernward. "Do Politics Have Artefacts?" *Social Studies of Science* 29 (1994): 411–431.

Johnson, Jim (Bruno Latour). "Mixing Humans and Non-Humans Together: The Sociology of a Door Closer." *Social Problems* 35 (1988): 298–310.

Jones, Kayla Brooke. "Toward Sustainable Community: Assessing Progress at Dancing Rabbit Ecovillage." Master's thesis, University of North Texas, August 2014. http://digital.library.unt.edu/ark:/67531/metadc700019/m1/3/. Accessed November 15, 2016.

Kanter, Rosabeth Moss. *Commitment and Community: Communes and Utopias in Sociological Perspective*. Cambridge, MA: Harvard University Press, 1972.

Kasper, Debbie V. S. "Contextualizing Social Practices: Insights into Social Change." In *Putting Sustainability into Practice: Advances and Applications of Social Practice Theories*, edited by Emily Huddart Kennedy, Maurie J. Cohen, and Naomi Krogman, 25–46. Cheltanham, UK: Edward Elgar, 2016.

———."Ecological Habitus: Toward a Better Understanding of Socioecological Relations." *Organization and Environment* 22 (2009): 311–326.

———. "Redefining Community in the Ecovillage." *Human Ecology Review* 15 (2008): 12–24.

Kennedy, Emily Huddart, Maurie J. Cohen, and Naomi T. Krogman, eds. *Putting Sustainability into Practice: Applications and Advances in Research on Sustainable Consumption*. Cheltenham, UK: Edward Elgar, 2015.

Kennedy, Emily, Harvey Krahn, and Naomi T. Krogman. "Taking Social Practice Theories on the Road: A Mixed-Methods Case Study of Sustainable Transportation." In *Innovations in Sustainable Consumption: New Economics, Socio-Technical Transitions, and Social Practices*, edited by M. J. Cohen, H. S. Brown, and P. J. Vergragt, 252–276. Cheltenham and Northampton, MA: Edward Elgar, 2013.

Kinkade, Kathleen. *Is It Utopia Yet? An Insider's View of the Twin Oaks Community in Its Twenty-sixth Year.* Louisa, VA: Twin Oaks Publishing, 1994.

———. *A Walden Two Experiment: The First Five Years of Twin Oaks Community.* Scranton, PA: William Morrow, 1974.

Kirk, Andrew. *Counterculture Green: The Whole Earth Catalog and American Environmentalism.* Lawrence: University of Kansas Press, 2001.

Kleiman, Jordan Benson. "The Appropriate Technology Movement in American Political Culture." PhD diss., University of Rochester, 2001.

Kline, R., and Trevor Pinch. "Users as Agents of Technological Change: The Social Construction of the Automobile in the Rural United States." *Technology and Culture* 37 (1996): 763–795.

Kozeny, Geoph. *Visions of Utopia: Experiments in Sustainable Culture* [motion picture], Volume 1. Rutledge, MO: Community Catalyst Project, Fellowship for Intentional Community, 2004.

———. *Visions of Utopia: Experiments in Sustainable Culture* [motion picture], Volume 2. Rutledge, MO: Community Catalyst Project, Fellowship for Intentional Community, 2009.

Kroes, Peter, Andrew Light, Steven A. Moore, and Pieter Vermass, *Philosophy of Design: From Engineering to Architecture.* Berlin: Springer, 2007.

Labay, D. C., and T. C. Kinnear. "Exploring the Consumer Decision Process in the Adoption of Solar Energy Systems." *Journal of Consumer Research* 8 (1981): 271–278.

Latour, Bruno. "On Recalling ANT." In *Actor Network Theory and After,* edited by John Law and John Hassard, 15–25. Oxford: Blackwell Publishers, 1999.

———. *Pandora's Hope: Essays on the Reality of Science Studies.* Cambridge, MA: Harvard University Press, 1999.

———. *Pasteurization of France.* Cambridge, MA: Harvard University Press, 1988.

———. "Technology Is Society Made Durable." In *A Sociology of Monsters: Essays on Power, Technology, and Domination,* edited by John Law, 103–131. London: Routledge, 1991.

———. *We Have Never Been Modern.* Cambridge, MA: Harvard University Press, 1993.

Leopold, Aldo. *A Sand Country Almanac.* 1949; rpt. New York: Oxford University Press, 1966.

Levitt, Steven D., and John A. List. "*Homo Economicus* Evolves." *Science* 319 (2008): 909–910.

Litfin, Karen. *Eco-Villages: Lessons for Sustainable Community.* Malden, MA: Polity Press, 2013.

———. "Gaia Theory: Intimations for Global Environmental Politics." In *Handbook of Global Environmental Economics,* edited by Peter Dauvergne, 502–518. Cheltenham, UK: Edward Elgar, 2005.

———. "Sovereignty in World Ecopolitics." *Mershon International Studies Review* 41 (1997): 167–204.

Living Energy Farm website. www.livingenergyfarm.org. Accessed January 2, 2013.

Lorenzen, Janet A. "Going Green: The Process of Lifestyle Change." *Sociological Forum* 27 (2012): 94–116.

Lovins, Amory. "Energy Strategy: The Road Not Taken?" *Foreign Affairs* 55 (1976): 65–96.

———. *Soft Energy Paths: Toward a Durable Peace.* New York: Harper & Row, 1977.

———. "Soft Energy Technologies." *Annual Review of Energy* 3 (1978): 477–517.

Luke, Timothy W. "Ephemeralization as Environmentalism: Rereading R. Buckminster Fuller's *Operating Manual for Spaceship Earth.*" *Organization & Environment* 23 (2010): 354–362.

Lurie, Julia. "California's Drought Is So Bad that Thousands Are Living without Running Water." *Mother Jones*, July 31, 2015. http://www.motherjones.com/environment/2015/07/drought-5000-californians-dont-have-running-water. Accessed September 22, 2015.

Macnaughton, P., and J. Urry. *Contested Natures*. Cambridge: Polity Press, 1998.

Madre, L. D. *Midwifing Death: Returning to the Arms of the Ancient Mother.* Austin, TX: Plain View Press, 2005.

Maniates, Michael, and John M. Meyer, eds. *The Environmental Politics of Sacrifice.* Cambridge, MA: MIT Press: 2010.

Marcuse, Herbert. *An Essay on Liberation*. Boston: Beacon Press, 1969.

———. *One-Dimensional Man: Studies in the Ideology of Advanced Industrial Society.* Boston: Beacon Press, 1964.

———. "Some Social Implications of Modern Technology." *Studies in Philosophy and Social Sciences* 9 (1941): 138–162.

Marx, Karl. *Capital: Volume 1.* 1887; rpt. New York: International Publishers. 1967.

———. *The Class Struggles in France, 1848–1850.* 1850; rpt. New York: International Publishers: New York, 1964.

———. *Early Writings*. New York: Vintage, 1974.

———. *Economic and Philosophic Manuscripts of 1844*, edited by D. J. Struik. London: Lawrence and Wishart, 1970.

———. *The Poverty of Philosophy*. London: Martin Lawrence Limited, 1847.

Marx, Leo, "Technology: The Emergence of a Hazardous Concept." *Technology and Culture* (2010): 561–577.

Massey, Douglas S. "American Apartheid: Segregation and the Making of the Underclass." *American Journal of Sociology* (1990): 329–357.

Mauss, Marcel. *Seasonal Variations of the Eskimo: A Study in Social Morphology*. London: Routledge & Kegan Paul, 1979.

———. "Techniques of the Body." [1934]. *Economy & Society* 2 (1973): 70–88.

McDonough, William. *The Hannover Principles: Design for Sustainability*. Charlottesville, VA: William McDonough, 1992.

McDonough, William, and Michael Braungart. *Cradle to Cradle: Remaking the Way Things Work*. New York: North Point Press, 2002.

McKibben, Bill. *Deep Economy: The Wealth of Communities and the Durable Future*. New York: Holt, 2007.

Meadows, Donnella H., Jorgen Randers, and Dennis L. Meadows. *The Limits to Growth: A Report for the Club of Rome's Project on the Predicament of Mankind*. New York: New American Library, 1974.

Merkel, Jim. *Radical Simplicity: Small Footprints on a Finite Earth*. Gabriola Island, BC: New Society Publishers, 2003.

Merton, Robert K. *On Theoretical Sociology*. New York: The Free Press, 1967.

Meyer, John. *Engaging the Everyday: Environmental Social Criticism and the Resonance Dilemma*. Cambridge, MA: MIT Press, 2015.

Midgley, D. F., and G. R. Downing. "Innovativeness: The Concept and Its Measurement." *Journal of Consumer Research* 4 (1978): 229–242.

Mills, C. Wright. "Situated Actions and Vocabularies of Motive." *American Sociological Review* 5 (1940): 904–913.

Miner, Horace. "Body Ritual among the Nacirema." *American Anthropologist* 58 (1956): 503–507.

Misa, T. J. "The Compelling Tangle of Modernity and Technology." In *Modernity and Technology*, edited by T. J. Misa, P. Brey, and A. Feenberg, 1–30. Cambridge, MA: MIT Press, 2003.

Mitcham, C. "The Concept of Sustainable Development: Its Origins and Ambivalence." *Technology in Society* (1995): 311–326.

Mont, Oksana. "Institutionalisation of Sustainable Consumption Patterns Based on Shared Use." *Ecological Economics* 50 (2004): 135–153.

Moore, Steven A., and Barbara B. Wilson. *Questioning Architectural Judgment: The Problem of Codes in the United States*. London: Routledge 2014.

Mumford, Lewis. *The City in History: Its Origin, Its Transformations, and Its Prospects*. New York: Harcourt, Brace, 1961.

———. *Technics and Civilization*. New York: Harcourt, Brace, 1934.

Murphy, H. M., Edward A. McBean, and Khosrow Farahbakhsh. "Appropriate Technology—A Comprehensive Approach for Water and Sanitation in the Developing World." *Technology in Society* 31 (2009): 158–167.

Nader, Ralph. *Unsafe at Any Speed: The Designed-In Dangers of the American Automobile*. New York: Grossman, 1965.

Nye, David E. *Consuming Power: A Social History of American Energies*. Cambridge, MA: MIT Press, 1998.

———. *Electrifying America: Social Meanings of a New Technology, 1880–1940*. Cambridge, MA: MIT Press, 1997.

Ollman, Bertell. *Alienation: Marx's Conception of Man in a Capitalist Society*. Cambridge: Cambridge University Press, 1971.

Ostlund, L. E. "Perceived Innovation Attributes as Predictors of Innovativeness." *Journal of Consumer Research* 1 (1974): 23–29.

Park, Robert E., and Ernest W. Burgess. *Introduction to the Science of Sociology*. Chicago: University of Chicago Press, 1921.

Park, Robert E., Ernest W. Burgess, and Roderick D. McKenzie. *The City*. Chicago: University of Chicago Press, 1925.

Pearce, Joshua M. "The Case for Open Source Appropriate Technology." *Environment, Development and Sustainability* 14 (2012): 425–431.

Pollan, Michael. *In Defense of Food: An Eater's Manifesto*. New York: Penguin Press, 2008.

———. *Omnivore's Dilemma: A Natural History of Four Meals*. New York: Penguin Press, 2006.

Pred, A. "Philadelphia" (book review). *American Historical Review* 87 (1982): 268–270.

President's Commission on Critical Infrastructure Protection. *Critical Foundations: Protecting America's Infrastructures*. Washington, DC: US Government Printing Office, 1997.

Princen, Thomas. *The Logic of Sufficiency*. Cambridge, MA: MIT Press, 2005.

Pursell, Carroll. "The Rise and Fall of the Appropriate Technology Movement in the United States, 1965–1985." *Technology and Culture* 34 (1993): 629–637.

Ravitz, Jessica. "Utopia: It's Complicated." CNN. http://www.cnn.com/interactive/2015/09/us/communes-american-story/. Accessed September 16, 2015.

Reckwitz, Andreas. "Toward a Theory of Social Practice: A Development in Culturalist Thinking." *European Journal of Social Theory* 5 (2002): 243–263.

Reynolds, Michael. *Comfort in Any Climate*. Taos, NM: Solar Survival Press, 2001.

———. *A Coming of Wizards: A Manual of Human Potential*. Taos, NM: High Mesa Foundation, 1989.

———. *Earthship Volume 1: How to Build Your Own*. Taos, NM: Solar Survival Press, 1990.

———. *Earthship Volume 2: Systems and Components*. Taos, NM: Solar Survival Press, 1990.

———. *Earthship Volume 3: Evolution Beyond Economics*. Taos, NM: Solar Survival Press, 1993.

———. *Journey: Part One*. Taos, NM: Earthship Biotecture, 2008.

Robertson, T. S. *Innovative Behavior and Communication*. New York: Holt, Rinehart, and Winston, 1971.

Rome, Adam. *The Bulldozer in the Countryside: Suburban Sprawl and the Rise of American Environmentalism*. New York: Cambridge University Press, 2001.

Rose, M. H. "Urban Gas and Electric Systems and Social Change, 1900–1940." In *Technology and the Rise of the Networked City in Europe and America*, edited by Joel A. Tarr and Gabriel Dupuy, 229–245. Philadelphia: Temple University Press, 1988.

Rudolph, Richard, and Scott Ridley. *Power Struggle: The Hundred-Year War Over Electricity*. New York: Harper & Row, 1986.

Sale, Kirkpatrick. *Human Scale*. CITY?, India: New Catalyst Books, 2007.

Salzman, James. "Thirst: A Short History of Drinking Water." *Yale Journal of Law and the Humanities* 18 (2006): 94–121.

Sandhill Farm Community website. www.sandhillfarm.org. Accessed January 4, 2013.

Sanne, Christer. "Willing Consumers—Or Locked-In? Policies for a Sustainable Consumption." *Ecological Economics* 42 (2002): 273–287.

Schatzberg, Eric. "Technik Comes to America: Changing Meanings of Technology before 1930." *Technology and Culture* 47 (2006): 486–512.

———. *Wings of Wood, Wings of Metal: Culture and Technical Choice in American Airplane Materials, 1914–1945*. Princeton, NJ: Princeton University Press, 1999.

Schatzki, Theodore. *Social Practices: A Wittgensteinian Approach to Human Activity and the Social*. New York: Cambridge University Press, 1996.

Schelly, Chelsea. *Crafting Collectivity: American Rainbow Gatherings and Alternative Forms of Community*. Boulder, CO: Paradigm Publishers, 2014; reprinted in 2016 by Routledge.

———. "Everyday Household Practice in Alternative Residential Dwellings: The Non-Environmental Motivations for Environmental Behavior." In *The Greening of Everyday Life: Challenging Practices, Imagining Possibilities*, edited by John Meyer and Jens Kersten, 265–280. New York: Oxford University Press, 2016.

———. "How Policy Frameworks Shape Environmental Practice: Three Cases of Alternative Dwelling." In *Putting Sustainability into Practice: Advances and Applications of Social Practice Theories*, edited by Emily Huddart Kennedy, Maurie J. Cohen, and Naomi Krogman, 195–203. Cheltenham, UK: Edward Elgar, 2016.

———. "Residential Solar Electricity Adoption: What Motivates, and What Matters? A Case Study of Early Adopters." *Energy Research and Social Science* 2 (2014): 183–191.

———. "Testing Residential Solar Thermal Adoption." *Environment and Behavior* 42 (2010): 151–170.

———. "Transitioning to Renewable Sources of Electricity: Motivations, Policy, and Potential." In *Controversies in Science and Technology*, vol. 4, edited by Daniel Lee Kleinman, Karen Cloud-Hansen, and Jo Handelsman, 62–72. New York: Oxford University Press, 2014.

Schlosberg, David. "The Pluralist Imagination." In *Oxford Handbook of Political Theory*, edited by John S. Dryzek, Bonnie Honig, and Anne Phillips. 2008. doi: 10.1093/oxfordhb/9780199548439.003.0007.

Schnaiberg, Allan. *The Environment: From Surplus to Scarcity*. New York: Oxford University Press, 1980.

———. "Redistributive versus Distributive Politics: Social Equity Limits in Environmental and Appropriate Technology Movements." *Social Inquiry* 53 (1983): 200–215.

Schnaiberg, Allan, and Kenneth A. Gould. *Environment and Society: The Enduring Conflict*. New York: St. Martin's Press, 1994.

Schor, Juliet. *Plenitude: The New Economics of True Wealth*. New York: Penguin, 2010.

Schumacher, E. F. *Small Is Beautiful: Economics as if People Mattered*. New York: Perennial, 1973.

Schwartz, Shalom H. 1977. "Normative Influences on Altruism." In *Advances in Experimental Social Psychology*, edited by L. Berkowitz, 221–279. New York: Academic Press.

Sennet, Richard. *The Craftsman*. New Haven, CT: Yale University Press, 2008.

Sewell, William Jr. "A Theory of Structure: Duality, Agency, and Transformation." *American Journal of Sociology* 98 (1992): 1–29.

Sheridan, Alan. *Foucault: The Will to Truth*. London: Routledge, 1990.

Shilling, Chris. "Embodiment, Experience, and Theory: In Defence of the Sociological Tradition." *Sociological Review* 49 (2001): 327–344.

———. "Sociology and the Body: Classical Traditions and New Agendas." *Sociological Review* 55 (2007): 1–18.

Shilling, Chris, and Philip A. Mellor. "Cultures of Embodied Experience: Technology, Religion, and Body Pedagogics." *Sociological Review* 55 (2007): 531–549.

Shove, Elizabeth. "Beyond the ABC: Climate Change Policy and Theories of Social Change." *Environment and Planning A* 42 (2010): 1273–1285.

———. *Comfort, Cleanliness, and Convenience: The Social Organization of Normality*. New York: Berg, 2003.

———. "Efficiency and Consumption: Technology and Practice." *Energy & Environment* 15 (2004): 1053–1065.

———. "Energy Transitions in Practice: The Case of Global Indoor Climate Change." In *Governing the Energy Transition: Reality, Illusion or Necessity?* edited by G. Verbong and D. Loorbach, 51–74. New York: Routledge, 2012.

———. "On the Difference between Chalk and Cheese: A Response to Whitmarch et al.'s Comments on 'Beyond the ABC: Climate Change Policy and Theories of Social Change.'" *Environment and Planning A* 43 (2011): 262–264.

Shove, Elizabeth, M. Pantzar, and M. Watson. *The Dynamics of Social Practice: Everyday Life and How It Changes*. Thousand Oaks, CA: Sage, 2012.

Shove, Elizabeth, and Gordon Walker. "What Is Energy For? Social Practice and Energy Demand." *Theory, Culture, and Society* 31 (2014): 41–58.

Shwom, Rachael, and Janet A. Lorenzen. "Changing Household Consumption to Address Climate Change: Social Scientific Insights and Challenges." *WIREs Climate Change* 3 (2012): 379–395.

Siu, Paul S. P. *The Chinese Laundryman: A Study of Social Isolation*. New York: New York University Press, 1987.

Skinner, B. F. *Walden Two*. New York: Macmillan, 1948.

Slater, Philip. *Microcosm: Structural, Psychological, and Religious Evolution in Groups*. New York: Wiley, 1966.

Spaargaren, Gert. "Sustainable Consumption: A Theoretical and Environmental Policy Perspective." *Society & Natural Resources* 16 (2003): 687–701.

———. "Theories of Practices: Agency, Technology, and Culture: Exploring the Relevance of Practice Theories for the Governance of Sustainable Consumption Practices in the New World-Order." *Global Environmental Change* 21 (2011): 813–822.

Spaargaren, Gert, S. Martens, and T. Beckers. 2006. "Sustainable Technologies and Everyday Life." In *User Behaviour and Technology Development: Shaping Sustainable Relations between Consumers and Technologies*, edited by P. P. Verbeek and A. Slob, 107–118. Dordrecht, the Netherlands: Springer.

Stern, Paul C. *Environmentally Significant Consumption: Research Directions.* Washington, DC: National Academy Press, 1997.

———. "What Psychology Knows About Energy Conservation." *American Psychologist* 47 (1992): 1224–1232.

Strauss, Anselm L. "A Social World Perspective." In *Creating Sociological Awareness,* edited by Anselm Strauss, 233–244. 1978; rpt. New Brunswick, NJ: Transaction Press, 1990.

Swan Conservation Trust, http://swantrust.org/. Accessed April 6, 2016.

Swidler, Ann. *Organization with Authority: Dilemmas of Social Control in Free Schools.* Cambridge, MA: Harvard University Press, 1979.

Tarr, Joel A. "Sewerage and the Development of the Networked City in the United States, 1850–1930." In *Technology and the Rise of the Networked City in Europe and America,* edited by Joel A Tarr and Gabriel Dupuy, 159–185. Philadelphia: Temple University Press, 1988.

Tarr, Joel A., and Gabriel Dupuy, eds. *Technology and the Rise of the Networked City in Europe and America.* Philadelphia: Temple University Press, 1988.

Telfer, Kevin. "Earth Mover (A profile of Mike Reynolds)." *Architects' Journal* (2003): 18–19.

Thies, Clifford F. "The Success of American Communes." *Southern Economic Journal* 67 (2010): 186–199.

Thomas, W. I. *W. I. Thomas on Social Organization and Social Personality: Selected Papers.* Chicago: University of Chicago Press, 1966.

Thomas, W. I., and Florian Znaniecki. *The Polish Peasant in Europe and America.* Edited by Eli Zaretsky. Champaign: University of Illinois Press, 1926.

Thomas Edison to Edward Hibbard Johnson, 1886. From the Thomas Edison papers at Rutgers University. Special Collections Series, Vail Papers (1885–1888, 1900): [ME004; TAEM 148:5].

Thoreau, Henry David. *Walden, or Life in the Woods: And, 'On the Duty of Civil Disobedience.'* New York: New American Library, 1999.

Tocqueville, Alexis de. *Democracy in America.* [1835]. Translated by Arthur Goldhammer. New York: Penguin, 2004.

Toulmin, Stephen. *Cosmopolis: The Hidden Agenda of Modernity.* Chicago: University of Chicago Press, 1992.

Trotsky, Leon. *The Permanent Revolution.* Translated by Max Schachtman. New York: Pioneer Publishers, 1931.

Truman's Point Four Program, June 24, 1949. In *Documents of American History,* 7th ed., edited by Henry Steele Commager, 558–559. New York: Appleton, 1963.

Twin Oaks Community. "Life in a Feminist Ecovillage." Community Pamphlet, 2004.

Twin Oaks Community website. www.twinoaks.org. Accessed January 4, 2013.

Unruh, Gregory C. "Escaping Carbon Lock-in." *Energy Policy* 30 (2002): 317–325.

———. "Understanding Carbon Lock-in." *Energy Policy* 12 (2000): 817–830.

Urry, John. "Consuming the Planet to Excess." *Theory, Culture & Society* 27 (2010): 191–213.

Usselman, S. W. "From Novelty to Utility: George Westinghouse and the Business of Innovation during the Age of Edison." *Business History Review* 66 (1992): 251–304.

Vaisey, Stephen. "Motivation and Justification." *American Journal of Sociology* 114 (2009): 1675–1715.

Wakefield, R. A., and P. Stafford. "Appropriate Technology: What It Is and Where It's Going." *The Futurist* (April 1977): 72–77.

Wann, David. *Simple Prosperity: Finding Real Wealth in a Sustainable Lifestyle.* New York: St. Martin's Press, 2007.

Warde, Alan. "Consumption and Theories of Practice." *Journal of Consumer Culture* 5 (2005): 131–153.

Weber, Max. *Economy and Society: An Outline of Interpretive Sociology*. Translated by G. Roth and C. Wittich 1935; rpt. New York: Bedminster, 1968.

———. *From Max Weber: Essays in Sociology*, edited by H. H. Gerth and C. Wright Mills. 1946; rpt. New York: Oxford University Press, 1958.

———. *The Protestant Ethic and the Spirit of Capitalism*. 1904; rpt. New York: Scribner, 1930 [1904].

———. "The Social Psychology of the World Religions." In *From Max Weber: Essays in Sociology*, edited by H. H. Gerth and C. Wright Mills, 267–301. 1946; rpt. New York: Oxford University Press, 1958.

Westen, Anthony. *Mobilizing the Green Revolution: An Exuberant Manifesto*. Gabriola Island, BC: New Society Publishers 2012.

Whitmarch, Lorraine, Saffron O'Neill, and Irene Lorenzoni. "Climate Change or Social Change? Debate Within, Amongst, and Beyond Disciplines." *Environment and Planning A* 43 (2010): 258–261.

Wiese, A. "The House I Live In: Race, Class, and African American Suburban Dreams in the Postwar United States." In *The New Suburban History*, edited by Thomas J. Sugrue and Kevin M. Kruse, 99–119. Chicago: University of Chicago Press, 2006.

Wilson, Charlie, and Hadi Dowlatabadi. "Models of Decision Making and Residential Energy Use." *Annual Review of Environment and Resources* 32 (2007): 169–203.

Wilson, Charlie, and Tim Chatterton. "Multiple Models to Inform Climate Change Policy: A Pragmatic Response to the 'Beyond the ABC's' Debate." *Environment and Planning A* 43 (2011): 2781–2787.

Winner, Langdon. "Do Artifacts Have Politics?" *Daedalus* 109 (1980): 121–136.

———. *The Whale and the Reactor: A Search for Limits in the Age of High Technology*. Chicago: University of Chicago Press, 1986.

Zablocki, Benjamin. *The Joyful Community*. Baltimore: Penguin, 1971.

Zaltman G., and N. Lin. "On the Nature of Innovations." *American Behavioral Scientist* 14 (1971): 651–673.

Index

About the Author

CHELSEA SCHELLY is an assistant professor of sociology in the Department of Social Sciences at Michigan Technological University. She received her PhD from the Department of Sociology at the University of Wisconsin–Madison and is the author of *Crafting Collectivity*.

MADELEINE'S QUEST

JULIEN EVANS

STEEMROK PUBLISHING

Also by Julien Evans:
Chalk and Cheese
The Sommerville Case
The Damocles Plot
Flight 935 Do You Read

Non-fiction:
How Airliners Fly
Handling Light Aircraft

Author's Notes

The chemistry and physics details in the story are generally accurate, although the descriptions of the results of laboratory variations in reaction parameters such as pressure, temperature and catalyst performance may differ from reality. When heated to high temperature in the presence of a suitable catalyst, carbon dioxide (CO_2) will react with methane (CH_4) to produce syngas (carbon monoxide (CO) and hydrogen (H_2)). The image on the cover of this book includes pictorial representations of the molecules of these substances. Currently (2015) the concentration of carbon dioxide in the atmosphere is approximately 400 parts per million.

The Christ the Saviour cathedral in Moscow, destroyed by Stalin, was eventually rebuilt after the collapse of the USSR

All the characters in this story are fictitious, with the following exceptions:

Gerard d'Erlanger, a director of the pre-war British Airways, who helped to set up the Air Transport Auxiliary

Amy Johnson, pioneer aviatrix, who died while serving with the ATA

Jacqueline Cochran, American pioneer aviatrix

Ranald Porteous, Chief Test Pilot at Auster Aircraft Ltd

Freddie Laker, proprietor of Laker Airways

Barbara Jane Harrison, BOAC stewardess who died trying to save the lives of her passengers during the accident to Boeing 707 G-ARWE at London Heathrow airport on 8 April 1968

Thanks are due to those contributors to the PPRuNe website who offered useful information concerning the operation of the BOAC B707

CHAPTER 1

"Thank you for gracing us with your presence, Luke." Professor Madeleine Maunsell didn't bother to hide her sarcasm. "Your monthly report seems to have gone astray as well."

The post grad researcher and the chemistry scientist looked at each other, Luke Stirling's normally self-assured (some would say smug) expression temporarily expunged by his mentor's scorn. Joe Curtis, Luke's lab-coated colleague, was avoiding the embarrassment by absently fiddling with a cathode ray tube control panel on the workbench apparatus a few feet away.

"Sorry, Professor. My car wouldn't start so I had to walk."

"There are buses. You come in from Abingdon, don't you?"

"Sunningwell. Sometimes walking's quicker than waiting for a bus. I got a lift for the last bit."

Madeleine glanced at the clock on the wall then back at the errant researcher.

She shook her head. "An hour late, and not for the first time recently."

"Sorry, Professor."

"Report?"

"Done the rough version. I'll type it tonight and bring the fair copy tomorrow."

"No, you'll do it today. Handwritten, if necessary. In blood, if necessary. On my desk, three o'clock at the latest."

"Okay."

"You did some scans of the nickel, didn't you? Are we still on nickel or did we decide to change?"

"No, it's still nickel. You said don't change the catalyst while we're still varying surface characteristics. Then we'll try ruthenium—that's what you suggested."

"Right. What did the scans show? We're on Steady State, aren't we? Was there any carbon poisoning?"

Luke looked across at Joe, and raised his eyebrows.

"We don't think so, Professor," said Joe, taking the question. "Based on yield rates anyhow. We think the SEM piccies confirm that. The resolution's quite good."

"It would be good to know it's serving its purpose," sighed Madeleine. The new Scanning Electron Microscope bought from Minnesota Research Inc had eaten significantly into the Department's 1967 budget and July was still a fortnight away.

The scientist turned again to Luke.

"Will there be scans attached to your report?"

"Yes, Professor."

"Good."

Madeleine turned to leave but was intercepted by Xiu Ying Song, another of the post grad students working in the lab. Daughter of a Hong Kong Chinese father and a Scottish mother, she was happy that everyone anglicised her first name to "Susan". She had heard that Luke sometimes referred to her as "Ying Tong" when talking about her to other people but she was not offended. She knew he was not a malicious character, even if his sense of tact was noticeable by its absence. It was all the fault of the radio comedy group calling themselves "The Goons". Their "Ying Tong Song" had been a popular ditty a decade previously.

"Excuse me, Professor."

"Yes, Susan?"

"We want to increase the temperature another twenty degrees. Is that okay? We want to see if the conversion rate gradient changes. We're think we're on a plateau now."

"Where are you now, temp wise?"

"Nine fifty Kelvin."

"What about safety regs?"

"Mr Syerman says we can go up to ten twenty with the current protection. Any higher and we'd have to put a screen in."

"Pressure still one bar?"

"One point three."

"Okay. Does it affect what James is doing?"

"No, Professor. The UV chamber is okay up to ten seventy. Mr Syerman checked with the manufacturer."

Madeleine smiled at her young researcher. "That's fine, Susan." She looked across the lab to the workbench Xiu Ying shared with James Ibsley.

"Another absentee?"

Xiu Ying followed her glance.

"James has gone out to . . . to . . ."

"Buy doughnuts?" offered Madeleine with raised eyebrows.

Xiu Ying blushed but did not respond.

"We take turns, Professor," came Luke's voice from the other side of the lab. "The bakery's just over the road, as you know. You have to go early or they're all gone. Jim'll be back in two shakes."

The chemistry professor shook her head. "What sort of set-up am I running here? Half my researchers missing half the time."

"A happy one, Professor," called Luke. "Mid-morning doughnuts keep our morale high."

"Not to mention our calorie intake," laughed Joe Curtis.

Madeleine smiled. "They are nice doughnuts, I'll give you that."

"I can nip over and get you one," offered Luke, seeking atonement for his late arrival.

"Thank you, Luke, I'll give it a miss this time. Besides, I don't want to drag you away from writing your report, knowing how much you enjoy doing them."

"My favourite activity."

"Three o'clock."

"Should be able to manage—"

"Without fail."

"Yes, Professor."

A final glance round her fiefdom, then Madeleine walked out into the corridor, heading for her office a few doors down. Seconds later James Ibsley came into the lab carrying the day's doughnut supply in a paper bag, slightly out of breath as he had raced from the bakery and come up the stairs rather than wait for the lift.

"We were lucky," said James in his antipodean twang. "They only had a few left."

"Madam Methane was here a moment ago, asking where you were."

James wrinkled his brow. "Surely she knows about doughnut patrol by now."

"You know what she's like when she's in a mood," said Joe. "She's already had a go at The Duke."

"What for?"

"Couldn't get my stupid car going," said Luke. "Needs a new battery."

"Well, you will drive a Dagenham Dustbin," said James.

"Excuse me. My Ford Prefect is the pride of British engineering . . . well, it used to be, anyhow."

The discussion about Luke's suspect mode of transport petered out and the researchers returned to their respective tasks.

Laboratory 2C was ensconced on the second floor within the Oxford University Organic Chemistry Gas Research Faculty, located in the city centre just across from New College, on the other side of Holywell Street. The entrance to the faculty could be found in Jowett Walk.

The main apparatus in Lab 2C was a complex arrangement of tubes, retorts and cables adorning a large cylindrical chamber set horizontally on the long workbench where Xiu Ying and James Ibsley were stationed. Several gas cylinders populated the floor space adjacent to the bench, black for carbon dioxide and orange for methane.

On the opposite side of the lab sat the less bulky kit used by Luke and Joe, two nondescript grey metal cabinets the size of television sets connected by rubber tubes to other gas cylinders and by electrical cables to a pair of cathode ray tube monitors. In a separate area stood several microscopes and a set of electric balance scales. On one of the mysterious grey cabinets was sellotaped a photo of film star Catherine Deneuve torn from a newspaper, while the other sported a similar image of Kathy Kirby, the pop singer. The two items of apparatus were Thermal Conductivity Detectors, also known as katharometers and the photos served the purpose of differentiating between them. Thus the post grads referred the to left TCD as "Deneuve" while its sister was "Kirby".

For their joint PhD thesis, Luke and Joe had agreed on the title "Methods for Assessing and Observing the Effect of Catalytic Variations on the Reformation of Methane by Carbon Dioxide". Within three years, if all went to plan, they would be able to style themselves Doctor Stirling and Doctor Curtis.

By contrast, Xiu Ying and James Ibsley were working on "Comparison of Steady State and Intermittent Periodic Gas Flows on the Reformation of Methane by Carbon Dioxide". Like their colleagues, the two researchers would attain Doctor of Philosophy status if the project was completed successfully.

Joe Curtis had been a student at Trinity, his performance during his exams earning him a well deserved First Class degree in chemistry, with Honours. His colleague Luke was a graduate of Cambridge. After his Finals, Luke was borderline First-Upper Second. A viva went his way when the assessment board grilled him on Gibbs Free Energy, one of the subjects he had by good fortune chosen to revise thoroughly. He was invited by Professor Maunsell into her team because at the interview she had been impressed with his knowledge of splitting of d-orbital energy levels in crystal fields. The tiny hint of smugness she was prepared to overlook.

Participating in the same graduation ceremony as Joe Curtis the previous summer, Xiu Ying was another Oxford chemistry graduate with First Class Honours. Now the two of them were friends rather than the mere acquaintances who had attended the same organic chemistry lectures.

James Ibsley had graduated from Aukland University, New Zealand. Research opportunities being somewhat limited in his homeland, James had looked further afield. Besides Oxford, interesting offers came from two American Universities, both of which tempted him with bigger grants and better fringe benefits. But his New Zealand girlfriend had landed a job as a dancer in "You Must Be Joking!", a successful London West End musical, and James decided to let his heart rule his head. The girlfriend lived in a flat in Fulham with two fellow performers and James spent as much time with her as he could.

The project the four young researchers were devoting their efforts to was the brainchild of Professor Maunsell, who was renowned in scientific circles for her work on how valence electrons in transition metal catalysts affected organic chemistry processes. In her filing cabinet at home were documents congratulating her on the distinctions she had achieved during her academic career. Unlike some other recipients of these prestigious awards, Madeleine felt no need to parade her certificates in frames on the walls of her house or her office.

Besides overseeing various research projects, more mundane tasks for Madeleine included lectures and tutorials for undergraduate chemistry students at the University. Both Xiu Ying Song and Joe Curtis had benefitted from the esoteric erudition of the lady who was now their personal mentor.

The aims of the project in Lab 2C were simple. Firstly, the idea was to find the best method of forcing methane to react with carbon dioxide to generate carbon monoxide and hydrogen. The second part of the project was to determine whether the method could be scaled up to reduce the large quantities of carbon dioxide escaping to the atmosphere from burning fossil fuels, with the resulting products used in other industrial processes. The clear benefit would be a reduction in the millions of tonnes of carbon dioxide released into the atmosphere every year by the world's power stations. The apparatus in Lab 2C allowed the researchers to adjust and measure variables affecting the reactive process, including temperature, pressure, electromagnetic radiation and catalysts.

"Madam Methane" was the more polite nickname by which the researchers referred to their mentor. Talking to his mates in the pub after a few bevvies, Luke would sometimes come out with "Mrs Cowfart", even though it had been pointed out to him many times that methane gas from livestock emanated primarily from the other end of the animal. But Luke acknowledged to anyone who would listen to his slurred opinions that the professor was "a great lady who would have been really fanciable twenty years ago."

* * * * *

Luke's mentor entered her office, sat at her desk and kicked off her shoes. She wondered whether she had been too grumpy with the youngsters. They weren't a bad bunch. Luke could be irritating at times but the four of them approached their tasks with enthusiasm. And Luke occasionally sparkled with a touch of genius. Occasionally his insight into d-orbital chemistry matched her own. The Duke, they called him. He wasn't a bad lad really.

Through the open windows wafted a gentle summer breeze carrying the shouts of students wandering along Jowett Walk two storeys below. Demob happy, thought Madeleine. The exams were over and the boys and girls who had endured them were celebrating accordingly. Within a week the city's population would dramatically dwindle as the students went home for the vacation. Madeleine would sometimes admit to her friends that these days she preferred the quieter Oxford of summer. Must be old age! In less than five years time she would reluctantly arrive at her half century.

"Can I get you anything?" Madeleine's secretary put her head round the door. "You look like you could use a cup of tea."

"Tea . . ." smiled the professor. "Man's greatest invention. Thanks, Verity. I wouldn't say no."

The secretary wrinkled her nose. "No, Prosecco. That wins the prize. Tea comes a close second though."

Madeleine laughed. "You're twenty years younger than me! You'll be a tea jenny one day!"

Verity disappeared and Madeleine sat back in her chair, thoughtful for a moment. Her eyes strayed to the picture frame on her desk, which held three photos. In the first her brother Nigel was sitting in the captain's seat of a parked Boeing 707, twisting round to grin at the camera. Outside the cockpit could be seen the tail of another 707 in BOAC livery. The photo below was of a group of young women in dark military uniform standing in front of a Spitfire. Madeleine was near the centre of the group.

The colour photo alongside the other two featured Madeleine and her parents, all grinning. It had been taken at an award ceremony. The certificate she was holding had just been presented to her by the President of the Royal Society.

Among the papers on the professor's desk was the letter that had arrived that morning. She picked it up with a sigh. She'd already read it twice.

CHAPTER 2

**Science Research Council,
22, Whitehall,
London SW1.**

**Tel: 01-944 4073 (3 lines)
Telegram: "Sciresco, London"
Telex: 827742**

June 21st, 1967

Our ref: JD/MM

Prof. M. Maunsell,
Organic Chemistry Gas Research Faculty,
University of Oxford,
Jowett Walk,
Oxford.

Dear Professor Maunsell,

I refer to our recent telephone conversation regarding future funding of the Methane Reformation Project. Since then the S.R.C. has had further discussions with Sir Melvyn Hocklan, the Chairman of the Allocation of Resources Committee at the Ministry of Technology. The upshot of the discussions is that funding for the M.R. Project is still under review, but it is more likely than not that the Committee will decide that funding will be discontinued at the end of this calendar year. The decision will be made not later than September 30th.

As the S.R.C. General Project Manager I am sorry to have to bring you these bad tidings. As you probably know, the Chancellor of the Exchequer has impressed upon all public sector departments the need to find economies where possible and this stricture has been applied to the S.R.C. In the case of M.R. Sir Melvyn mentioned several areas of concern, which he will include in a formal communication to me and which of course I shall forward to you as soon as I receive it. Two of the points raised by the Committee are set out herewith.

Firstly, it is noted that the Reformation Process is endothermic and requires temperatures in the region of 730 degrees Celsius before significant yields are obtained. If the energy required is furnished by the burning of fossil fuels then the overall reduction in CO2 emissions is significantly reduced, and may even be negative, in which case the benefits are restricted to the uses that can be made of the carbon monoxide and hydrogen (syngas) generated. As you know, these chemicals can be manufactured industrially by processes that are not expensive. If the syngas is used as a fuel then again one of the combustion products will be CO2.

Secondly, it is noted that currently the atmosphere contains 320 parts per million (ppm) of CO2, averaged over the whole planet. The mechanism by which solar radiation reradiated from the earth's surface increases vibrational and rotational motion in the molecules of the gas by I.R. absorption and the ensuing transfer of enthalpy to other atmospheric gases will be known to you. It has been estimated that the physical properties of the atmosphere and meteorological parameters would not be significantly altered by re-radiation of I.R. by CO2 until the concentration of the gas reached levels of approximately 400 ppm. Using current data, this level of concentration is unlikely to be reached within the next hundred years, by which time it is likely that alternatives to fossil fuels will have been developed.

As you know, the current Chairman was appointed some time after the M.R. Project funding agreement was reached. In my opinion the previous incumbent was impressed with your presentation when you originally applied for funding. Perhaps a personal interview with you might persuade the current Chairman—and indeed the whole Committee—to revisit the decision to withdraw funding if that becomes their position. If you would like me to arrange this please write or phone—

"Tea for the lady."

Verity had come back in. She put the cup and saucer on Madeleine's desk, noting that the professor was still holding the letter.

"What's going to happen, Maddie?"

The professor dropped the letter back on her desk. "I wish I knew."

"The Faculty wouldn't get rid of staff, would it, if they cancelled MR?"

"No, definitely not." Madeleine smiled. "Your job's safe, Verity. As long as I have any say in the matter, anyway. We're doing other things as well, don't forget."

"That's a comfort to hear."

"And I still have spotty undergraduates to indoctrinate with the creed of organic chemistry."

"But MR was your baby, wasn't it?"

"Wrong tense, Verity! *Is* my baby, if you please! It's not dead yet."

"Sorry!"

"It's the four post grads I feel sorry for. If we lose MR then we'd have to send them packing. The SRC grants are funding them. They'd have to find something else for their PhDs and someone else to fund them."

"That would be a shame. They're a nice lot. Even Luke when he's behaving himself."

"Well, I'm not giving in yet." Madeleine picked up the letter again. "John Davis has told me to go and nag the committee. It's worth a shot."

"You show 'em, Professor!"

"Strike while the iron's hot, eh? Verity, will you call Mr Davis and fix an appointment? The number's on this letter. Let me check my diary and see if there's a suitable—"

The phone rang in Verity's office next door. She pressed a switch on the intercom box on the professor's desk and picked up the receiver.

"Professor Maunsell's office. Who's calling, please?" She covered the mouthpiece and said quietly, "Prudence".

Madeleine held out her hand to take the receiver, releasing her secretary with a nod.

"Dizzy! How the devil are you?"

Verity gave her boss a smile and left the office.

"Bearing up," came the lilting female voice in the receiver. "What about you?"

"Fighting Whitehall bureaucrats," said Madeleine ruefully.

"And winning, I hope."

"Too soon to say, Dizz."

"Are you still good for Saturday?"

"Should be. Remind me what the plan is."

"Subject to weather we'll head for Sandown. If it's too crosswindy there we'll make it Bembridge."

"If it's Bembridge we could have lunch at the Crab and Lobster."

"Of course."

"Is Pam coming?"

"She says yes."

"What time for departure?"

"I'll do the preps myself. If you and Pammy roll up at Booker at, say, ten thirty, we should get airborne about eleven."

"Flight time?"

"Forty minutes with zero wind."

"Sounds OK."

Both Prudence Dismore-Whyte and Parminder Chopra had been wartime colleagues of Madeleine Maunsell and both of them could be seen in the Spitfire photo on Madeleine's desk. The ladies in the picture were all members of the Air Transport Auxiliary group, set up by airline director Gerard d'Erlanger in 1939 as an organisation of non-combatant pilots whose task was to ferry military aircraft from their places of manufacture to their operational bases. It would not be true to say that the females in the ATA came from all walks of life. The upper classes and aristocracy were well represented and a few had been presented to King George VI as debutantes before the exigencies of war brought an end to the ritual of the "coming out" ceremonies. The impecunious and those lacking good connections were generally not to be found amongst the ATA pilots of either sex. Many of them had paid for their own expensive training before the war started.

Lying just to the west of Maidenhead in Berkshire, White Waltham airfield was the ATA's main base. The feathers of the genteel British pilots stationed there had been ruffled somewhat by the arrival of the less-retiring American female cohort, led by the indomitable Jacqueline Cochran. It took a while for the native pilots to establish a rapport with their transatlantic colleagues. As the organisation grew so pilots arrived from the British colonies and indeed from many nations outside the Empire.

Many firm friendships were established among the White Waltham pilots, deriving partly from common interests and background and partly from the need for mutual support during their arduous and often dangerous work. Few of the ATA pilots had been trained in instrument flying or use of radio and their deadliest enemy was the fickle British weather. In 1941 pioneer aviatrix Amy Johnson got lost in bad weather while ferrying a transport from Blackpool to Oxford and was drowned after bailing out over the Thames Estuary. A temporary gloom settled on White Waltham when the news came through, but soon the sadness was dismissed with a metaphorical shrug and a laconic "too bad—let's get on with the job".

Madeleine's two closest friends were typical of the disparate group of fliers. Prudence Dismore-Whyte, known to everyone as "Dizzy", came from a well-to-do family who owned farmland in Derbyshire's Peak District. In her teenaged youth she had been a champion speed and figure skater who was expected to represent Great Britain in the 1940 Winter Olympics, had the war not intervened. Prudence's first serious boyfriend was a fellow athlete whom she had met while skiing in Kitzbühel. William had a pilot's licence and--more impressively—his own Miles Falcon, a sleek monoplane. Prudence needed no persuading to take up his offer of a joyride and even before they had landed she knew that flying was the thing for her. Her parents did not discourage her and her father offered to pay for lessons, although he drew the line at advancing funds so she could buy her own machine.

Soon Prudence had qualified for her own licence and after building up experience on cross-country flights she found employment as an instructor at Stapleford Tawney, an airfield in Essex. William entered his Falcon for an air race and Prudence was all set to be his copilot. But she was unwell on the day the race started and William found a replacement. On the first take-off the Falcon was seen to enter a very tight turn. Suddenly its nose dropped and it fell to the ground, bursting into flame on impact. Both pilots perished. The watching experts considered that William had pulled too hard on the stick and stalled the wings under excessive 'g' loading. At the outbreak of war Prudence was still getting over her grief. Thinking that some sort of positive activity would help to restore her frame of mind she applied to join the ATA and was accepted.

Completing the trio of female pilot friends based at White Waltham was Parminder Chopra. Parminder's father was a wealthy Indian businessman who had owned a franchise selling imported Vauxhall cars in Bombay. His eldest daughter learned to drive a car at the age of 14, which made her father very proud. But pride gave way to doubt when Parminder announced two years later that she wanted to learn to fly. Manoj Chopra told his daughter that flying was not the sort of activity a young girl of good caste should get involved with. But Parminder's strength of character was a match for her father's and she soloed in a De Havilland Moth before her eighteenth birthday. Her father's pride was restored and indeed reinforced the day she took him on a pleasure flight from the local airfield at Juhu. From the spectators' enclosure her mother watched the event apprehensively and offered a silent prayer of gratitude when the little biplane landed safely.

Parminder learned of the existence of the ATA through her aviation contacts and persuaded her parents to let her go to England to join the Empire war effort. She easily passed the skill test designed to filter out those not up to the job and eventually found herself ensconced at White Waltham, where the general discrimination against female pilots was for her compounded by her dark complexion and foreign name. No wilting flower, she either ignored the barbs or retaliated with the wit for which she soon became well known. To make things easier for herself and her hosts she adopted the name "Pamela".

Of the three White Waltham friends only Prudence had followed a career in aviation after the war. She resumed instructing, now at Booker airfield, near High Wycombe, and then progressed to instructing instructors. Skilled at aerobatics, she was often hired for stunt flying scenes in films and television programmes. Parminder had held a Private Pilot's Licence for a few years but had then let it lapse as other interests and raising a family took up her time and resources. Madeleine of course went back to the academic career that the war had interrupted.

Parminder was the only one of the three who had married. She had first met her future husband during a spell of duty at the ATA ferry pool at Hamble airfield, where he was an operations officer. Ignoring racist comments and criticism from his colleagues, Rick Collins started to take the Indian girl sailing whenever she had free days at Hamble and love had blossomed. Now Mr and Mrs Collins owned an office cleaning business in Aylesbury. Parminder's father offered to help pay for enrolment of their three children at a good English public school but she had courteously declined, remembering the snobbery which had sometimes tainted the atmosphere at White Waltham. The local school was the best place for the kids to meet normal people, she told her father.

* * * * *

The planned trip to the Isle of Wight followed the pattern that had been established not long after the war. Every few weeks Prudence would hire one of the Booker planes for the day and the three friends would fly to another airfield for an extended lunch and catch up. In the single-engined Piper Cherokee they now used Prudence would sit in the copilot's right seat so that Parminder could take the controls for the outbound flight and Madeleine for the return, or vice versa. The only difference between these outings and similar friends' lunches all over the country was that no alcohol was consumed. All three had witnessed events during the war when pilots under the influence had embarrassed themselves. Or worse.

Old habits die hard. Like most pilots (and ex-pilots), Parminder and Madeleine were both intensely self-critical and intensely competitive. Sometimes these attributes mutated into parody. The unspoken challenge was: who could do the smoothest landing?

"Nice landing, Pam."

"No, it was hopeless. I should cut off my hands."

"Nice landing, Maddie."

"Pathetic. Even worse than my usual crashes."

"Okay, Maddie," the professor now heard in the phone receiver. "See you on Saturday."

"Thunder, lightning and rain."

"Hurly-burly."

The snippets from Shakespeare's Macbeth were Parminder's invention as a sign-off amongst the friends between themselves after a White Waltham wag had dubbed them the Three Witches.

Barely had Madeleine replaced the receiver when the intercom buzzed again.

"I've got John Davis on the line, Maddie. He wants to arrange a date with you to talk about funding."

"Put him through, please, Verity."

CHAPTER 3

Nigel Nixon swung his maroon Triumph Vitesse convertible off the A4 just west of Maidenhead and headed towards Pinkneys Green. With the car's soft top retracted, the pilot's light brown hair rippled in the warm slipstream. Ray-Ban sunglasses shielded his eyes from the fierce July sun. After half a mile he turned right again into the driveway of a large Edwardian building. Partially obscured by ivy on one wall was a sign saying Fearnville Nursing Home for the Elderly.

Having parked the car, Nigel made his way indoors and stopped at the reception desk where a matronly figure was seated apparently tackling the crossword in a newspaper.

"Hello, Mary," grinned the pilot, removing his sunglasses. "How is she today?"

The receptionist looked up and peered over her reading glasses at the visitor. "Not too bad, Mr Nixon. A bit grumpy, but not more than usual. The fine weather's helped a bit. She's in the garden."

"No problems recently?"

Mrs Shrigley sighed. "There was an escape attempt a week or so ago."

"Oh dear, not again."

"We're used to it. She's not the only offender. Luckily our security doors stop them in their tracks."

"Is she lucid today?"

"Seems to be. Is Professor Maunsell coming in?"

The pilot looked at his watch. "Yes, should be here soon."

"Why don't you go through, Mr Nixon—"

"Nigel."

"Yes. I'll bring a nice cup of tea out for you, and one for Dorothy too. Would you like a chocolate biscuit?"

"Thank you, Mary."

Dorothy Nixon's wheelchair was facing the sun but the Fearnville staff had thoughtfully erected parasols over the heads of those residents who preferred to sit in the shade, including Nigel's mother. There were a dozen or so residents in the garden and one or two of the nursing staff fussing around them, efficient in their crisp, light blue uniforms.

"Hello, Mum," said the pilot, pecking Dorothy on the cheek.

"You're late," came the muted reply. The old lady did not turn her head to look at her son.

"No, I'm not," countered Nigel. "It's ten thirty. We always come between ten and eleven. You know that."

"Well, nobody told me."

Nigel grabbed a nearby chair and positioned it in front of his mother to bring himself into her line of sight. Now she fixed him with gimlet eyes.

"Where's Madeleine?"

"She'll be here soon, Mum."

"She cares even less about me than you do."

"That's not true, Mum."

"Yes it is. She's only your half-sister, you know. Have I told you that before?"

"Mum, I've known that for thirty years."

"She's the child of your father's tart. That's why she doesn't care about me."

"That's nonsense, Mum, and you know it. She does care about you even though she's not your child. You should be grateful for the help she gives us. There's no obligation. She does it out of the goodness of her heart."

A snort of dissent from Dorothy.

"And," continued her son, "if you don't treat her more respectfully she won't come and visit you any more."

"As if I cared."

Nigel sighed. They had been through this exchange, or variations of it, many times before. Sometimes he was the baddie and sometimes it was Madeleine. The pilot and his sister knew it was pointless trying to defend themselves when Dorothy was in this sort of mood.

Nigel forced his face into a smile. "So, what's new? Are you enjoying the nice weather?"

"It's too hot for me."

The pilot looked at his watch again. It would be easier when his sister showed up and shared the load. Neither of them liked solo visits and standard procedure was for the two of them to meet at Fearnville and spend an hour or so with Nigel's mother. Occasionally they summoned up the fortitude to have lunch with her, trying to ignore the barbs and criticisms. Because Nigel's rest days varied from week to week the easiest procedure was for the chemistry professor to align her availability with Nigel's when possible. The siblings made sure that Dorothy saw the two of them on average once per week.

Fearnville had been chosen because it was roughly halfway between Nigel's house in Epsom and Madeleine's in Woodstock, a few miles northwest of Oxford, and because the staff were trained in dealing with recalcitrant elderly residents. Dorothy had divorced Henry Maunsell, her first husband, around the time of Madeleine's birth. Her second husband had died two years ago and the widowed Dorothy became used to living alone and fending for herself, with help from Nigel and Madeleine when circumstances dictated. But a year after her husband's death she had had a minor stroke which impaired her mobility and, more significantly, degraded her mental faculties. She became forgetful and sometimes confused. Madeleine arranged for carers to visit Dorothy's house to prepare meals and help her with washing and bathing and domestic duties but Dorothy's progressively irascible behaviour led to the carers withdrawing their services, citing their client's rudeness and lack of co-operation.

When discussing his mother Nigel and Madeleine would always come to the same conclusion. They had to cut her a bit of slack because she was getting old and old people sometimes went strange. And because she never used to be as difficult as she was now, which might be the result of the stroke. And because, when all was said and done, she was Nigel's mother.

Sometimes brother and sister would comment on the irony of the situation. Dorothy had granted her son Power of Attorney to help her manage her affairs but had then objected when he used those powers to sell her house in order to pay the care home fees, which depleted her assets at the horrendous rate of £48 per week. Her occupational and state pensions met only a small fraction of the cost. So this vast torrent of cash was paying to keep Dorothy in an institution which she hated. Well, perhaps hatred was too strong a word. Dislike ranging from mild to intense was probably a more accurate assessment. An unspoken worry was what would happen if Dorothy's funds ran out before she shuffled off her mortal coil. Enquiries indicated that state provision would be nowhere near as bountiful. At 78 years of age it could be that there were several years of life left in her yet. Another irony—what was the point of longevity if she was getting more and more miserable as time went by?

Today's routine would be different for Nigel. After his duty was done he would follow his sister's car northwest along the A423 to Oxford rather than homewards. But as they approached the city centre he would take the A34 ring road round the city and continue northwards towards the airfield at Kidlington, midway between Oxford and Woodstock. There was a vague plan that he might drop in at Madeleine's later on for a cup of tea if she was home early from the lab. On the other hand if they managed to do all their catching up while they were at Fearnville he might head for home after Kidlington.

"Here's your tea, Dot," came a female voice. A nurse was walking towards them carefully balancing two teacups. Nigel took them from her with a smile and placed them on the little table beside his mother's wheelchair.

"I'm not Dot, I'm Dorothy," said Nigel's mother with a scowl. "How many times do I have to tell you?"

"Sorry, I forgot!"

Another of his mother's whims, thought Nigel. Most of the time she was quite happy with "Dot" but occasionally the grumpy circuits would cut in and she would object. Nigel mouthed his own apology to the nurse, who just shrugged with a wry smile.

"Oh, your sister phoned," said the nurse now. "She can't make it today. Can you phone her back? You can use the phone in reception if you want."

"I'll do it now," said Nigel.

"Are you going?" asked Dorothy. "You've only just got here."

"No, I'm going to phone Maddie. She can't get here today."

"Why not?"

"I don't know, Mum. That's why I've got to phone her. To find out."

Nigel followed the nurse back indoors, grateful for a temporary respite from his mother's scorn but dreading the prospect of having to deal with her thereafter without support until he could decently get away.

"I'm sorry, Nige. It's a mess here," were his sister's words when he got through. "What state is she in?"

"Her usual miserable self. You're in the dog house for going AWOL."

"Bugger! We've got a bunch of Russians chemists here looking at our set up. Prof Henstridge was supposed to be showing them around but he's off sick so I'm filling in. Are you staying at Fearnville for lunch?"

"Not if I can get away with it. I'm supposed to be going to Kidlington this afternoon so I've got a real excuse if I need it. I'd prefer to have lunch there."

"Do you want to meet later?"

"Oxford or Woodstock?"

"Probably Oxford. I'll be in the lab till about six, catching up on stuff I would have done earlier. Is that too late?"

"Okay. These Kidlington sermons usually take a couple of hours. I should be free about four."

"Alright, come to the lab when you've finished. We could have dinner if you want."

"I'll think about it. I'll check with Jackie. Not sure if she's got a plan for tonight."

"Well, good luck anyway."

"With the Kidlington Kids or with my mother?"

A laugh down the line. "Both."

"Twenty pimply cadets are a lot easier to handle than you-know-who."

"I don't doubt it."

The conversation over, Nigel went back to the garden with heavy heart. He was intercepted by the carer who had brought them tea.

"She's fallen asleep. Do you want me to wake her?"

The pilot thought for a moment. He was tempted to scarper but he knew that when she woke up Dorothy would give the Fearnville staff hell for letting him go. He sighed.

"Yes, that's probably for the best. She'll only take it out on you lot if I do a runner."

"You're very kind. Let's do it together then."

*   *   *   *   *

Hooray! Nigel rejoined the A423 and put his foot down. The Vitesse responded with a satisfying roar of engine.

Hooray! DCO, as they used to say in the Air Force. Duty Carried Out. He'd managed to endure an hour and a quarter with The Misery and now his mood lifted at the thought that the rest of his day would be more pleasurable. In around an hour if the traffic wasn't too bad he would arrive at Kidlington airfield, where he would be addressing a group of cadets training to be airline pilots.

The airfield's catering facilities were adequate rather than epicurean but it would be nice to have a spot of lunch with a few of the instructors, one or two of whom he knew from his Air Force days.

The Kids were actually post graduate university students who had successfully passed the rigorous selection procedure for sponsored flying training by the two state airlines, British European Airways and British Overseas Airways Corporation. The training to airline pilot standard took approximately twelve months, after which they would convert onto their first airliner type as junior copilots.

Nigel was one of the captains who volunteered for these lecture duties. The idea was to give the cadets an insight into how their careers would progress after qualifying and to answer any questions they might throw at him. So this particular afternoon he might be describing the set-up by which they would eventually be pounding the circuit in a BOAC Boeing 707 at Prestwick airport at the same time as his sister was entertaining her Russian visitors.

The Kidlington Kids were not short on self-motivation. The young men knew they were lucky to have been selected and they applied themselves with vigour to their chosen profession. The course was an intensive mix of practical flying and ground subjects and the standards demanded were high. Around a fifth of the cadets would be chopped during the course for failing to make the grade. Similar courses were run at Hamble, on the south coast, the same airfield from which First Officer Madeleine Maunsell and her chums delivered warplanes to their operational bases a quarter of a century previously.

Nearly there. As Nigel turned into the airfield entrance he noted a Twin Comanche climbing away sluggishly with noticeable sideslip. Obviously a practice engine failure on take-off. He could imagine the instructor's exhortations to his student. *Keep it straight, laddie! Watch your airspeed! Identify the dead engine!*

CHAPTER 4

The "Trout" was fortunate in that the mass exodus of students for the summer vacation was partly offset by increased tourist numbers. Some of the pub's customers turned up after visiting the nearby city of Oxford and some dropped in on their way to Blenheim Palace, a few miles to the north.

On this July evening there were a few young people supping their drinks at outside tables alongside the weir interrupting the flow of the River Thames but none were undergrads. At one table sat half a dozen pilot cadets from Kidlington discussing the relative merits of England cricketers Basil D'Oliveira and Geoffrey Boycott. Two more cadets, their training sponsored by Olympic Airways, were chatting in Greek.

At another table sat Luke Stirling, Joe Curtis, Xiu Ying Song and Gillian Bradshaw, Luke's girlfriend. The mood was sombre and no words had been spoken since Joe's remark, "Not looking too good then."

Gill was a biology student who had just sat her finals. In a week or so Harold MacMillan, the Chancellor of Oxford University, would present her with her degree, an upper second. Two months from now she would start gainful employment with a pharmaceutical firm in Bristol, where her parents lived. They were pleased that their daughter had acquitted herself well at university and that they would now see more of her, but less than thrilled when she told them that at weekends she would be driving to Oxford to stay with her boyfriend. Nor were they impressed when she announced her intention to join the Labour Party, with the ultimate ambition of getting herself elected as an MP. On the other hand they liked Luke, who would before too long become Dr Stirling. He had been going out with Gill for three years and seemed to genuinely care for her. As yet Mr and Mrs Bradshaw were unaware of the calamity that might befall their daughter's boyfriend. Luke and his fellow

23

researchers had only just discovered the grim news themselves and the pub meet was to discuss whether there was anything they could do to avert disaster.

Now Xiu Ying restarted the discussion. "Well, the Prof said there was a faint chance we could carry on."

"She didn't tell us what 'faint' meant, though," said Joe.

"Well, she said she was going to see the people in London who make these decisions, didn't she," said Luke. "Perhaps we should offer to go with her, to back her up."

"I'd go along with that," said Joe.

"I don't know what I'd do if the project gets wound up," said Xiu Ying. A sad smile settled on her lips. "Perhaps I should swap my chemistry degree for a law degree. Then I could go and work for my father in Hong Kong."

"It won't be the end of the world," said Luke. "We've all got good degrees. If the worst comes to the worst we could find research projects in the States. They're crying out for Oxbridge graduates over there. Look at Jimmy—he was offered a place at MIT, wasn't he?"

Joe nodded. "And Berkeley. But he graduated from Aukland."

"Same thing. They'd snap us up."

"I'm not sure I want to live in the States," said Gill quietly.

"Look, we're jumping the gun on this," sighed Luke. "Mrs Cowfart might be able to persuade the men with the money bags to keep it coming."

"Isn't your boyfriend American, Suzie?" asked Joe.

"Yes, from Missouri."

"Post grad here isn't he?"

"Yes, history of languages."

"Is he going home after he gets his PhD?"

"Probably."

"What about you? Will you go with him?"

Xiu Ying sighed. "I don't know."

"Jimmy'll never leave the UK," said Luke. "Not while his bird is prancing round on a stage in London."

"The main question is—are we wasting our time, anyway," said Joe. "I mean, who gives a toss about CO2 in the atmosphere? If it was really important there'd be loads of other people looking into it besides us. The Prof seems fixated on it and I love the work we're doing but—" Joe's comment petered out in a shrug.

"I like it 'cos it's an intellectual challenge," said Xiu Ying, "but you have to ask yourself if there's any practical value."

"But that's true for loads of research projects," offered Luke. "Nothing wrong with that. Sometimes the benefits of research aren't appreciated until some time afterwards."

"That's where the Americans do better than us," said Xiu Ying. "They've got a more long term outlook. My boyfriend says the British won't put money into a project unless there's a return on investment in five minutes."

"So who's funding your boyfriend's research?" asked Joe.

"The language faculty at St Louis University."

"Wonder if they'd send over a few dollars for a Brit project reforming methane with CO_2," joked Luke.

"Well, I like your idea of offering to support the Prof when she goes to see the SRC," said Joe.

"Who are they?" asked Gill.

"Science Research Council. All our dosh comes from them."

The tranquillity of the warm summer evening was mildly disturbed by the gentle buzz of an aircraft passing overhead. The single-engined trainer was heading north, towards the airfield at Kidlington. None of the methane group paid any attention. By contrast all eight of the cadet pilots looked up. A discussion started as to whether the plane was positioning for a straight in on zero three or left base for two seven. The conversation in Greek at the adjoining table was perhaps the two Olympic cadets asking themselves the same question.

Xiu Ying turned her head, half-listening to the pilots. A thought had been triggered.

"Didn't the Prof use to fly planes in the war?" she wondered now.

"That's right," said Luke. "There's a photo on her desk. They took planes to their bases from the factories or something like that. According to the RSM they flew all sorts—fighters and bombers and everything else."

"Regimental Sergeant Major" William Syerman was the Safety Officer at Gas Research. The humorous tag came from his ex-military background. Eight years of army service as an explosives expert and bomb disposal officer had started during the war and culminated in his promotion to Warrant Officer in the Devonshire Regiment. His expertise in handling hazardous materials qualified him more than adequately for the post of Safety Officer in the laboratory. Luke and James took great delight in taunting him with unmilitary behaviour just to hear him bellow his opprobrium. Disregarding the students' intellectual prowess, the RSM would chastise them as if they were dim-witted schoolchildren.

"She never married, did she, the Prof?" asked Xiu Ying now.

"Don't think so," said Joe. "I heard somewhere that she had a boyfriend during the war but he got killed."

"Yeah, I heard something like that," added Luke. "Or maybe she's a lesbian. She lives with another woman, doesn't she?"

"She doesn't look like a lesbian, though," said Joe.

"So what's a lesbian supposed to look like?" asked Gill.

"I dunno . . . big jaw . . . moustache . . ."

"She's not a bad looker," said Luke, "for someone pushing fifty."

"She's got more brains than you," said Gill, disapproval in her tone.

"Debatable," said Luke playfully. "I had to correct her on the ruthenium d-orbital energy levels, remember."

"She's a world expert on transition metal catalysts, mate," said Joe, "as well you know. You don't even come close."

"So why did she go from flying planes to organic chemistry?" asked Xiu Ying.

"The RSM told me her brother is an airline pilot," said Luke. "Perhaps it was a family thing. The Flying Cowfarts."

"I think our big problem is endothermicity," said Joe, continuing the trend of non sequiturs. "If we can reduce temp and boost electromag input the process would be more efficient, specially if we use solar photo-voltaic."

"PV costs a fortune, though," said Luke. "Even if we can zap more CO_2 that way. You don't get enough UV from solar, we all know that. And the power for your UV generator will push out CO_2 somewhere along the line."

"Aren't we switching to ruthenium soon?" asked Xiu Ying.

"Yep. Madame Methane says we can do it any time we want. Personally I don't think it'll do much better than nickel." Luke put his tongue in his cheek. "I had to explain to her that the electron flux won't be high enough—"

"You prat!" laughed Joe, flicking a beer mat at his fellow researcher. "That's what she told you!"

"So," said Xiu Ying, "what will the Prof say to the SRC guys if they ask her where the improvements are coming from?"

"I dunno," said Luke. "Maybe if she walked in wearing a see-through blouse and no bra and a mini-skirt and black fishnet tights—"

Gill leant forward to cover Luke's mouth with her hand. "You're right, Joe," she said with a laugh. "He is a prat!"

CHAPTER 5

From a distance they were a lonely speck in a vast panorama of tranquil sea and cloudless sky, a dark blue and silver glint hovering six miles above the ocean's smooth surface under the late morning sun blazing down almost directly above them.

Close up the speck grew into a sleek Boeing 707-436 arrowing north-eastwards through the Pacific skies at 82 per cent of the speed of sound, equivalent to 550 miles per hour. Since the jet airliner's take-off from Honolulu two hours earlier the fiery combustion chambers in its four Rolls-Royce Conway engines had already converted more than 12 tons of kerosene into carbon dioxide and water vapour. In the exhaust gases streaming from the engines the water vapour froze instantly into ice crystals to form the dazzling white contrails marking the airliner's progress through the cold upper atmosphere. Although cynics had it that the Conway was merely a device for turning fuel directly into smoke and noise—with a little residual thrust—the British bypass engine would in fact pull any aircraft it was attached to through the sky while burning fewer pounds of fuel per mile than the older turbojets.

On the 707's swept back fin the golden Speedbird logo confirmed the owner of this technical marvel as the British Overseas Airways Corporation. The forward fuselage bore the four-letter BOAC acronym in the same gold leaf.

Closer still, an observer peering in through the cockpit windows would note five men comfortably ensconced inside. In the forward left seat Captain Nigel Nixon was munching a smoked salmon sandwich, his gaze casually sweeping the skies outside, with occasional glances at the instrument panel in front of him. On his lap was a folded copy of the *Hawaiian Chronicle* newspaper, on top of which lay the Voyage Report the pilot had been adding entries to.

The BOAC Operations Manual had no instructions regarding extraneous reading matter in the cockpit. Some captains forbade any such material, which made for long hours of tedium when normal conversation had exhausted the range of topics the crew members might want to bring up. Nigel's approach was more lenient. Reading papers or magazines or books was permitted between top of climb and top of descent, with his prior permission. A front seat pilot could not avail himself of this privilege if his colleague's seat was vacant, and Nigel was fastidious in applying his own rule to himself. Thus, when Senior First Officer Rodney Smith had gone back to the navigator's station, his captain had duly put down the paper he had been reading.

Rodney was now standing by the table at the navigator's station where Second Officer John Cooper was seated with pencil and ruler in his hands. Between the captain's seat and the navigator's, First Officer Mike Archer dozed in the observer's seat.

"Too easy, isn't it?" Rodney said to John. "When the Loran's working well, I mean. What do you reckon for a track error?"

The younger man positioned the ruler on the chart taped to the table.

"Four miles right."

"Accuracy?"

"About the same."

"Wind component?"

"Plus eighteen . . . seven better than forecast."

"So . . . next ETA?"

"Er . . . " John looked down at his log. " . . . two minutes ahead of plan."

"What shall we tell the captain?"

"Alter heading one degree left?" suggested John tentatively.

"He'll give you a funny look if you say that," laughed Senior Engineer Officer Frank Armitage at the flight engineer's panel on their right. "He'll think you're taking the piss."

"What do you mean?" asked the youngest crew member.

"Tell him to go five left then four right." It was an old gag and Frank quickly apologised for inflicting it on them.

"So no change, then," said John, working out what was being suggested.

"Yeah," said Rodney, "tell him the error and new ETA but tell him to maintain heading."

"Okay."

"No . . . tell you what . . . Frank, will you tell Nigel? John and I will Up Periscope and take a sun line. Good practice for him."

The flight was outside VHF radio communications range and so the crew had removed their headsets for comfort. If Oceanic Control wanted to talk to them they would hear the Selcal alert tone from the high frequency receiver. For normal conversation the crewmen had to raise their voices slightly above the rush of slipstream past the windows.

When Second Officer John Cooper wasn't immersed in his work he would reflect on how well life had turned out for him. This was the first time he'd ever been outside the UK. Two years previously he had left school, aged eighteen, to start his flying course at Hamble, his only previous experience a few trips in a glider with the Combined Cadet Force at school. Now he held a commercial pilot's licence, with instrument rating, and was calculating the position of a Boeing 707 carrying one hundred and twelve fare-paying passengers from Honolulu to Los Angeles. In First Class, charming all who spoke to her, was famous Hollywood film actress Nell O'Neill. And it was up to John to tell the captain what heading to set on the autopilot to make sure that Nell and the other hundred and eleven arrived safely in LA.

Actually, as John well appreciated, final responsibility for navigation rested with Rodney, who was training the newcomer in various techniques. Overall the trip would last two weeks. London Heathrow to New York Kennedy (two night stopover), Kennedy to LA (two night stopover), LA to Honolulu (three nights), then back home on the same routing. A Sydney-based crew had taken over the outbound aircraft and another Sydney crew had brought 707 Foxtrot Delta into Honolulu, where Captain Nixon and his crew had boarded.

John was not allowed to fly the jetliner himself, nor even to touch the controls. The closest he had come to achieving the ultimate goal was when Captain Nixon had allowed him once or twice to sit in the copilot's seat and monitor what the autopilot was doing during the cruise.

Whenever John and his chums languishing at the bottom of the seniority list asked about likely dates for starting their conversion courses the responses and rumours were so wild in their variation as to be meaningless. No matter. Patience was the key. Training for the Flight Navigator Licence was interesting enough. And the chaps he was crewed up with for his first trip seemed like a good bunch. No one had told him off when he shot the wrong star on the periscope sextant during the night flight from LA to Honolulu and found the cocked hat he drew on the chart where the position lines intersected was ninety-two miles wide. Mortified by his error he expected a torrent of negative judgement from the others. Ninety-two miles! How could he have made such a mistake! But Rod put a hand on his shoulder and gave him a smile. "We've all done it," he said. "Have another go."

Second only in importance to the rumours about when the new recruits would actually start 707 Ground School (Lecture 1:

Electrics) were the stories they'd picked up about some of the captains they'd eventually be flying with. Captain Gilmore was supposed to be a grumpy martinet who would not reply to any crew member failing to address him as "Captain Gilmore, sir" and whose largesse when it came to giving away take-offs and landings was notable for its infrequency. In comparison, Captain Hayling allegedly vacated his seat not long after the undercarriage was retracted, swapped uniform shoes for carpet slippers and let his two copilots run the show until the halfway down the descent, when he'd don his shoes and slide into his seat for the approach. As often as not, so the rumour went, he would turn to the copilot in the right seat with a wink and a "Show me how to land it, fellah!"

Discreet enquiries before departure from Heathrow had elicited the information that "Soapy" Nixon was one of the good guys. The nickname derived from his frequent exhortation to the crew members he was training that they must follow SOPs, the acronym for Standard Operating Procedures. Generous, too, thought John. The first landing had been done by the captain himself at JFK and the second by First Officer Mike Archer at LAX where yellow smog had cut the vis to barely a mile. Their night arrival into Honolulu coincided with a rain squall sending a gusty wind lashing across the wet runway, which concentrated all their minds.

"You happy?" the captain had asked Rod, who was flying the approach.

"I am if you are," came the reply.

"Your landing, Rodders. Good luck."

"Amen," muttered Frank Armitage, watching like a hawk from the flight engineer's seat.

Another whim the new boys had to familiarise themselves with was the correct form of address when speaking to more senior pilots, which again varied in formality according to personal preference. Captain Nixon's requirements, as relayed to John by Rod at the initial briefing at Heathrow, was that "sir" or "Captain" would be expected whenever the crew were in uniform, both in the air and on the ground. In civvies, off duty, "sir" would do until the captain specifically allowed "Nigel". Never "Nige."

John quickly noticed that different rules applied to Frank Armitage, who addressed his captain randomly as "Skipper" or "Skip" or "Nigel", regardless of where they were and what they were wearing. The junior pilot also heard "Nige" a few times from the flight engineer's lips, which didn't seem to cause offence. Perhaps the familiarity was allowed because, according to Rod, the two older men had been fellow crew members in Lancaster bombers during the war. After a few drinks, John was informed, you could sometimes get them to relate some of their wartime experiences.

*   *   *   *   *

The Second Officer retracted the sextant, the pressurised cockpit air hissing as it escaped past the instrument seal until the hatch was securely closed again. He stood down from the stool to find the captain standing by the nav table, where Rod was seated.

"Having fun?"

"Yes, sir. Just doing another sun line. It's not far off overhead so Rod says it'll be a good test of accuracy."

"Good stuff. I might have a go myself later, see if I can remember how to do it."

"Yes, sir."

"Work out your position line later, John. Go and sit in my seat for a bit. Mike's up there right side. Rod will look after the nav—give you a break. You know how to tweak the heading if needed?"

"Yes, sir."

"Mike'll keep an eye on you—if he doesn't fall asleep again."

"Okay, sir."

"Tell him to let you do a couple of turns, say twenty degrees either way, just so you can see the response. Go left first to get us closer to track."

"Okay, sir."

As the Second Officer moved forwards, marvelling that he was going to be allowed to steer the 707, even if via the electrical intermediary of the autopilot, Nigel turned to Rodney.

"How's he doing?"

"Not bad. Still a bit nervous after the muddle outbound."

"I've done that more than once."

"Me too."

The captain nodded. "They're so young. What is he? Twenty? Twenty-one?"

From the engineer's panel the moustachio'd face of Frank Armitage turned towards them.

"Just remind me, Skip, how old were you and me when we were bombing shit out of the Jerries in 478?"

"At least twenty-three, Frank!"

"There you are, then."

"What was 478?" asked Rod.

"Best damned Lanc in Bomber Command," said Frank. "No matter how badly Nigel abused it it always brought us home."

"Even though I had the most useless flight engineer!" joked the captain.

"I think you mean the best flight engineer."

All three men momentarily looked forward, aware that the Boeing was rolling into a left bank. Nothing was said while they waited for the wings to level again. After half a minute a roll right brought the aircraft back to its original heading.

"Someone's having fun," said Frank.

The captain grunted and tapped his finger on the Loran set. "I'll watch the kit for a bit if you want a break, Rodders."

"Thanks, sir. I'll go back and check with the Chief, see what booze we want to liberate from the bar for later."

"Good thinking."

The cockpit door closed behind Rod and Nigel sat himself in the observer's seat, casually watching the instrument panels in front of the two younger pilots. Then he turned to look at the engineer's panel. Frank was reading a magazine about antique clocks.

"You still making those things, Frankie?"

"Yep. Selling 'em too. Have to. BOAC don't pay me enough."

"My heart bleeds."

"You want to buy one?"

"I've heard about your prices. I'd have to sell my car!"

"You still got that Vitesse?"

"Uh-huh. Nice motor. What are you driving these days?"

"Mark Ten Jag."

"They're paying you too much!"

"That'll be the day."

There was a minute's silence. Looking forward, Nigel could see the two copilots chatting to each other. He sighed and muttered to himself: "Well, I'd better see if I can get anything out of the Loran." He raised his voice to speak to the engineer. "How are we doing on fuel?"

Frank reached out for his clipboard and ran a pen down one of the columns on his log. "We're fat, Skipper. Three and a bit tons over diversion minima."

"Groovy . . . as the young people say. San Diego fuel gives us Palm Springs as well, doesn't it?"

"Yeah . . . more or less. Loads of fuel." Frank jerked a thumb at the Loran screen. "Assuming you don't get us lost."

"Bloody cheek!"

The engineer rubbed his moustache. "Sad to hear about Maddie, Nige. Give her my love when you next see her. Is there anything you can do to sort things for her?"

The captain shook his head. "Not really. As I told you the other day, I don't even understand what she's supposed to be doing. Something about stopping carbon dioxide getting into the atmosphere."

"Well, I have to say that seems a pretty pointless exercise to me. But she's a bleeding professor, isn't she . . . there must be something else she can turn her talents to."

"You would have thought so. I think she's worried about her post grad students. Their grants would be cancelled."

"She should have stuck with flying, Nigel. Everyone said she was a bloody good operator on the four-fans. Better than some of the blokes, they said."

"So I heard."

"Apparently she was a whizz on instruments. Wasn't it you who taught her the basics?"

"Yes. They were losing too many people who couldn't handle crap weather. I didn't want her ending up in a smoking hole in the ground."

"I was never uncomfortable the time I flew with her." The engineer's mouth curled into a mischievous smile. "You on the other hand . . ."

"Watch it, matey!"

Both men laughed and Nigel turned his attention to the Loran, fiddling with the control panel, watching the trace on the screen.

"Did she never think about civvy flying, Nige?" asked Frank, picking up the conversation. "She could have hacked it. No doubt about it."

"Difficult choice for her, she always said. She liked flying but she knew the chances of finding a job were remote. No one wanted female airline pilots in the post war years."

"A few of the gals got in," observed Frank.

"Yeah, that's true," agreed the captain, still watching the Loran screen. "But Maddie loved chemistry too. So aviation's loss was science's gain. She's still pally with a couple of the ATA girls, though. That posh bird . . . what's her name?"

"Dizzy, wasn't that it?"

"Yeah, and the Indian girl."

"I remember her. A stunner."

"Yeah. Dizzy is an instructor at Booker. The three of them do trips together sometimes."

"Three women in a plane," laughed Frank. "I bet it's quiet!" He paused briefly. "Has Maddie got a fellah?"

"Not that I know of," said the captain. He thought for a moment. "We don't talk about these things."

"She had a boyfriend at White Waltham, didn't she?"

"She was engaged. But her fiancé bought it a couple of weeks before they were due to get hitched."

"I remember vaguely, now I come to think of it. Mossie pilot, wasn't he?"

"Yes, pathfinder on 139 Squadron. Shot down over Dortmund."

Again the two men fell silent. The engineer ran his eyes over his panel, flicked a few switches, checking the 707's generators and inverters, then leant back in his seat and picked up his magazine again.

"I know where we are," beamed Nigel, looking up from the chart where he'd constructed a fix from the two position lines he'd found on the Loran.

"Halfway to nowhere?"

"Over the Pacific Ocean!"

"No need to watch out for flak then!"

CHAPTER 6

The 12th century parish church of All Souls guarded the western outskirts of Woodstock, a ten minute walk from Madeleine's house. In fine weather many of the parishioners would make their way to the nearby Blenheim Palace Park after services for a stroll along the lake. In the eighteenth century Capability Brown had designed the landscaped gardens for the Duke of Marlborough. It had been the architect's idea to create the lake by damming the River Glyme.

Four members of the All Souls choir were now seated round the dinner table in Madeleine's house. Their discussion had strayed from the best musical settings of the Eucharist to the latest idiosyncrasies of Benedict Avery, the dour choirmaster at All Souls. Madeleine's guests were Jean Coster and Daphne Edwards, two of the choir's altos, and Matthew Claridge, bass. A fifth place at table had been laid for Toni Clark, Madeleine's lodger, but she had phoned to say she would be late home from work.

Daphne took a last mouthful of spaghetti bolognese and laid her fork on her plate. The others had already finished.

"All I'm saying is it's not very democratic. He didn't ask our opinion—he just foisted it on us."

"He's the choirmaster. He can do what he likes," said Matthew, a barrel-chested man who could comfortably sustain a bottom D note with considerable power. "He's always behaved that way—ever since he joined the choir, anyway."

"Nobody actually likes his compositions but no one's prepared to tell him face to face."

Jean sighed. "That's because we know how he'll respond."

"Sulk like a child," said Daphne. "That's what usually happens when he doesn't get his way."

"Toni says she quite likes his setting," said Madeleine.

"She would, wouldn't she," scoffed Daphne. "The modern Bohemian Woman," she added mockingly.

"Who doesn't believe in God," added Jean. "You would think that's a vital prerequisite for a singer in a church choir, wouldn't you?"

"Well, I'm not so sure myself," said Madeleine. "About God, I mean. But I love religious music. Except for Benedict's setting of the Eucharist, of course. I agree with the rest of you on that."

"The Sanctus—it sounds like a cat being tortured. There's absolutely no harmony between the sopranos and the altos. I hate those bits where we're only a semitone apart as we go down the scale."

Matthew took out his pipe. "May I?"

The others murmured approval and watched as the bass singer stuffed the pipe bowl with aromatic tobacco and held a match to it.

"What's that peculiar society Toni's a member of?" asked Daphne. "Some sort of atheist group."

Madeleine smiled. "The Atheist Christian Society, or something like that."

"Ridiculous!" scoffed Jean. "How can you have such a thing?"

"She's tried to explain it to me," said Madeleine. "The idea is that you can follow the Christian code without believing in Jesus Christ or God."

Matthew blew out a cloud of smoke. "Well, maybe she's right, there," he smiled, "in the opposite case, anyway. There are some so-called Christians in our congregation who could do with a spot of retraining!" He carefully avoided eye contact with the other diners as he spoke.

"Including Benedict Avery!" laughed Madeleine to block any danger that Daphne or Jean might ask him who he was referring to.

"Look, I don't mind having a word with him," said Matthew, following Madeleine's lead by moving away from tricky ground. "I'll try something diplomatic. I'll say we find it difficult to do it justice."

"As I said before, it's about time we did Harwood again," said Daphne. "We haven't done it for ages."

"My favourite, I must agree," said Madeleine, with Matthew nodding his concurrence.

The meeting she had called had achieved its purpose, thought Madeleine as she cleared away the main course detritus. If Matthew was happy to tell Benedict that the choir didn't want to do his Eucharist setting then it would save her having to take on the task. It wasn't that she deliberately avoided confrontation—she could stand her ground when necessary—but the choirmaster's misogynistic proclivities were well known so the bass singer was likely to make more headway.

What could not be disputed was the choir's reputation, which had soared since Mr Avery took charge. Despite the previous incumbent's competence steadily diminishing in inverse ratio to his consumption of alcohol, he had been retained in post because the vicar hadn't had the heart to dismiss him. Matters came to a head when a hymn book hurled during Friday practice had injured one of the boys. The choirmaster was sent packing.

It didn't take long to realise that the new man was not all sweetness and light. The first insult was his demand that the choir practise most of the hymns for the following Sunday. *Hymns! We don't need to practise hymns! We do the Eucharist and the anthem for Evensong! That's what we've always done!*

Mr Avery also unsettled some of the choir members by having them sing their parts without the accompaniment of the others. So for example he'd single out the tenors and ask them to do their bit on their own, testing their pitch control. The boy sopranos were used to following instructions from figures in authority without challenge and most of the men were similarly untroubled. But three or four of the altos and tenors voiced their disapproval at this further questioning of ability, especially Jean Coster and Daphne Edwards, two of the most senior choir members.

Initially the only compliment that could be paid to the new man was that at least his admonitions were untainted by the whiff of alcohol. But before long parishioners began to remark on how much better the choir was beginning to sound. Then they noticed that the surplices of the boy choristers were now adorned with medals hanging from ribbons round their necks. It turned out that Mr Avery had enrolled them in the Royal School of Church Music.

There was some controversy over the medals and the ranks they denoted. The newly appointed Head Chorister's medal was chrome plated and dangled from a red ribbon. For the Deputy the medal was unplated pewter. The other ranks wore blue ribbons, navy blue for the Seniors (four boys) and sky blue for the Juniors (three boys). Mr Avery insisted that the Head and Deputy should be selected by audition. But parents of other boys quickly objected. To fend off internecine warfare the vicar intervened and instructed his choirmaster that appointment would be in accordance with length of service. Mr Avery grumpily backed down. Harmony was restored.

"Hello, everyone, sorry I'm late. Couldn't get away," came the voice from the doorway. It was Toni, Madeleine's lodger. In contrast to the sensible clothes of the other ladies Toni was wearing a multicoloured kaftan and pale green leggings. Her short cut chestnut hair was streaked with red. "Any scoff left for me?"

"Toni! Good to see you!" greeted Daphne with a beam with Jean enthusiastically nodding agreement. "We were just talking about you, weren't we, Maddie?"

"There's plenty of spag bol left," said the chemistry professor to her lodger. "And a bit of salad, too, if you want a few vitamins."

* * * * *

It was through the church choir that Madeleine and Antoinette first met. At the start of practice one Friday a couple of years previously the new alto ("call me Toni") had been introduced to the others as a new resident of Woodstock. She raised a few eyebrows with her unconventional clothes and exuberant opinions but happily conformed when the vicar asked her to consider quieter attire for the Sunday services. "That's cool, Rev!" came the reply.

Some time afterwards Madeleine happened to bump into Toni at the shops in Woodstock and the two of them decided to have a chat over a cup of coffee at the Cafe Marcello. Toni mentioned that she might have to leave the choir because she'd split up with her boyfriend and the house they owned together would have to be sold. Madeleine made sympathetic noises but Toni dismissed the matter with a wave of her hand. Anyway, she had continued, it would be impossible to buy her ex-lover out because the rent she could realistically charge a lodger would not cover the extra mortgage and more than one lodger would be too claustrophobic. She was thinking she'd move out into rented accommodation herself, somewhere nearer where she worked, and stick the equity she'd accumulated into a building society.

Her job, apparently, was with the Emcro film production company based in Oxford. "Originally they called themselves 'Empire Chromatic'," Toni would explain to anyone interested enough to ask. Toni herself was an assistant producer. "I even get a name check on some of the stuff we churn out," she would say. "Half a second of fame as my name rolls up the screen."

Madeleine had come up with an alternative solution to Toni's accommodation problems.

"I live on my own in a largish house. If you lodged with me you could rent your house out to two or three lodgers or a family. That would cover your mortgage, wouldn't it? We could start off with a trial period to see if we're compatible and then take it from there."

And so the deal was struck. The landlady found the tenant easy to get on with despite their differing life styles and tastes. At thirty-eight, Toni was seven years younger that Madeleine and her variable working hours a contrast to the professor's regular nine-to-five routine.

After initial wariness they gradually became friends and no longer reticent about personal matters. One evening they polished off a bottle of Chardonnay and, unusually for Madeleine, started a second. The professor even accepted one of her lodger's roll-your-own cigarettes, a rare indulgence for her.

"So, is there a man in your life, Maddie?" asked Toni, wreathed in smoke.

The older woman shook her head. "No."

"What about in the past?"

Madeleine narrowed her eyes. "I nearly got married in 1944 but my fiancé . . . died."

"Sorry."

"He was killed in the war."

"And since then?"

Madeleine thought for a moment. "There have been one or two . . . liaisons. Nothing earth-shattering. No, that's not true . . . there was this American scientist a few years ago . . ." She sighed. "These days I seem to manage without . . ." Her voice tailed away and for a minute or two the pair of them sat silently, puffing at their cigarettes.

Madeleine looked at her lodger. "How about you, Toni? Have you got . . . anyone?"

A shake of the head. "No, not just now. As you know, I split up with Clive—my boyfriend."

"Had you been together long?"

"Three years . . . maybe four."

"Was it mutual, the break-up?"

"I went to bed with one of his friends."

"That would explain it."

Toni blew out smoke. "Well, yes and no. The agreement was we were free to shag other people if we wanted to."

Madeleine nodded, unable to think of a suitable comment. It was not an area of behaviour she was familiar with.

"Clive was jealous, I suppose. Maybe I shouldn't have done it, knowing they were good friends. But he fucked one of my friends a while back and I wasn't too bothered. I watched him do it."

"What do you mean?"

"It was a threesome."

"A what?"

Toni smiled. "Clive and me and—"

Madeleine saw the light and held up her hand. "No further information required."

"I like sex, Maddie."

"Yes, I got that impression."

"But I'm not a nympho. I'm selective about who I get naked with."

Madeleine sipped at her wine. She hesitated, as if undecided about pursuing the conversation.

"Like I said, I'm not addicted to it," continued Toni, ending the pause. "I can go for months without sex. It tends to be an impulsive thing for me."

The chemistry professor's curiosity finally overcame her reticence. "Threesome? How does that work? Is it always two women and one man or can it be the opposite? Do the two women . . . er . . . do they . . . I mean . . ."

Toni laughed. "Anything goes! Two girls can have fun if they're in the mood. With or without a man." She registered the raised eyebrows on her landlady's face. "Don't worry, Maddie, I won't make a pass at you!"

"You wouldn't get very far!"

"I know that. You can sense these things."

"Unknown territory for me," muttered Madeleine. "Girls, I mean."

"It's usually blokes for me. I'm not really a lesbian."

Madeleine shook her head. "Well, this is a strange conversation. I didn't think I'd be discussing sexual proclivities with my lodger."

Toni topped up the wine glasses. "I haven't been to bed with anyone since Clive and I broke up. No one I've met recently has floated my boat. Anyway, occasional celibacy isn't such a bad policy. You can relax a bit instead of looking around for excitement all the time. Sort of calming."

"I think you just get out of the habit of sex, don't you," said Madeleine. "It's ages since I've looked at a man and thought, 'come to bed with me'. Perhaps I'm too old for that sort of thing now."

"I like your brother," said Toni. "He could have his wicked way with me if he wanted."

Madeleine put a smile on her face. "He's already taken."

"He's got family, hasn't he?"

"Sort of. He's divorced, but he lives with a woman and her daughter. She's a divorcee too. Jackie, she's called, short for Jacqueline. She's quite a bit younger than Nigel and her daughter's about twelve, I think."

"You'll have met her, then?"

"Uh-huh. She seems okay. Ex-stewardess."

"In BOAC, like Nigel?"

"Yes. Does some sort of ground job at the airport now. If you're a stewardess the rules say you have to stop flying when you get married. A bit unfair because the ruling doesn't apply to the stewards or pilots."

"Well, married women are going to produce babies, aren't they?"

"Suppose they're in a job where they want to carry on working?" said Madeleine. "What's to stop them having nannies?"

"Not on this planet, Maddie! Unless you're in the upper classes or you've got loads of money!"

"It's also cynical commercialism, Toni. You'll attract male passengers if they think the stewardesses are available, so the thinking goes."

A pause.

"You were a pilot in the war, weren't you?"

"Yes."

"How did the men take it? Women encroaching on their territory?"

"Some were happy, others weren't."

"What was the biggest plane you flew?"

"Lancaster, same as Nigel."

"Those big bombers? Like the Dambusters?"

"Yes."

"How come you both got into flying?"

"I'll tell you about it one day."

Toni slowly shook her head in wonderment. "You're incredible. You flew big planes and now you're an expert on . . . what is it?"

"Catalysts in organic chemistry."

"Amazing. I don't even know what that means."

Toni reached out to top up their wine glasses again but Madeleine put a hand over hers.

"Nigel's your half-brother, isn't he?"

"That's right. We've both got the same father."

"I'm an only child," said Toni. "When I was a kid I desperately wanted a brother or sister but it never happened."

"Well, we grew up apart," said Madeleine. "My dad split up from Nigel's mum before I was born. So I was brought up by my mum and Nigel by his mum. We didn't meet until we were teenagers, and that was by chance."

"No reconciliation between the two families, then?"

Madeleine shook her head. "Only Nigel and me, not our parents."

"But you both see your dad, don't you?"

"Yes, when we can. He and my mum live in France."

"Oh, yes, I remember you telling me. Provence, isn't it?"

"That's right. Nigel sees them more than me, 'cos he gets free airline tickets. I get out there once or twice a year. We all spent Christmas together there last year, which was nice."

"So, Nigel's mum and step-dad. You don't have anything to do with them?"

Madeleine picked up the bottle, hesitated a moment and then poured an inch of wine into both glasses. "His dad died two years ago. His mum's in a nursing home near Maidenhead. Nigel and I go to see her once a week but it's hard work . . . she's going senile and getting quite bitter and twisted. She doesn't like me because I'm the offspring of her first husband and sometimes she says nasty things."

"So why bother, if all you get is grief?"

"A sense of duty really, relieving Nigel of some of the burden. You have to remember she's a widowed old lady who's gradually losing her faculties. She wasn't quite so horrible when she was younger."

"I was lucky, I think," said Toni. "My parents have always loved me, despite what they call my unconventional lifestyle. They were very upset when they found out I'd had sex with another girl. But they said whatever I got up to I would always be their daughter and that parents' love was unconditional."

* * * * *

After Jean and Daphne and Matthew had left Madeleine and Toni began clearing up and washing the dishes.

"They don't really like me, do they?" asked the lodger quietly.

"Who?" responded her landlady, though she already knew the answer.

"The other choir members."

"Nonsense," said Madeleine. "It's just . . . certain individuals who . . . think you're a bit . . ."

"Weird?"

"I was going to say 'unusual'."

"I'm thick skinned, as you know, Madds, but if there's bad feeling—"

"Stop it! You're well liked and well respected in the choir. To be frank, the fault lies with . . . those individuals, not with you."

"Jean and Daphne aren't in my fan club."

Madeleine dropped a couple of plates into the bowl. "They're a different generation, Toni, that's all. If you don't fit the standard pattern you're suspect."

"You mean married, two kids, nice little house and all the rest of it."

Madeleine smiled. "Something like that."

"Well, it's just not me. I don't want to be tied down to a domestic life. I value my freedom. Although at times I wish I'd had a kid or two. I like children."

"I'm happy in my rut, I suppose." Madeleine paused, then sighed wistfully. "Kids . . . yes, I regret not having kids. And I think I would have been happy as Terry's wife."

"Terry?"

"My fiancé."

"Oh, yes."

For a moment the only sound was Toni stacking crockery as she dried the items. "Everyone likes you, Madds," she said, restarting the conversation, "even if you don't fit the pattern."

"I think that's because I'm . . . unthreatening. Alright, no hubbie, no kids but I dress like them, talk like them."

"You're nothing like them. You're . . . interesting. And you've got brains. More than anyone else in the choir, I reckon."

"I wish Daphne wouldn't call me 'Professor' at choir practice. I have asked several times."

"That's just snobbery," said Toni. "People think people with titles are automatically superior. It's hypocritical."

"In what way?"

"They're very humble when they're saying their prayers but some of them don't like the idea of poor people or uneducated people or non-white people in their middle-class church."

"Sadly, I think you're right there," agreed Madeleine.

"Jean thinks I'm beyond the pale because I'm an atheist."

"Has she said so?"

"Not directly, Madds," said Toni, picking up a fresh dry tea towel. "That's not her way. Sarky comments is more her style."

Madeleine picked up the bowl and emptied it. "There'll always be differences in any group of people. The best part about the choir is that the whole is greater than the sum of its parts. So you don't see eye to eye with Jean and Daphne and we all think Benedict is . . ."

"A pain in the backside."

Madeleine laughed. "I was going to say 'difficult'."

"Ever the diplomat, Maddie."

The clearing up complete, Toni announced she was off to bed.

"Sleep well," smiled her landlady.

Toni turned as she got to the door. "Oh, I'll be away for a fortnight next month. I just found out today."

"Anywhere nice?"

"San Francisco . . . we're thinking about doing a film about this new hippie movement and San Fran seems to be the centre of it all. I'm going out there with a colleague for a recce."

"You lucky thing! Any room in your entourage for an old chemistry professor?"

"You're welcome to come along, Maddie, but you'd have to pay your own way."

The landlady shook her head sadly. "Too much going on, Toni. Maybe the next time if you give me more notice. Nigel can sometimes get me discounted fares. I'd love to go to the States for a holiday rather than just to give chemistry lectures."

"It's great, this hippie thing. It's going to stop wars and spread love around the world."

"I'm all for that," laughed Madeleine.

"John Lennon's right . . . All You Need Is Love."

"I agree."

"Maybe I'll find a nice American while I'm working there. He'll shag me rotten for two weeks and then we'll split up, never to meet again."

"Uh-huh."

"You're smiling, Madds. Are you thinking you'd like a piece of the action as well?"

"No comment."

CHAPTER 7

Verity Abbott found Bill Syerman in Lab 2C, tinkering with the large cylindrical apparatus on the workbench which was usually the domain of Xiu Ying Song and James Ibsley. Apart from the safety officer the lab was empty because the four post grads had gone to London with the professor.

"Having fun, Bill?"

"Not the word I'd chose, Verity."

"What are you doing?"

"Replacing the heater element in the mixer chamber. They can crank it up a few more degrees if they want."

The professor's secretary sighed. "Do you actually know what goes on in here, Bill? I know a bit about the chemicals and so on because I have to type out Madeleine's reports. But what's actually happening in the apparatus?"

"I probably don't know much more than you, V." Bill smiled at her. "Let's have a cup of tea and compare notes."

"That's why I came to find you. The kettle's not working in my office. Could you have a look at it?"

"Sure."

"I was going over the road to get a doughnut. Want one?"

"I'll never say 'no' to an Oakley's doughnut."

The kettle malfunction was no more than a wire coming adrift inside the plug. Bill commented to Verity that someone should have noticed that the cord grip was loose and Verity nodded, pretending to understand what he was talking about. Her husband, Gerald, took care of that sort of thing at home. Perhaps, she thought, she should learn about these things herself.

Back in the lab, Bill resumed his work on the mixer chamber, pausing every now and then for a mouthful of doughnut and a swig of tea. Having completed her office duties for the day, Verity brought her own mug into the lab to chat with Bill.

"How's the family?" she asked.

"All fine, thanks," replied the safety officer. "Kids driving the memsahib mad of course. Bloody school holidays!"

"Going away at all?" The Organic Chemistry Gas labs would shut down for the last two weeks in August for the annual summer break.

"Yes. We've hired a caravan in Bognor. How about you?"

"Gerald and I are off to the Lake District. Lots of walking. Let's hope the weather stays fine."

For a minute or two there was only the sound of Bill humming to himself as he fastened the protective cover on the mixer chamber.

"There we are, all ship-shape and Bristol fashion," he said, laying his screwdriver on the bench and wiping his hands on a cloth. "That should keep the hooligans happy for a while."

"I wonder how they're getting on in London," said Verity.

"I'm not hopeful," said Bill. "I know the Prof's good at making her case but I've heard other people saying there's not much point to what they're doing. There must be better things for the SRC to spend their money on. For a start they could send it to the MOD to stop the rundown of the armed forces."

It was a frequent complaint of ex-Warrant Officer William Syerman that the continual cutting of the Ministry of Defence budget by the Labour government of Harold Wilson was a strategic error. "Look, we've just had the Six-Day War in the Middle East for God's sake," he would say. "In a troubled world we should be strengthening our military, not weakening it."

"I know Susan is worried about what's happening in Hong Kong," said Verity. "The riots and demos."

Bill responded through a mouthful of doughnut. "Bloody communist agitators! The Chinese government's behind it, of course."

"Susan phones home every other day. She says her father says the situation isn't as bad as the newsreels are saying. But she thinks he's understating it to stop her worrying."

"Bloody lefty students waving their Little Red Books. Lock the bastards up and throw away the key—that's what I would do."

"How's it going to end? They're planting bombs now, the demonstrators."

Bill nodded, sipping his tea. "Yeah, I know. The MOD called me a day or two ago, wanted my advice."

Verity raised her eyebrows. "Why?"

"I was in bomb disposal during my days with the Devonshires, as I think I've told you before."

"Yes, I remember."

"So, it seems the HK Police asked the Foreign Office how to deal with the IEDs. The FO johnnies asked the MOD."

"IEDs?"

"Improvised explosive devices . . . home made bombs."

"It's just horrible. Killing ordinary people who just want to live their lives in peace."

"Spot on, Verity! They're clever bastards—most of the devices are fakes but they still cause panic, which is the intention, of course."

"You're not going to Hong Kong yourself, are you, Bill?"

The safety officer smiled. "No, I'm retired, V, remember? They're sending out a few serving officers but they're also asking some of us old hands with experience how we would deal with the problem. I was a trainer, you see."

"Weren't you ever frightened, Bill, when you had to deactivate the bombs?"

"And how! I had a scare or two over the years. One of my buddies got blown up when we were in Aden."

"Was he killed?"

"There wasn't much of him left to put in the coffin."

"That's awful! How did you cope with it?"

"We used to joke about it. One of my mates used to eat half a tube of Smarties before a job and told everyone they were his 'brave pills'."

"What's going to happen in Aden, Bill?"

"Wilson's pulling the troops out. The locals can carry on killing each other after we've left, if they want to."

"Is that a good thing, us leaving?" asked Verity.

The safety officer scratched his chin. "It looks like running away, doesn't it? I appreciate this country can't afford to keep all the overseas bases going but I just think the socialists are cutting too much too quickly."

"They've kept us out of Vietnam though. Isn't that a good thing?"

Bill picked up the screwdriver and tapped it on the bench. "On balance, yes," he said slowly. "It's probably an unwinnable war. There are millions of Chinese commies ready to march south if they think the Yanks are coming out on top."

Verity sighed. "What a world we live in, eh, Bill."

"Look on the bright side, V. Five years ago we were on the verge of nuclear war, remember?"

"Yes, it was frightening."

"Yeah. We all let out a sigh of relief when Khruschev backed down."

"So are the Russians not such a threat these days?"

"They're not our friends, V. I was unhappy when the Prof had to show those Russian scientists round the labs. I don't think we should be sharing what might be sensitive information."

"It was officially an 'Exchange of Information on Catalysis' according to the briefing we got."

"I just hope it wasn't all one way, Verity. What did we get in return? Plus, we don't want another Nene."

"What do you mean?"

"The Nene was a Rolls-Royce jet engine, as invented by Frank Whittle. We gave the Reds a few of them and they promptly made their own copies and stuck them in their MiG-15s and shot down our guys in Korea."

Verity shook her head sadly. "As I said, what a world we live in."

"Puts worrying about our budget into perspective, doesn't it?"

"That's one way of looking at it." Madeleine's secretary pointed to the apparatus on the work bench. "But it's a shame they're thinking of winding up this project because there's not enough funding."

"That scanning microscope cost a fortune. That used up a load of dosh." Bill tapped the top of the mixer chamber with his screwdriver. "And this thing . . . thirty-seven pounds, that new element cost, just so they can heat the gases up another twenty degrees. That sort of money would pay for a week in a seaside caravan for a family of four."

"So, I know they use methane and carbon dioxide," said Verity. She pointed to the apparatus. "They mix them in that thing, don't they?"

"Yes, that's the idea."

"And they produce hydrogen and carbon monoxide, as far as I can make out from Maddie's reports."

"Correct."

"Any idea why?"

"My understanding is that the products are useful but the main reason is to cut down the carbon dioxide pumped out by the power stations. They're looking to react it with methane to remove it from the exhaust gases."

"Why . . . is it poisonous or something?"

"No, I don't think so. I think it's something to do with trapping the sun's heat but I'm not sure how that happens."

"I gather Susan and James do the mixing of the gases and the other two do the analysis of the end products."

"That's about it, V. I think they're trying different conditions to get the best conversion yield. They can vary the temperature in the heat chamber and they can switch on UV light in there too—I think that helps."

"The bit that I don't understand is what the catalyst does."

Bill puffed out his cheeks. "All I know is it speeds up the reaction. Beyond that it's a closed book to me."

"Valence electrons," said Verity mysteriously.

"Pardon?"

"Valence electrons. They spend a lot of time talking about them. There's not a Methane Reformation report that leaves my typewriter without a fair sprinkling of valence electrons in the text."

"That's something to do with curtains, isn't it?"

Verity laughed. "That's 'valance', Bill. We're talking about 'valence' with an 'e'."

"We're talking Double Dutch I reckon," came the weary reply.

"One report I had to type was about infra-red radiation," said Verity. "The earth gives out radiation when the sun heats it and the CO2 reflects it back from the sky. Is that what you were talking about?"

The safety officer screwed up his eyes. "Yes, I think that's the main reason for the research. But maybe they're wasting their time and money. I haven't heard anyone else talking about it being a problem. You don't hear about it in the papers or on the telly."

Verity sighed. "So I suppose that's why the funding is being shut off."

"Well, it's not like we're going to lose the Prof. She's doing lots of other work and taking lectures and tutorials. And she's already said our jobs are safe."

"Thank goodness!" muttered Verity.

"For sale, one scanning electron microscope, almost new, one previous owner. That would bring in a quid or two. They could give us a pay rise."

"What about the post grads, though, Bill. What'll happen to them?"

"They can go out and get proper jobs. Do something useful for a change."

"Oh, come on, Bill. They've got brains. I've heard them talking about opportunities in America. Our loss would be America's gain."

"Yeah, they're okay, I suppose." The safety officer grinned. "Tell you what, though. That Luke could do with a year or two in the army. That'd sort him out!"

CHAPTER 8

The Government buildings in Whitehall were impressive from the outside but, lacking air conditioning, oppressive for their occupants in the July humidity. Room 303 at number 22 was no exception, and the large fan on the desk seemed merely to circulate the uncomfortable warmth.

Behind the desk, with their chair backs close to the oak-panelled wall, sat John Davis, General Project Manager at the Science Research Council, and next to him a heavily jowled middle-aged man whom Madeleine had not seen for several years. Professor Alfred Grove was standing in for Sir Melvyn Hocklan, the Chairman of the Allocation of Resources Committee. John Davis had introduced him as a member of the Spending Review Board. It transpired that all expenditure sanctioned by the Committee now had to be approved by the Review Board, acting on behalf of the Treasury. Madeleine forced a smile as she shook his hand but memories of academic rivalry dampened her spirit. She had previously won an award which Professor Grove's supporters thought should rightfully have gone to him. His cold eyes now belied his smile and suggested that the slight still rankled.

Next to Madeleine on the other side of the desk, seated line abreast, were Luke Stirling, Xiu Ying Song and James Ibsley. Joe Curtis was attending a family reunion and Madeleine had reassured him that he was not letting the side down by his absence. She had been pleasantly touched by the offer of the students to support her when she was putting the case for continuation of the Methane Reformation project.

The meeting started with Madeleine summarising the progress of the project, writing points on a flip chart in felt-tip and occasionally bringing in the post grads to add detail. It was disappointing to detect lack of encouragement from Professor Grove, who listened with stern detachment. As if to compensate, John Davis threw in the occasional friendly smile and nod of the head.

The task completed, Madeleine thanked the two men opposite for their attention and sat down again.

There was a long pause which accentuated the stuffy atmosphere in the room and discomforted all those present apart from Professor Grove, who seemed to Madeleine to be perversely enjoying the moment. Revenge for the award she had snatched away from him all those years ago? Her old rival looked directly at her for a moment or two and took a deep breath.

"Well, Madeleine, thank you for your presentation. It clearly means a lot to you and I'm impressed with the dedication of you and your team." There was another pause and it was obvious that a "but" would follow. A wasp flew in through an open window and haphazardly buzzed round the room. James Ibsley waved it away with a flick of his hand. Madeleine sensed that Mr Davis was uncomfortable, as if he too knew what was coming next, which of course was to be expected.

"But," continued Professor Grove, "with the current squeeze on Government spending we have to be sure that funds are directed to where they will be most productive and I'm afraid the Methane Reformation project doesn't fit into that category."

Madeleine nodded. "I'm curious, Professor Grove—"

"Alfie."

"Thank you. I'm curious as to how the decision has been made. I understood that the Government allocated funds to the SRC and the Allocation Committee decided how those funds are distributed. Is it Sir Melvyn who has cut us off?"

Professor Grove furrowed his brow, evidently considering how to frame his answer. "We asked Sir Melvyn to talk to us at the Review Board because we were reviewing the SRC budget in light of reduced public spending. We stressed that prioritisation would be expected to be firmly applied and less . . . critical . . . projects decelerated or in extreme cases abandoned. Sir Melvyn agreed that MR fell into the latter category, I'm afraid, Madeleine."

"So was it your decision or Sir Melvyn's?" She saw Professor Grove stiffen and realised her question might be a veiled attack. To restore civility she added: "What I mean is . . . was it the Allocation Committee or the Treasury? If I want to appeal against the decision, who should I be talking to?"

"It might be possible to organise a compromise," said Professor Grove, ignoring the question. "If you can find private sponsors to fund the bulk of the research then the Review Board would not object if the SRC allocated top-up resources."

"What do you mean by 'bulk'?"

"We would need to discuss it."

"Rough figure?"

"Based on you current expenditure around 75 percent would probably meet our requirements."

Madeleine shook her head. "Impossible," she muttered.

"Excuse me. May I make a comment?" All eyes turned to Luke Stirling.

"Go ahead, young man."

"I just wanted to say that we think we've been really privileged to do this work for Prof Maunsell. She's been an inspiration to us and even if you shut us down it won't have been a waste of time."

Xiu Ying and James nodded their agreement.

"We're at the forefront of catalytic research," continued Luke. "We probably know more about the effects of variations in surface topography than anyone else in the world."

"That's a very bold statement to make," said Professor Grove, eyes narrowed. He turned to Madeleine. "Is that true?" he asked.

"I would say that we're among the front runners."

"We get enquiries all the time," said Luke. "Other people follow us. And we've just started using tungsten carbide doped with palladium, which Prof Maunsell suggested might speed up removal of carbon deposits during the periodic operation. We're getting positive results. No one else is doing that yet."

Madeleine turned to look at her student, mildly and pleasantly surprised. It was Luke himself who had floated the palladium variation and she had agreed after he had impressed her with his knowledge of the transition metal's properties. He now knew more about the subject than she did. Exasperating though he was at times, Luke could switch on the charm when required. His chivalry was a welcome antidote to the afternoon's grim proceedings.

For a moment silence descended. The wasp had relaunched itself and hovered towards John Davis, who brushed it aside.

"I think you've backed the wrong horse," said Professor Grove eventually, looking at Madeleine.

"What do you mean?"

"The catalysis stuff is excellent. It's acknowledged you're a leading light. I know a bit about it myself, of course." For a moment his eyes twinkled and a hint of a smile played around his lips. The award Madeleine had beaten him to was for exploring the catalytic properties of transition elements. "But stopping CO_2 getting into the atmosphere—it's just pointless."

"The total is increasing every year. One day it'll be significant."

"Not while you and I are alive and kicking. Or our children or grandchildren for that matter. Where are we now? Three hundred parts per million?"

"Three twenty."

Professor Grove snorted. "The experts say you'd have to get up to four hundred to make any difference to the atmosphere."

"Granted," said Madeleine wearily. "But isn't it a good idea to have the technology in place for when it's needed?"

The Review Board man shook his head. "The problem won't arise. There'll be other energy sources available long before we get to four hundred. Nuclear, solar, even wind and tide. They'll be cheaper than coal and oil."

Madeleine shrugged. She had no answer.

"So, don't worry about it. You backed the wrong horse this time. I know you've got plenty of other things going on so you'll have more time for them."

John Davis cleared his throat. "What will you post grads do if the project is wound up?" he asked.

Luke and James both began to answer but Luke indicated to his colleague that he should go ahead.

"We've looked at other options," said James. "It's most likely we'd split up. I'll try to stay in the UK because my girlfriend works in London. I think Joe's looking at the States. I don't think he'd have a problem finding something."

"I might go back to Hong Kong," said Xiu Ying. "I'm sure I can get a teaching position, maybe even at the University. My Dad might be able to help me."

John Davis turned to Luke. "What about you, young man?"

"Not sure," came the reply. "The States sounds good to me but my girlfriend might not want that so I'll have to talk it over with her."

Professor Grove stood up abruptly. "Well, I think we're just about done here," he said, his voice gruff again. His eyes scanned the others. "Let me say once more I've been impressed with your work and I'm sorry it might be coming to an end. You can rest assured that full funding is guaranteed until the end of October, when the situation will be reviewed. If you can find a sponsor to take over, all well and good. As I said, we're happy to let the SRC make a contribution. Whatever happens, I wish you youngsters well and hope things work out okay for you."

He walked round the desk and towards the door then stopped and turned. "You too, Madeleine. All the best."

"Thank you," responded the chemistry scientist quietly.

Professor Grove nodded. "Perhaps we'll meet again in happier circumstances."

"Perhaps."

The atmosphere brightened after the Review Board member had left the room. John Davis was effusive with his apologies.

"Don't mention it," said Madeleine. "There was nothing you could do."

"Are you going back to Oxford now?"

"I thought I might look at the shops first. The others are going . . . where is it?"

"We're going to Battersea Fun Fair," grinned Xiu Ying. "We asked the Prof to come with us but she won't."

"Think I'm a bit old for roller coasters and candy floss," Madeleine said ruefully.

John Davis looked at his watch. "I was going to walk down to the Embankment for lunch. Would you lot like to join me? There's a nice cafe in the Victoria Gardens—the Siena."

The chemistry scientist surveyed her charges. "May I be excused?"

"You go with Mr Davis," said Luke. "We'll go straight to Battersea if you don't mind."

Madeleine smiled at the Project Manager. "It sounds nice. Count me in."

* * * * *

"So you were at Bletchley during the war?" queried Madeleine as John Davis raised his hand to attract the waiter's attention. They had just decided that a refill of coffee would round off a pleasant lunch.

"Yes."

"In what role?"

John smiled. "Guessing the undecoded content of the enemy signals to help the analysts choose the rotor combinations to set up in the bombe. So it was more my knowledge of German than my maths degree which was useful to them . . . oops!"

The chemistry scientist noted his sudden concern. "What is it?"

"I hope you're not a Russian spy!"

"Nyet!"

"Well," continued John with a grin, "we were told we weren't supposed to talk about our work to outsiders. Daft, isn't it? The war finished two decades ago and yet we're still bound by the Official Secrets Act."

Madeleine nodded. "Perhaps they're still using decoding procedures these days that you were using back then and they don't want potential enemies to find out."

"More likely the bureaucrats love the power it gives them to control other people."

"I'm very familiar with that," laughed Madeleine. "It's the same in academia."

The waiter showed up and topped up their cups.

"Where did you do your maths degree?" asked Madeleine, picking up the conversation.

"York . . . how about you?"

"New College, Oxford. Just across the road from the Chemistry Gas labs where I work."

"Did I hear you were a pilot during the war?"

"Yes, delivering aircraft to their bases." Madeleine gave the SRC manager a brief summary of the White Waltham set-up.

"That's very impressive, I must say."

"A lot of it was mundane, really."

"Did you have any scary moments?"

Madeleine smiled. It was the most asked question when people found out about her background.

"There were times when it was big relief when the wheels touched the ground."

"So after the war the chemistry was more important to you than the flying? You didn't want to carry on?"

"It was a close run thing, John. I loved both. But there weren't many opportunities for female pilots . . . or for the men either, actually. There were loads of demobbed pilots looking for work but not many jobs available."

"It was much the same at Bletchley. There wasn't a great demand for cryptologists."

"What did you do?"

John took a sip of coffee and licked his lips. "Started teacher training but soon realised it wasn't for me." He paused for a moment and frowned. "Sounds silly now but as a boy I wanted to be a footballer."

Madeleine raised her eyebrows in surprise. "Really?"

"I was good at football at school. Had a trial with Chelsea as a teenager. But everyone told me I would be wasting my intellectual abilities kicking a leather ball round a field and they pointed out that professional footballers don't earn much. So it was off to York to read maths."

"Where did the German come in?"

"My mother was from Elsenach, in Thuringia. I was more or less bilingual." John laughed. "She thought it funny that one of the men I worked with was called 'Turing'."

"Ah, yes, Alan Turing."

There was a silence as they both remembered the brilliant codebreaking mathematician who had been driven to suicide after prosecution for homosexual activity.

The chemistry professor restarted the conversation. "So you gave up teacher training . . . what happened next?"

"Very boring. I applied for a post as a statistician. Wound up as a civil servant, which is where I am today. Middle class job, earning a middle class salary, father of two middle class children who are also doing middle class jobs."

"Nothing wrong with that. Does your wife work?"

"Yes . . . as far as I know she's still working. Haven't seen her for a while."

"Oh . . ?"

"We're divorced."

"Oh."

"Wasn't acrimonious. Just drifted apart."

There was a pause.

"What about the children? Where do they live?"

"My daughter lives in Basingstoke and my son in Swindon. Both married."

"Any grandkids?"

"Not yet . . . how about you?"

"I'm not that old, John!"

"Sorry, I meant children rather than children's children."

"You're forgiven!"

"You're very kind!"

They both laughed and then Madeleine looked across at her companion. "No children."

"Didn't you want them? Sorry, that's a bit personal."

"I never married, John. My fiancé was killed during the war and I haven't met anyone to take his place."

John nodded slowly and Madeleine sensed he was struggling to know what to say next. She put on a bright smile. "What are your hobbies, then?"

Her companion returned the smile. "I compile crosswords for the newspapers."

"Interesting."

"And I do cartoons for them, too, and for other publications."

"Also interesting."

"And I play bridge."

"So do I."

"In a club?"

"Uh-huh . . . you?"

"Yes . . . Teddington, where I live."

Madeleine nodded. "If you find yourself in the Woodstock area come and play with us."

"Very kind. I'll make a note."

CHAPTER 9

"Pretty good, Maddie . . . like you've never been away!"

"No, I let the nose drop while we were inverted. Lost points for that."

"Rubbish! Do another one . . . do a four-point hesitation."

"Here we go!"

In the clear blue August sky the Chipmunk trainer accelerated in a shallow dive until the airspeed needle wound up to 120 knots. In the front cockpit Madeleine eased the stick back until the nose was a few degrees above the horizon. Her left hand firmly held the throttle fully open to ensure the Gipsy Major engine was delivering its full one hundred and forty horsepower.

Madeleine flicked the stick right and then centred it again. The aircraft responded with a rapid roll to the right which abruptly stopped with the wings at right angles to the horizon. Her left foot moved forward so the rudder would hold the nose up. Another flick rolled the aircraft upside down. Madeleine centred the rudder, now pushing the stick forward to keep the nose in the correct position, the pilots hanging in their straps. Two more flicks of ninety degrees brought them back to level flight. The chemistry professor inched the throttle back to cruise power.

"Perfect!" came the comment over the intercom from Prudence Dismore-Whyte in the rear cockpit. "Better than I can do them!"

"Over-rotated a bit on the third flick."

"But you instantly corrected. You haven't lost it, Madds!"

"Nice of you to say so, Dizzy."

"Let's try a stall turn left and right."

"On the Tiger, left was more difficult than right, wasn't it? Gyroscopic forces from the prop if I remember correctly. Or was it the other way round?"

"Do one and you'll find out!"

Madeleine pulled the Chipmunk round in a steep turn to check there were no other aircraft around that might get in their way. She asked her instructor to confirm they were still clear of regulated airspace.

"You tell me! What's that town just to the east?"

"Basingstoke?"

"Reckon so. We just need to keep clear of Aldermaston and the TMA." Aldermaston was the nuclear research establishment a few miles southwest of Reading and the London Terminal Manoeuvring Area protected airliners approaching and departing London Heathrow airport.

The original plan was a standard Cherokee outing from Booker, destination Goodwood, where the three ex-ATA pilots would land for lunch. But Parminder Collins had called Dizzy a few days previously to cry off, telling the instructor she needed to support her daughter at her Pony Club Gymkhana. By coincidence the Wycombe Flying Club had recently acquired a de Havilland Chipmunk training aircraft for aerobatic instruction so Dizzy had suggested to Madeleine that the two-seater should be their mount for the trip to Goodwood. En route Madeleine would get a chance to try some of the manoeuvres she and the other ATA pilots practised during their wartime flying on Tiger Moths and other types. The instructor's briefing to her friend had been brief and concise.

"The Chippie is just like the Tiger. Just do everything twenty knots faster."

The first stall turn was a shambles. Madeleine left it too late to rudder the nose round and down. The Chipmunk hung motionless for a moment, nose pointing vertically upwards, then started to slide backwards. Instinctively both pilots grabbed the stick tightly to prevent reverse airflow slamming the control surfaces against their stops. Then the nose abruptly dropped, the negative 'g' forces throwing both pilots against their shoulder straps, and the little aircraft headed vertically downwards.

"Sorry!" said Madeleine as Dizzy pulled the Chipmunk out of the dive.

"Let's try it again, Maddie. You can always cheat a bit, you know."

"What do you mean?"

"As you pull up, wind on a few degrees of bank in the direction you're going to yaw at the top. If you're doing a display no one will notice you're not exactly vertical as you go up and it'll help the nose round."

The reprise was successful and after another stall turn in the opposite direction the instructor suggested they had a go at a Porteous Loop.

"What's that?" asked Madeleine. "I don't remember those."

"Invented by Ranald Porteous, the Auster display pilot. At the top of the loop it's full pro-spin control inputs. You do one turn of the spin then recover to complete the loop. I'll do one as a demo then you can have a go."

After another ten minutes frolicking in the summer sky Dizzy told Madeleine to reset her direction gyro and turn southeast towards their destination. Silhouetted against the midday sunlight the South Downs slid slowly towards the Chipmunk at a sedate 90 knots and soon the pilots could see Selsey Bill in the further distance and the silver waters of the Channel beyond.

They flew a standard overhead join at Goodwood airfield, Dizzy pointing out the orientation of the three grass runways and then leaving Madeleine to her own devices as they let down on the dead side and joined the circuit crosswind.

"Half flap, seventy knots on base leg, full flap sixty on finals," reminded Dizzy. "Try to three point her, Madds, just like the Tiger or the Spit. If you wheel her on like a Lancaster she'll bounce to kingdom come."

"Not bad!" came from the rear cockpit after touchdown.

"A little bounce."

"Not higher than the width of a gnat's you-know-what." In their ATA days the part of the gnat's anatomy Dizzy was referring to would have been specified more crudely.

They taxied onto the visiting aircraft apron and shut down, happy to get out of their flying suits and let the sea breeze refresh them. A bowser growled towards the Chipmunk and Dizzy supervised the refuelling while Madeleine tidied the cockpits.

"Time for a well-earned lunch," said the instructor with a grin.

"Which I'll pay for."

"If you insist!"

* * * * *

After their meal the two pilots wandered over to the area set aside for deck chairs and settled themselves comfortably. The temperature had dropped a degree or two and the chemistry professor and the flying instructor congratulated themselves on choosing a perfect day for their trip to the coast. A Cessna and an Auster were practising circuits, their landings observed with critical eyes by the two women.

"Ouch!"

"Held off a bit high."

"That's more like it."

"Oh, dear!"

Madeleine turned to her friend. "Shame Pam couldn't make it."

"Three in a Chippie would be a tight fit!"

"How old's her daughter?"

Dizzy narrowed her eyes. "Seventeen . . . eighteen, maybe. Have you met her?"

"Don't think so."

"She's got twin boys as well."

"Yes, I remember, a bit younger, aren't they?"

"They're a lovely family. If I were ever to have kids I'd want them to be like Pammy's kids."

"We've left it a bit late, Dizz," said Madeleine ruefully.

"I think I'd be useless as a mother," said the instructor. "Too impatient. I'd wallop them if they were misbehaving."

"How's Norman?"

"Well, thank you."

Norman Aldridge, Deputy Manager at South Bucks Aero Engineering, was Dizzy's partner. Like her and Madeleine he had never married. SBAE was also based at Booker so the instructor and the engineer could frequently share transport when their work schedules coincided.

"Give him my regards."

"Will do . . . what about you, Madds? Have you got a man hidden away somewhere you haven't told us about?"

"Nope."

"Do you ever get lonely?"

"No . . . I don't think so. Anyway, I'm too selfish to have some one else intruding into my life."

"I doubt that."

"I've got my friends and I've got Nigel and my Mum and Dad."

"And how is Captain Gorgeous?"

"He's fine, thanks. Swanning around in a 707 somewhere no doubt."

"Or sitting by a hotel swimming pool with beautiful stewardesses."

"Or putting the world to rights in a bar with grumpy old pilots and flight engineers."

"I'm glad you and me and Parminder keep in touch. I've got so many happy memories of the Waltham days."

Madeleine nodded. "A great bunch of people . . . well, most of them, anyway."

"Some of them were decidedly odd. Remember that Squadron Leader? We called him 'Stuck on Transmit' 'cos he was always telling you things you'd already heard a thousand times."

"And the chap who used to refer to people as 'it' if they annoyed him."

"Yes, Dave somebody . . . 'it came wandering in and demanded a cup of coffee . . . well, I told it it would have to sort it out itself . . .'"

"Who was that dreadful American woman? Used to treat the ground crew like servants. 'Foghorn', we called her."

"Name escapes me now," said Dizzy. "But I remember they called you 'Miss Iffy'."

"I think Pammy started that one."

"Because you were instrument qualified, wasn't it?"

"Not officially. But we were losing people in bad weather, weren't we? As you'll remember, we lost Amy that way. Nigel unofficially taught me the basics of instrument flying. So the girl doing IF became Miss Iffy."

"And some wag turned it into Mississiffy, as I recall."

"And that became 'Sternwheeler'."

"That's right, I remember now. Childish, weren't we?"

"You said it! But I didn't mind the 'Iffy' name. I think it encouraged others to get instrument time."

"Including me, when I saw how useful it was. I did a couple of hours under the hood with Nigel, too."

"Yes, looking back, my brother probably helped to save a few lives."

"I'm sure you've told me before, but which one of you started flying first?"

Madeleine glossed over her father's divorce from Nigel's mother and reminded Dizzy that she had first met her half-brother at a table tennis tournament organised for church youth clubs. Not knowing they were related, they soon became friends and began doing things together when circumstances allowed. Living over thirty miles apart the chances did not arise very often. The easiest option was to take separate trains into Northampton, telling their respective parents they were meeting friends, which was the truth when they could persuade others to join them. Sometimes Madeleine would catch the bus to the airfield at Sywell, where most Saturdays Nigel worked at the flying club, cleaning the aircraft, sweeping the hangars and running and fetching as required. He didn't get paid but sometimes one of the local pilots would take him up for a brief flight round the local area. One or two of them would let him handle the controls.

Narrating the story to Dizzy, Madeleine could still picture the momentous day when a grinning Nigel pointed to a biplane a little larger than the others with a three-seat enclosed cabin in front of the pilot's cockpit. The aircraft was a Fox Moth, he told her, and the pilot had asked him if he and his friend would like to come for a flight in it. She was hooked, she told Dizzy, from the moment the wheels left the ground.

Not long afterwards Nigel met her again at Sywell and she could tell from his expression something was wrong. He took her hand and led her to the restaurant, where he sat her down and went to order a pot of tea. Madeleine waited for him to come back to the table, dreading what he was going to tell her.

"There's no other way to say this," said Nigel.

"What is it?"

"You and I are related."

"How do you know?"

"You're my sister . . . my half-sister."

It was quite a shock, Madeleine told Dizzy. It seemed that Nigel had found a document at home referring to "Nigel Maunsell". When he asked his mother why his name had changed she said quite openly that her first husband was named Henry Maunsell and so that was the name recorded on her son's birth certificate and other documentation. After Dorothy married Cyril Nixon her son's name changed when Cyril officially adopted him. Then Nigel mentioned to his mother that one of his female friends happened to have the same name and wasn't it a strange coincidence. To his alarm, Dorothy flew into a temper and told him that his new friend was almost certainly the bastard child of his father. Nigel's mother forbade him to meet her again.

Nigel was angry because he had planned to ask Madeleine if she would be his girlfriend and now that was impossible. She told him that she would have agreed to go out with him but there was no reason why they couldn't be friends. Continuing to defy his mother's prohibition Nigel arranged another meeting. He asked Madeleine to ask Henry if he would come too. The reunion brought plenty of tears and hugging and father and son both agreed to keep in touch. They also agreed that Dorothy would not be party to this arrangement.

The half-siblings both started flying lessons at the same time, Madeleine told Dizzy, hers paid for by Henry when he realised she couldn't be talked out of it and his in lieu of payment for his work at the flying club, with a parental contribution when further funding was required. Nigel was careful to make sure that his mother never knew he was still meeting his half-sister and Henry, fearful that the top-up would instantly cease if the liaisons were discovered.

They both learnt the art of flying quickly, pretending there was no element of competition between them. They both soloed on the same day, one after the other, Nigel after nine hours and fifteen minutes of instruction and Madeleine after nine hours and ten minutes. As soon as he finished school Nigel successfully applied to join the Air Force. His sister successfully applied to join the ATA four years later, interrupting her degree course.

"Where you became Miss Iffy," repeated Dizzy now.

"I've got a different nick-name now."

"Go on."

"Madam Methane."

"Because of your research, I suppose."

"I believe an alternative is 'Mrs Cowfart', though I haven't heard that with my own ears."

Dizzy enquired about the progress of the MR project and Madeleine related the sorry tale of its probable demise, the flying instructor nodding her head in commiseration.

"What are the chances of finding a sponsor?"

"One or two American companies have shown a bit of interest."

"So they're taking this carbon dioxide problem seriously, then?"

The chemistry scientist shook her head. "No, they're not fussed about that but they like what we're doing on catalyst research. It's big business if you can speed up industrial chemical processes."

"So there's some hope you can carry on?"

"They would want us to relocate to the States. I don't want to do that and the post grads aren't that keen either."

"Shame. Isn't there anyone on your side?"

"The SRC Project Manager. He makes sympathetic noises which I think are genuine and he says he's trying to get the Treasury to sanction funding for continuation."

"Good for him."

"Yes, but there are powerful forces lined up against him. I'm meeting him again this week to discuss tactics."

"I'm glad I stuck to flying, Maddie. I'd hate to be stuck in an office arguing about funding."

Madeleine smiled. "Well, actually, we're meeting in Oxford for dinner and then the theatre."

Dizzy raised her eyebrows. "Madeleine Maunsell, you're a dark horse. You *have* got a new man . . . and you weren't going to tell us."

"Hardly a new man, Dizz. We've been on a couple of dates and we mainly talk about finding money for MR. That's all there is to it."

"So far."

"So far."

CHAPTER 10

"The jewels of our father, with washed eyes Cordelia leaves you. I know you what you are, and, like a sister, am most loath to call your faults as they are named. Love well our father. To your professed bosoms I commit him. But yet, alas, stood I within his grace, I would prefer him to a better place. So, farewell to you both."

Jacqueline Holmbeck frowned and ran her finger down the page. "Sorry, Emm, lost the place. Too busy admiring your performance."

"It's Regan's line, Mum. 'Prescribe not us our duty' or something like that."

"Found it!"

"Actually can we start again?"

"Okay, but I'll need to go to the shops before they shut or you and Nigel will have no tea."

"Just one more run through, then."

Jackie, seated on the sofa, flicked through her Penguin "King Lear" to find the right page. Her daughter, Emma, walked to and fro in front of her, muttering some of her lines. There were twenty-six years between them, although more than once Jackie had been pleased to hear people ask her if she was Emma's elder sister. Both were slender, with long blonde hair, high cheekbones and green eyes.

"Is Daddy coming to the performance?" asked Emma.

"He said he would."

"Shame Nigel can't make it."

"Probably for the best though, darling. The two of them prefer not to be in each other's company, as you know."

"Maybe Nigel would be allowed to come to one of the rehearsals."

"Maybe."

"What about Auntie Maddie? Does she know about it?"

"Yes, I think Nigel's told her."

"When's he supposed to be back?"

Jackie looked at her watch. "Any time, assuming he's not stuck in traffic."

"I wonder if he'll bring me another baseball cap. My friends think they're really cool."

"He never comes home empty handed, darling."

Thursday evenings were quite relaxed at Nigel Nixon's house in Epsom because Emma's homework load was light and didn't include maths, which her mother couldn't help with if Nigel wasn't at home. The English teachers at Mounthill Grammar School were in the habit of letting pupils off their homework if they were taking part in the autumn term Drama Competition, the idea being that they would use the time to practise their lines. In fact it was rumoured that some pupils had signed up for the competition purely to avoid English homework.

Emma Holmbeck thought she would have a go "just for a laugh", as she told her mother. When the casting was under way for the Second Year Team B production of "King Lear", Act One, Scene One she was offered the part of Cordelia without the faintest idea of the character. Jackie knew the Bard's work a little and told her daughter she had the best female part. Surprisingly, Emma had immersed herself in the task and said she would have been happy playing one of the other sisters as she thought they were more realistic characters because they had the imperfections Cordelia lacked—being a baddie was more fun, in Emma's words.

Nigel showed up halfway through the rerun, dropping his suitcase and his flight bag in the hall on his way to the lounge. He was in civilian clothes, his uniform packed away in his case.

"My two favourite ladies," he greeted them with a grin.

"Good trip?" asked Emma.

"Yes, thanks."

"How was Chicago?" asked Jackie.

"Quite pleasant, actually. Just like an English summer at this time of year."

"I remember trips there in the winter. The coldest place on the planet."

"Ah, the good old days on the Strat!"

"I always preferred the Connie."

"I thought you flew on Comets, Mum," said Emma. "The ones that kept crashing."

"That was later," said Jackie. "They used Stratocruisers and Constellations on the Atlantic runs."

"O'Hare was hectic," said Nigel. "Some people reckon it's the busiest airport in the world. Non-stop yak on the RT." Nigel's flight had arrived at Heathrow in the early hours and he had caught up on his sleep in the BOAC crew dormitory located just outside the airport boundary before driving home.

"I'm off to the shops, dear. Lamb chops for tea?"

"We'll need mint sauce, Mum."

"Noted. I'll check the larder."

"If I make you a cup of tea, Nigel," said Emma, "will you run through 'Lear' with me?"

"Can I be the king?"

"You'll be everybody apart from Cordelia."

Nigel Nixon and Jackie Holmbeck had been a couple for half a dozen years, both divorcees. There were no offspring to complicate the separation of the airline pilot and Denise, his first wife, who had found herself a man who wasn't away from home so much and who shared her interest in painting. In fact the two had originally met at the Epsom Artists' Society. Nigel did not challenge her accusation of insufficient attentiveness, nor did he comment on Denise's selfishness in breaking up her new partner's marriage.

Jackie had originally met Nigel on the Comet fleet before she herself was married. They flew together a few times and grew fond of each other although the friendship never strayed beyond innocence. At that time she was an "A" girl stewardess and he a First Officer on the elite fleet, the first jets to grace the world's airways. Etched in both their memories was a traumatic moment during a stopover in Cairo. The entire crew were enjoying an evening out at the Al Khal restaurant when the manager came over and summoned the captain to his office. The captain reappeared grim-faced a few moments later and broke the dreadful news to his crew that Comet Yoke Peter had come down in the Mediterranean near Elba and it looked like their were no survivors. A second similar Comet disaster three months later forced the authorities to ground the entire fleet pending investigation. The cause was found to be explosive decompression in the airliners' cabins arising from metal fatigue, a phenomenon not fully understood at the time, exacerbated by design and manufacturing flaws. BOAC Comet services would not resume for four years, during which time Nigel had been promoted to command on the piston-engined Argonaut fleet and Jackie had met Andrew Holmbeck, married him and given birth to their daughter.

Jackie devoted herself exclusively to bringing up Emma for the first two year's of the little girl's life and then applied for work as a ground girl with her former employer, hiring a child minder to cover her working shifts, three days a week at a check-in desk in the makeshift terminal buildings on Heathrow Airport's North Side. Soon after Emma started school her mother added a fourth duty day each week, looking after the company's passengers who now departed from the newly opened Oceanic terminal in the airport's Central Area.

Jackie was upset and hurt when she learnt her husband had found another woman, a secretary at the office where he worked as a solicitor. But then she realised she had been deluding herself when she told herself she still loved him. She asked for compassionate leave to sort herself out. The split was more acrimonious than Nigel's from his erstwhile wife because Andrew and Jackie could not agree terms on his rights of access to their daughter, whose custody had been awarded to her mother. Eventually Jackie realised her vindictive lack of cooperation was hurting Emma so she gave ground and a settlement was reached. She promised Emma that she would stay off work as long as needed to make sure her daughter didn't feel neglected. Emma, who seemed to withstand the upheaval better than her mother, sensed that Jackie's unhappiness was compounded by not seeing her BOAC friends. She nagged her mother to start working again and gradually a sense of tranquillity returned, helped by Andrew having moved out of the marital home.

Soon afterwards Jackie used one of her concessionary staff travel tickets to take herself and her daughter to Barcelona for a short holiday. The check-in girl was a friend who discreetly upgraded them to first class and so a few hours later mother and daughter found themselves comfortably ensconced in the forward cabin of a Britannia turboprop cruising through French skies, the creature comforts they were plied with all the more enjoyable because they were free. The chief steward remembered Jackie from her days flying the line and so they had a good old catch-up, eagerly swapping gossip. While they were chatting in the galley the flight deck door opened and the captain appeared, wearing his uniform cap, as required by company regulations.

The pilot's eyes widened in surprise and then he grinned. "Jackie Holmbeck. Well I never."

"Hello, Nigel. I heard 'Captain Nixon' during the welcome announcement when we boarded and wondered if it was you."

"Why didn't you say something? You could have come up front for take-off."

"Last I heard of you, you were still flying Argonauts."

"They're all gone, as you know. Some sold to other companies, some turned into saucepans."

"How do you like the Brit?"

"Love it. Are you coming all the way to Accra with us?"

"Getting off in Barcelona."

"Shame."

The chief steward had left the galley to carry out various duties in the cabin and so the pilot and the ex-stewardess could cautiously explore more personal territory. When they discovered they were both divorced and neither was currently attached to a long term partner Nigel asked her for her phone number and said he would call her to arrange a date after she returned home.

Not long after they'd started going out Nigel asked Jackie if she would move in with him. After consulting her daughter, she agreed. Andrew accepted her suggestion that he acquire ownership of their house, compensating Jackie with a lump sum representing her share of the equity.

The animosity from Jackie's ex-husband towards her new partner had arisen when Emma innocently remarked to Andrew that it was "cool" having two Dads, even if she called one of them "Nigel" rather than "Dad".

"Let's get this straight, Emma," said Andrew with a scowl. "I am your father. Your only father. Nigel is not your father and never will be. Got it?"

"Okay," said Emma, less alarmed than her mother, who had heard the outburst. "But I see him more than you."

"More's the pity. But never forget that you are my daughter, not Nigel's."

"Mummy loves him. And so do I."

"More fool you. And her."

Emma suddenly got angry. "It's all your fault for leaving Mummy."

"No, it's not. It's her fault for not loving me."

"Stop it, you two!" pleaded Jackie, intervening. "We've been through this a million times."

With the passage of time the bitterness between the two men had receded somewhat, together with Andrew's resentment at seeing his daughter less than Nigel did. At Emma's prompting, Jackie had even managed to persuade them to meet each other to declare a formal truce.

Not long after the dust had settled she arranged a party for some airline friends. Emma briefly appeared before she went to bed and one of Jackie's guests asked her what she was going to be when she grew up. Would she work for BOAC, like her Mum?

"Yes," said Emma. "I would like to be a stewardess. But if I can't be a stewardess I suppose I'd have to be a pilot."

Cue much merriment among the assembled company, especially the cabin crew and pilots.

CHAPTER 11

For mid-October the weather was mild as Madeleine Maunsell and Toni Clark walked from Madeleine's house towards All Souls church for Friday evening choir practice. In contrast to the pale skinned chemistry professor the TV producer still glowed with her Californian sun tan. Under her duffel coat she was wearing one of her favoured kaftans. In her hair dangled strings of beads of various colours. Her landlady felt distinctly frumpy walking alongside her.

It had been an unusual day, thought Madeleine. Morale in the Gas Research lab was holding up as well as could be expected, thanks to a stay of execution for the Methane Reformation project. Funds had now been guaranteed until December with a possible extension to the end of January, thanks to intense lobbying on her behalf by John Davis. She had met the Science Research Council Project Manager a few times socially and found him good company. Twice he had partnered Madeleine at her Bridge Club in Woodstock, staying overnight in the spare bedroom.

A further bonus was that Luke and Joe had found a method of boosting the tungsten carbide catalyst effectiveness by changing the ratio of ruthenium to palladium in the doping and increasing the catalyst's surface area. Yield in the periodic process had improved by a shade under three percent and by a fraction over four percent in the steady state variation. Madeleine had lost no time in informing the SRC accordingly.

Earlier in the day Verity Abbott had put through a phone call to Madeleine from America. The caller was Maxwell Murdoch, an entrepreneur whose family had amassed their fortune constructing major engineering works, especially power stations, as Madeleine discovered later. Although the two of them had never communicated before by any means the American spoke with the directness associated with his countrymen, bypassing the hindrance of formality.

"Hey, Madeleine, how are you doing?"

"Very well thank you, Mr Murdoch."

"Max, please."

"Yes."

"Okay. I'll get to the point as this call will eat up all my dollars. I want to bring your Reformation project over to Wilmington. You and the whole caboodle—all your kit and all your people."

The chemistry professor took a moment to recover from the surprise. "Where did you say?"

"Wilmington, North Carolina. We'll pay your fare and sort you out with accommodation. Plus you'll get a fat paycheck."

"Why do you want us to work there?"

"It's what God wants."

"Pardon?"

"I got a message from the Lord, Madeleine. In a dream."

The chemistry scientist paused, wondering for a moment whether the American was joking. For a brief second she even entertained the idea that the whole thing was a prank dreamt up by her post grads for their amusement, probably at the instigation of Luke Stirling.

"Whaddya say, Madeleine?"

"What sort of message?" countered Madeleine, deciding to play along for the moment in case "Maxwell Murdoch" was indeed the genuine article.

"The Lord told me I gotta help him clean up the planet. So we gotta build more nuclear power stations and clean up coal-fired stations. That means trapping carbon dioxide before it gets out into the atmosphere. My people tell me you can do that sort of thing so that's why I'm offering you the deal. Whaddya say?"

"It's very sudden . . . Max. We would need to think about it."

"Yeah, that's okay. My people tell me you don't have close family so it wouldn't be difficult for you to come over. The students will follow you if we offer them enough dough."

"I do have friends," said Madeleine, reacting to the American's presumptuousness and apparent invasion of her privacy.

"Course you do, honey. Everyone's got friends. But we'll pay for them to come and visit with you as often as you like or you can go visit with them. It's not a problem."

"Couldn't we carry on doing the work here?" asked Madeleine, deciding the caller was genuine. "We could send the results to you in Willman . . . Willingham—"

"Wilmington. Well, I'm not so sure about that. It would kind of slow everything down."

"If we came out, how long would we be there?"

"As long as it takes. You'd get leaves of absence so you could go home to your friends."

"I'll need to think about it, Max. I'll tell the post grads what your plan is and see what they say."

"Great. I'll call you in a coupla days, see if it's gonna pan out. I hope we can cut a deal. Remember, we'll be doing the work of the Lord."

After ringing off, Madeleine set her secretary the task of researching Max Murdoch's background, which meant phoning contacts in various libraries. The professor walked into the MR lab and, unusually, found all four post grads working. She briefly smiled to herself, wondering if the students had set up a sophisticated lookout system to alert them to their mentor's imminent arrival to give them time to adopt the poses of dedicated researchers. Immediately she rebuked herself for her cynicism and reminded herself that never in the past had they bothered to prevent her seeing them less than fully occupied.

Luke and Joe were engrossed with "Deneuve", one of the Thermal Conductivity Detectors, staring at the trace on the cathode ray tube. On the other side of the lab Xiu Ying and James were checking the manometer measuring the pressure of the gases in the reaction chamber.

"Is this a good time to chat?" began Madeleine.

"Give us a couple of mins to record these readings, Professor," said Joe.

"We'll be free then, too," said Xiu Ying.

"I'll put the kettle on, then," said Madeleine. "Verity's out on a job so I'll make the tea."

"We've still got a couple of doughnuts left," said James. "We can divvy them up."

When the chemistry professor passed on the gist of the phone conversation she'd had with Maxwell Murdoch, James nodded.

"I've heard of this bloke somewhere. Stinking rich. Some sort of religious nut. There was a scandal, I seem to recall. Had an extramarital affair and then told everyone it was okay 'cos the Lord had forgiven him. My girlfriend heard about it because the wife was a well-known actress and dancer."

Melanie raised her eyebrows. "Interesting," she observed. "The thing is . . . do we want to decamp to North Carolina to do this work?"

"For me, it would depend on Gill, my girlfriend," said Luke. "I couldn't go if she didn't want me to."

"Same for me," said James. "I wouldn't want to leave England while Stephanie is working in the West End. The producer reckons 'You Must Be Joking' will run for at least another year before it runs out of steam. She won't leave that show until it closes unless a better offer comes along."

"So if the rest of us go, what will you do?"

"I have absolutely no idea. Teaching maybe. Or find another PhD project if there was anything available."

"What about you, Susan?" asked Madeleine. "Your boyfriend's American, isn't he? Would that make a difference?"

Xiu Ying smiled sadly. "He's not my boyfriend any more. We broke up."

The others looked at her in surprise. This was the first any of them had heard of it.

"What happened?" asked Joe. "Or don't you want to tell us?"

"He has got a new girlfriend."

"Then he's a fool," said Joe.

"Quite," added James.

"Have you got a new boyfriend?" asked Luke.

"No."

The diversion seemed to Madeleine to have run its course so after a respectful second or two she steered the conversation back to the original theme.

"What about you, Joe? Would the States appeal to you?"

Joe shrugged. "Possibly."

"What about Maria?" said Luke. "Would she go with you?"

Joe pouted. "No, I don't think so."

"Don't tell me you've broken up, as well?"

A pause. "Well, we're still going out but . . ."

Another pause. Luke turned to his mentor.

"Would you go, Professor?" he asked.

Madeleine smiled. "Like you lot, I've got mixed feelings. It would be nice not to have to worry about funding, wouldn't it? On the other hand I'd miss my friends and the other work I'm doing."

"Would this Murdoch geezer pay us a regular salary?" asked Joe. "I mean, if he's prepared to throw a load of cash at us . . ."

"His words were 'fat paycheck'," said Madeleine. "Figures weren't mentioned."

"Maybe it wouldn't work anyway," said Luke. "Okay, we can convert small amounts of CO_2 to CO but we don't know if it could be done on an industrial scale."

The chemistry scientist nodded. "I had the same thoughts, Luke. Theoretically the process could be scaled up, but there might be practical problems, such as the cost of the catalyst for a start. Even the well-stocked wallet of Max Murdoch might be overstrained buying a few tons of palladium."

"How much is he worth, anyway?" asked James.

"Verity's checking his background as we speak," came the reply.

"What about the religion thing?" asked Luke. "We wouldn't have to go to church or anything like that, would we?"

"I am a buddhist," said Xiu Ying. "I would not want to go to a Christian church."

"Alright," said Madeleine. "Even assuming the financial package was attractive and there were no religious . . . obligations . . . and partners did not object, I'm not sure we would want to carry on without you, James. We've all been a good team for a year now. It would be a shame to break it up."

"Nice of you to say so, Professor," said James with a smile as the others nodded their concurrence. "But I'm sure you could find a substitute."

"Yeah, but would he be able to play darts as well as you?" laughed Luke.

"Or she," added Xiu Ling. "Maybe we could find a lady chemistry post grad who could throw arrows at the wall."

"Women can't play darts," grinned Luke. "Well known fact."

"I beat you last time," said Xiu Ying.

"I deliberately let you win!"

"Bullshit!" said Joe. "She beat you fair and square. Mind you, you were probably half-pissed at the time."

"Right," said their mentor, "if I can drag you back to the main point . . . what am I going to tell Mr Murdoch?"

"It's a definite maybe," offered Joe.

"And a definite 'no' from me," added James.

"Just had a thought," said Luke. "Perhaps if we accepted Murdoch's offer the SRC would keep MR going here purely to stop the Yanks pinching our work."

"Good point," said the professor. "I'll mention it to John Davis next time I see him, see what he thinks."

"Something else to think about," said Joe. "Suppose we get established over there doing our thing and then the Lord tells Mr Murdoch to pull the plug. What would happen then?"

No one had an answer.

Verity Abbott came back from her detective work to find her boss rinsing the tea mugs in the secretary's office. She was about to take over when the phone rang. She picked up the receiver.

"It's your brother," she mouthed to Madeleine. "Shall I put him through to your office?"

"Hello, Maddie," came the voice in the earpiece as the professor sat herself down. "Are you well? You haven't been chopped by the Ministry yet?"

"Still here," replied his sister. "The SRC haven't closed us down yet but there's only funding for another few months. We might have to transplant ourselves to the US of A if we want to carry on."

Madeleine summarised the offer from Max Murdoch and her students' responses.

"Would you want to go across the pond to work?" asked Nigel.

"Mixed feelings, really," came the response. "I'd miss you and my friends."

"What about that Project Manager you've been dating?"

"What about him?"

"Are you . . . you know . .?"

"No, we're not."

"Well, maybe . . ."

"And maybe not. Shall we change the subject?"

A chuckle from the other end of the line. "Actually, the reason I called is because I got a call from Fearnville earlier today."

"Not another escape attempt, I hope?"

"No, she's in hospital . . . unconscious. They think she's had a stroke."

"Oh, sorry to hear that. What about tomorrow's visit, then. Is that off?"

"Yes. She's in Wycombe General. I'll pop along myself but there's no need for you to come."

"Are you sure?"

"The hospital says she's not responding to anything at the moment so there's no point in visiting her as far as you're concerned."

"Okay, but if she regains consciousness and you'd like me to help you deal with her I'll make myself available . . . not very well put, that, was it? 'Deal with her.' A bit heartless."

"Perfectly understandable, sis, the way she's treated you over the years."

After the call Madeleine buzzed Verity.

"What news on Mr Murdoch?"

"I'm coming in. I'll tell you all about it."

It turned out that Maxwell Murdoch IV was a big name in the States, owner and Chief Executive Officer of the giant Southern National Generation company. His great-grandfather had started the family business, building infrastructure to support the expanding railroad network as it brought supplies to the frontier edging progressively and violently westwards. Family lore had it that the original holder of the name was killed by a Pawnee arrow defending a locomotive depot against the red hordes near Tulsa, Oklahoma. Perusal of contemporary newspapers told a different story. With a few variations, the gist was that Maxwell Murdoch I met his end at the hands of the husband of the Pawnee wife he was caught *in flagrante* with in a cheap hotel. Some reports added that the wife also met her grisly end in the same manner.

Maxwell Murdoch II had exploited the demand of the burgeoning American industrial economy for electrical power plants at the start of the twentieth century, with his eldest son following in his footsteps.

The fourth bearer of the name showed himself possessed of the same fanatical business acumen of his predecessors seasoned with a flamboyance that some said strayed into wildness. Married three times, divorced twice, according to the sources Verity had consulted. She herself had unearthed a newspaper article about Mr Murdoch in the University's Bodleian Library archive. The article in the *Washington Post* included a photo of the 60-year-old entrepreneur, showing a portly man with a full mane of white hair and a moustache and goatee beard to match.

The article revealed that Maxwell Murdoch IV was a supporter of the Republican Party and a leading voice in the controversial Righteous Warriors of America organisation, whose policy on the Vietnam War was that it must be won by the anti-Communists at any price. Their creed was that God demanded nothing less, even if the enemy had to be tamed by tactical nuclear weapons. The Post also mentioned two assassination attempts on Mr Murdoch, in one of which a rifle bullet had clipped an ear.

"Sounds quite a character," said Toni now as they neared All Souls Church for choir practice. "It wouldn't be boring, working for him."

"It might be a bit too exciting for someone my age," said Madeleine. "But if I don't go, he may not want the post grads without me, so I'd be letting them down."

"When do you have to decide?"

"I can't imagine Maxwell Murdoch hanging around a long time waiting for an answer."

"Perhaps the Lord can guide you," said Toni. Her landlady threw her a quick glance to confirm her tenant's tongue was in her cheek.

"It would be more useful if we knew what his plan was for Nigel's mother," muttered Madeleine.

"Right," said Toni. "Let's go and sing his praises," waving to fellow choir member Daphne Edwards as they approached the church, "even if we don't believe in him."

"Amen."

CHAPTER 12

Choir practice was conducted in the church hall, a modest single storey building standing alongside the graveyard. Attendance this Friday evening was about average, with seven choirboys and Toni Clark singing treble, four altos, five tenors and four basses. The company included only four females since one of the altos was Albert Landred, who could manage around three quarters of the vocal range of the ladies. Two of the tenors were adolescents, ex-trebles whose voices had only recently broken and whose pitch control could therefore not be guaranteed. Choirmaster Benedict Avery did not hide his exasperation at the occasional departures from tonal accuracy.

Gregorio Allegri's "Miserere" had initially caused ructions for the choir of All Souls, Woodstock. Some of the choir members had been pressurising Mr Avery to let them add it to their repertoire for quite a while. His response was: "When you're good enough". Believing this ploy was merely a device for enforcing attendance at choir practice alto Daphne Edwards responded with her opinion that the choir had already achieved the necessary standard. Whereupon the choirmaster had modified his requirement to: "When I think you're good enough". Daphne and her alto colleague Jean Coster had taken to visiting the homes of the choirboys whose enthusiasm for the Friday practices was patchy to persuade their parents to persuade them to attend. But soon after rehearsals began for the 16th century chant the need for cajoling the less dedicated choirboys had faded when they realised they loved performing it.

This Friday evening the trebles didn't mind when Mr Avery required them to sing a few bars of the their line by themselves to verify accuracy, their confidence reinforced by Toni's contribution. By contrast Daphne and Jean bristled when the choirmaster expected the altos to subject themselves to the same discipline. When they asked Madeleine Maunsell to support their cause the chemistry professor diplomatically replied that she herself had no objection to the choirmaster's demand. Fellow alto Albert Landred, sitting quietly and expressionless at the end of the row, concurred with Madeleine. His unprepossessing demeanour suggested bank clerk or teacher. In fact he was one of England's most highly rated heart surgeons.

"Well, ladies, it's your choice," said Mr Avery now. "You can sing me your part or we can cancel the whole thing."

"That's blackmail," said Jean.

"Correct," said the choirmaster.

"It's demeaning," said Daphne. "You're challenging our professionalism. We've told you this before."

The young choristers watched the disagreement with the thinly disguised glee that always derives from seeing adults arguing with each other.

Bass Matthew Claridge spoke up. "Come on, Daphne. He's not asking you to give blood. It's just a bit of warbling. The boys have done it and we'll be doing it too. We all know you can sing like a lark so there's nothing to be frightened of."

"I'm not frightened of anything," said Daphne haughtily.

"Okay, we'll do it . . . under protest," conceded Jean, putting her hand on her colleague's arm to placate her and prevent further argument.

"Hallelujah!" muttered Mr Avery.

The altos successfully negotiated twenty odd bars of the work and the choirmaster smiled, almost without sarcasm.

"See, it didn't hurt, did it?"

"Nicely done," commented Toni, twisting round from the front row to address the altos.

"I'm sure we don't need your approval," said Jean. But sitting next to her Madeleine responded more firmly and pointedly: "Thank you, Toni."

"Right," said Mr Avery. "You tenors can have a go now."

The next fifteen minutes of rehearsal passed without any difficulty and the beautiful music soothed the souls of everyone present. During a pause Jean tapped Toni on the shoulder and apologised for her earlier rudeness.

"That's cool, baby," said the TV producer.

The cheerful arrival of the plump Mrs Partridge signified imminent tea break. It was the parishioner's duty to meet the catering needs of the singers, dispensing tea or coffee or soft drinks and biscuits.

As they sipped their tea seated a little apart from the others Toni asked Madeleine about John Davis. Typically she launched straight into personal questions without preamble.

"So come on, Madds, have you bedded him yet?"

"I see you've picked up American ways during your sojourn in California," observed the professor a touch ruefully. "No attempt at subtlety."

"Evading the question. Does that mean you have?"

Madeleine looked directly at her tenant. "No, I haven't. Both times he stayed over you were at home too, remember?"

"So if I hadn't been there you would have shagged him, is that what you mean?"

"No, Toni, that's not what I meant."

"Do you want to?"

"He hasn't suggested it."

"Another evasion, Maddie. What would you say if he did suggest it?"

"It would depend."

Toni laughed. "If you don't want him, I'll have him. He's quite dishy in a quiet, refined sort of way."

"Shall I tell him you fancy him?"

"If you're not interested I'll tell him myself. When's he next coming over?"

"Not sure. He's away at the moment and we haven't arranged anything for when he comes back . . ." Madeleine's response faded out because Matthew Claridge had come over for a chat.

"How was America, Toni? Don't think I've had the chance to talk to you since you got back."

"Fantastic, Matthew. It's all happening over there."

"Making a film, weren't you?"

"Yeah, the Hippie Movement."

"My son wants to go next summer, so he says. But I'd be worried. They're all taking drugs and riding huge motorbikes from what I can make out."

"How old's your son?"

"Seventeen . . . eighteen next year."

"Will you let him?"

"He says he can travel across the country for free. Apparently students can get jobs delivering cars when their owners are flying from one place to another. My son says he and a friend want to do New York to Los Angeles, which is a common route according to him. They don't get paid but the petrol is paid for and they're given so many days to complete the journey."

"Sounds like fun," said Toni.

"I told him I'll think about it."

"Well, the hippie thing might all be over by next summer. No one knows how long it's going to last or if it'll change. At the moment it's all good, well ninety percent anyway. Yes, there are some people hanging around dropping acid but they don't cause trouble and the odd spliff never hurt anyone. The mood is peace and love."

"And stop the Vietnam War."

"Yes. We filmed a few demos."

"But LBJ's not happy, judging by the news reports."

"The Government are worried 'cos they can't control what's going on. I just hope it doesn't get nasty. Some right-wingers are getting themselves agitated. That group Maddie was telling me about earlier—"

"The Righteous Warriors?" offered Madeleine.

"Yeah. Shame I didn't know about them while I was over there. I could have brought them into the film. I might insert a bit about them later if it's possible."

"When's the film coming out?" asked Matthew.

"It's being edited now. They're talking about the end of the year. Apparently the lawyers have to see the final cut before it's released—make sure no one can sue us for defamation or slander. They're very litigious, the Americans."

"What's it called?"

"Provisionally 'Flowers in Your Hair'. That's the title I want. The Emcro executives will have the final say."

"Anything else in the pipeline, Toni?" asked Matthew.

"Yes, we've started work on a new drama about the mining community in Northumberland between the wars. It'll be a class thing—there'll be a posh family involved too, servants and all the rest of it."

"Sounds interesting. TV or cinema?"

"TV series. The first episodes are being written."

"Workers against the toffs, then," chuckled the bass singer.

"Yes, but the main characters are two schoolgirls, daughters of miners. Everything hinges round them."

"I'll watch out for that." Matthew lowered his voice a little. "You've made your peace with Daphne and Jean, I hope? I noticed the little upset earlier."

"It was nothing. Storm in a teacup."

Madeleine also spoke quietly. "Daph told Toni off when she suggested that most of the hymns we sing are just Victorian pop music."

The bass laughed. "Haven't heard that one before. What do you mean, Toni?"

"Take a look through the English Hymnal. Half the hymns were written during the Victorian era. Snappy, simple tunes to which various people wrote lyrics, some artistic or poetic, but a lot of the words are trite, some just plain drivel."

"Don't tell our lady altos that!"

"Actually, Benedict said as much once during an unguarded moment," continued Toni. "But he thinks of the human voice as just an accompaniment to the music. He reckons you can turn anything into choral music, even a shopping list."

Madeleine smiled. "What about the Periodic Table? Could he set that to music?"

Matthew frowned. "Something to do with chemistry, isn't it? Seem to remember it from schooldays."

"Yes, it's the elements listed by increasing atomic number," said Madeleine.

"Don't tell me," said the bass, screwing up his eyes. "Hydrogen, helium, beryllium, boron, nitrogen, carbon—"

"Very good," laughed the professor, "except you missed out lithium and carbon comes before nitrogen."

"Can't imagine Jean and Daph singing along to that, somehow," chuckled Matthew. "They're not happy when their irreverence detectors activate. Sometimes the choristers deliberately sing daft words just to annoy them."

"Yes, I remember one of their joke hymns," said Madeleine. "We were doing 'Praise My Soul the King of Heaven'. What did the boys sing?—'Braise my vole, said Tim to Kevin'. Daphne didn't like that." Timothy Watts and Kevin Hall were two of the choristers.

"They wanted to throw me out of the choir, remember?" said Toni, "when they found out I was a non-believer."

"What's that group you belong to?" asked Matthew. "Atheists for God, something like that, isn't it?"

"The Secular Christian Society," said Toni. "We like J C's ideas as a social code but don't go for the Son of God stuff."

"Peace and love," offered Madeleine. "A bit like what's going on in San Francisco, judging by what you've been telling us."

"I agree," said Toni. "It's another form of socialism, just like the hippies."

"Jean and Daphne are staunch Tories," smiled Matthew. "If they suspected you were dallying with the comrades they really would chuck you out."

"Yes," said Madeleine, a little conspiratorially, "All Souls is very much a middle class church. No lefties or poor people allowed here."

"Yeah," said Toni. "Suppose a bloke of Middle Eastern appearance with long hair and a beard showed up."

"Riding on a donkey!" added Madeleine.

"They'd have a fit."

"Probably call the police."

"Maddie, you're just another subversive!" laughed Matthew. "I'm away before you corrupt me!" He ambled off to talk to other choir members.

"I'm also in trouble with Mrs Vicar," said Toni to her landlady.

"Don't tell me . . . hats!"

"Got a right bollocking last Sunday for not wearing one."

Yvonne Kenney, wife of the Reverend Godfrey Kenney, was known to be a stickler for correct dress in the place of worship. Ties for the gents and hats for the ladies.

"Standards must be maintained!" said Madeleine sarcastically.

"She's a snobby cow," said Toni, "but I'm thinking I might get my revenge."

"How?"

"Seduce Godfrey."

"What!"

"I know he fancies me. I might ask him for a bit of one-to-one religious instruction. Tell him he can try to convert me. You know— laying on of hands, and all that. And laying on of other parts, too."

"You wouldn't!"

"I might."

"You're awful!"

"That's me!"

Before Madeleine could think up a reply the choirmaster's voice sounded across the hall.

"Okay, everyone, it's back to work. You don't deserve it but I've decided we'll do Harwood this Sunday. We'll do the Agnus Dei now for practice. Let's see what a mess you make of that."

"Hooray," cheered several of the choir members.

CHAPTER 13

If you had to be stuck on your own somewhere away from home there could be worse places than Honolulu, Hawaii, thought Senior First Officer Rod Smith. Especially when you were a guest at the Holiday Inn, Waikiki Beach, at the expense of your employer, the British Overseas Airways Corporation. Especially when you could spend your time sunning yourself on the beach, drinking a cold Michelob and watching golden-skinned, bikini-clad girls splashing along the water's edge. Further out the huge rollers were sweeping surfers towards the shore. Rod could hear their shouts—whoops of delight from those who had successfully caught the wave and yells of frustration from the unfortunates who had wiped out.

In the nature of these things, Rod's enforced stopover had begun with a rostering problem back at Heathrow. Several crewmembers had succumbed to the early winter flu bug currently ravaging Britain, leaving the airline temporarily short of pilots, copilots more so than captains. Frantic swapping of crews by roster personnel had just about managed to keep the programme going as scheduled but at the cost of disrupted lives for those involved. Last minute changes meant that crews assembled and broke up more frequently than usual. Rod was one of those on leave who had volunteered to help out. He had arrived in Honolulu two days previously and would join a homeward bound crew the next day. There wasn't enough time to organise child care for his two kids which meant that his wife, Miranda, couldn't accompany him on a staff standby ticket. Besides which, as Miranda had said herself, for a three-night stopover, not to mention three days flying each way to get to Honolulu and back, it was hardly worth the effort of trying to sort everything out.

Rod looked at his watch. 1130. He would not be alone much longer. If he was running to time, Senior Engineer Officer Frank Armitage would soon be joining him. Frank had been travelling as a passenger on the 707 which Rod and his crew had brought in from

Los Angeles. A crew rearrangement left Rod on the Hawaiian island while the others flew back the next day with a different copilot.

Visiting the flight deck on the leg to Honolulu, Frank had told the crew he was heading for Hilo for a ten-day holiday with an old American friend. But during his time there he and the friend and the friend's son were going to fly over to Waikiki for two days so that the boy could practise surfing. The American was a Convair 640 captain on Hawaiian Airlines so deadheading between the islands was easy to arrange and cost free for himself and his friends and relatives.

Rod was about to pick up his copy of the *Pearl Dispatch* local newspaper when he spotted Frank walking along the beach towards him, accompanied by a man wearing a baseball cap who looked to be of a similar vintage to Frank and a slender youth whose age would have been mid-teens, thought the BOAC copilot.

"What ho, young Rodders!" called the flight engineer as the trio approached. He was wearing a straw panama hat more suited to a country parish vicar than a master of the technical systems of the Boeing jetliner.

Rod stood up and held out his hand. Frank made the introductions.

"This is Stan Hazzard, seniority number three on Hawaiian, so max respect, please!"

"Pleased to meet you, Stan."

"Sure, you too. This is my son, Carl."

"Champion surfer," said Frank.

"Not yet," said the boy, "but I'm working on it." He pointed out to sea at a group of brown-skinned local boys expertly riding a roller towards the beach. "Those kids are better than me and they surf on tea trays."

The four repaired to a nearby burger bar and Rod ordered three beers and a Coke. The group sat themselves down at a table in the shade.

"Just remind me how you two know each other," said Rod, looking at Frank and Stan.

"As I told you, we first met during the war," said Frank. "Stan was a B17 driver based at Alconbury. But the powers that be decided that some of the Yanks should get checked out on our heavies and vice versa. So Stan and some of his chums converted onto Lancs and some of our boys flew the B17. The idea was cross-fertilisation of operational techniques or some such nonsense. Stan was on 625 Squadron with me. We flew a few missions together."

"That's right," nodded the American. "Half a dozen or so, I guess."

"How did the Lanc compare to the Fortress?" asked Rod.

"We were impressed," said Stan. "Both types had their good points and bad points. The Lanc was faster and could carry more ordnance but was stuck in the lower levels and couldn't defend itself so well. The Seventeen could fly higher and bristled with guns but couldn't carry as much."

"Loads of guns, yes," said Frank, "but the Jerries quickly found the blind spots, so they were easy meat until the Mustang came in to protect them."

"True enough," agreed Stan, "but we were attacking in the daylight so it was easier for the bad guys to find us."

"Any nasty moments?" asked Rod.

"Plenty," said the American. "But at least we lived to tell the story. Plenty of our buddies didn't make it." He laid his hand on Frank's shoulder. "This man here saved me and my crew after the Essen raid."

"What happened?" asked the BOAC copilot.

"We got hit by flak and I got knocked out. Frank took over and flew the ship home. Next thing I knew I was in hospital with bandages all over."

Frank grinned. "I'm one of the few airmen in the world whose number of landings exceeds his number of take-offs—one and zero respectively."

"They gave you a medal, didn't they?"

"It didn't mean much. They dished them out with the rations."

Carl looked over, impressed. "Hey, that's cool. I remember Dad telling me about that before. So you're the one who landed the plane when Dad was unconscious?"

"Yessir. But it wasn't the best landing in the world. The port undercarriage collapsed."

"Jeez, man," said Stan. "You had one engine out, one misfiring and damaged flight controls!"

"I got a lot of stick from the rest of the crew."

"In jest!"

"I think so."

"They had to scrap that ship as I recall. Too badly damaged."

"Blame the Germans, not me!" said Frank, bringing grins from the others.

"I remember they brought us a replacement," said Stan. "The delivery pilot was a girl—that lovely Madeleine something-or-other. She was absolutely gorgeous. I met her a coupla times on the squadron and I fell for her."

"Is that Nigel Nixon's sister?" asked Rod. "She was in the ATA, wasn't she?"

"The very same," replied Frank.

"I remember asking her for a date," said Stan, "but she was already taken, she told me."

"She was engaged," said Frank. "But her fiancé went for a Burton on a Dortmund raid, as I recall."

"Did she find anyone else?"

"No, I don't think so," said the flight engineer. "No one permanent, anyway. She still goes by her maiden name."

There was a pause. Carl drained his Coke and stifled a burp. "Say, Dad, is it okay if I go and grab a board and shoot the curl?"

"Sure, son. I'll be catching a cab into town in a while, buy some fishing gear. I'll meet you back here at five. That suit you?"

"Swell." the youth stood up and offered his hand. "If I don't see you again it's sure been great meeting with you, Rod. Enjoy your stay."

"Do you ever meet Madeleine?" asked the American of Frank, picking up the earlier conversation after the youth had left.

A shake of the head. "No. I fly with her brother quite often. He's a skipper on the 707. He sometimes mentions her in conversation."

"What does she do? I mean—does she work?"

"She's a science professor at Oxford University. Runs a research facility."

"Wow, that's impressive. So she quit flying after the war?"

"Yes. I think she flies puddle-jumpers now, as a hobby."

"So you live in Hilo?" asked Rod, looking at the American.

"Yeah."

"Stan's got a lovely house on the sea front," said Frank. "With his own boathouse and jetty."

"Sounds good. And you fly for Hawaiian?"

"Uh, huh."

"The 640 will be just local flights, then," said Rod.

"Yeah, island hopping. It's okay. Pays the bills."

"No long haul?"

"Nope. There's talk of a coupla leased DC8s coming in to do flights to the mainland but I think it's just wishful thinking among some of the younger guys."

"Seniority number three—you'd be on the first conversion course, Stan," laughed Frank.

The American shook his head. "Not for me, thank you, sir. How do you think I shot up the seniority list? All the guys above me went off to fly bigger equipment on longer routes with other companies. I've done my share of droning over the oceans on the old piston Dougs. This suits me just fine. Anyway, I'm too old to convert onto the jets. They're way too fast for me. My brain works at propliner speed."

"Rubbish!" snorted Frank. "We had chaps converting onto the 707 and the VC10 who were the far side of fifty, much older than you."

A waiter came over and took their order for another three Michelobs.

"You're starting a command course soon, the rumour machine has it," said Frank to the copilot.

"Yes, in the New Year if all goes to plan."

"Just remember, Rodders, to treat your flight engineers with respect when you swap to the left seat."

The copilot laughed. "Of course. You know, I could never understand why some captains treat other people like dirt. No names and all that. They upset their first officers, their flight engineers, the cabin crew, the dispatchers, the ground engineers, you name it. I always thought what's the point of annoying people who could help you? It's the bleeding obvious really."

"I think it's more likely in pilots who got early promotions," offered Stan. "They forgot what it's like to be a cojo. I reckon the way to do it when you're dealing with other people is to ask yourself how you're being assessed. Are you coming across as a decent, reasonable guy or a complete asshole. Some guy once told me— whatever job you're in, ten percent of people are shits."

"There are one or two of the latter category in the Corporation," said Frank. "As Rod says, no names—no pack drill."

"I've been in the right seat nearly ten years," said Rod, "so I've had plenty of time to watch how captains do their stuff. You incorporate the good points into your own operation and make sure you never follow the bad examples."

Frank tapped his chin. "The last time you and I flew, Rod. That was with Soapy Nixon, wasn't it? We had that young kid doing his Flight Nav course with us. Soapy is one of the best, I would say."

The copilot nodded enthusiastically. "I agree. A joy to fly with. I did a trip with him just after the command course bid result was published. He gave me nearly all the take-offs and landings so I could get my hand in."

"Soapy?" queried Stan.

"He's unable to complete a sentence without including the word 'SOPs'," grinned Frank.

"As in Standard Operating Procedures?"

"Correct."

The American pilot chuckled. "I'm told the 707 handles just like the dear old Seventeen. All Boeing designs are the same, they say. You make a control input and half an hour later the ship responds."

"I love it," said Rod.

"Me too," agreed Frank. "But you pilots are so fickle. Suppose they offer you the VC10, Rodney. What then?"

The copilot screwed up his eyes. "Difficult one, Frank. The chaps on the Ten love it as much as we love the 707."

After a few more minutes aviation chat the American announced that he was headed Downtown for some fishing tackle.

"Seems like a nice chap," said Rod as the two Brits watched Stan climb into the taxi he had just waved down.

"Yeah, I'd say so," agreed Frank. "Not quite as brash as some of his countrymen."

"Nice that you kept in touch after the war."

"Yeah, we've been friends for over twenty years. He used to live in Cincinnati but after his divorce he met a Hawaiian girl and came to Hilo to live with her. Lovely girl. I get to stay with them once or twice a year. Go out fishing or sailing. Not a bad life."

"Does Carl live with them?"

"No, lives with his Mum in Cincinnati. But she's happy to let Carl come over during the school holidays whenever he wants to. Stan says he gets on better now with his ex than when they were married. Seems to happen a lot in the States. Couples divorce then carry on as friends."

"Sounds okay."

"I think he worries a bit about Carl," said Frank. "Stan wants him to join the military as an officer when he's old enough but the boy doesn't seem very keen. And back home Carl's been on a few anti-war demos, which Stan's not too happy about. I told him it's just the rebelliousness of youth and he'll grow out of it."

"He seemed okay to me," offered Rod. "Presentable appearance, good manners."

"Anyway," said Frank, "if the Vietnam war continues Carl will probably be called up whether he likes it or not."

"Poor sod!"

"He'll be fine, I told Stan. It's not the end of the world that he doesn't want to follow his Dad into the Air Force. Maybe he'll change his mind if he gets drafted. It's a generational thing, I think. Neither of my two boys are that way inclined either. They're both pleased National Service has finished."

"Bit of a Curate's Egg, National Service," said Rod. "A lot of people reckoned it was just a chance for psychopathic corporals to bully callow youths but I got into aviation through National Service. His Majesty's Government were kind enough to train me to fly his aeroplanes and pay me for doing it."

"It's all different now," said the flight engineer ruefully. "The idea of fighting for your country seems sort of old fashioned and unsavoury."

"All you need is love!" laughed Rod. "That's what we hear these days."

Frank smiled back. "Maybe the kids are right after all. God knows there's been enough killing the last fifty years or so. Anyway, Stan's an easygoing sort. He won't push Carl where he doesn't want to go."

"So he met Nigel Nixon's sister. Small world, aviation."

"You're right. I'm not surprised he fancied her. But she was a lovely lady, it has to be said. I bet there weren't many males who met her who didn't fancy her, me included. I wonder what she's like now—I haven't met her in years. It's funny, you know. Nigel always talks about her affectionately but when they were younger there was some sort of bust up for a while."

"What do you mean?"

The flight engineer took a swig of Michelob, apparently gathering his thoughts. "It might have been a family thing. I remember Nigel telling me about their parents splitting up around the time Maddie was born. But I'm thinking of White Waltham."

"What, the ATA?" queried Rod.

"Yes," said Frank with a nod. "Nigel used to bring a Tiger Moth into Waltham when he was on ground duties between tours on the Lanc. This aircraft had a retractable hood fitted over the rear seat so people could practise instrument flying. Not many of the ATA guys and gals had had formal IF training and we were losing them in bad weather so quite a few of them took up Nigel's offer of stick time under the hood."

"Did it work?"

"Certainly—later on it became compulsory. In fact Maddie and one or two others took on the instructing role. Useful after Nigel started his second Lanc tour, which meant he couldn't get to Waltham any more."

"So what was the upset between them?"

"It was when Nigel was training Maddie to instruct IF. They were doing night circuits. I was at Waltham on a temporary posting training ATA bods as flight engineers for ferry flights. Anyway, they finished the detail okay and we were expecting them to join us in the bar for a night cap but they didn't show up. No one thought anything about it but next morning at breakfast they were at loggerheads, not speaking to each other or anybody else."

"How strange."

"Quite, Rodders. Anyone asking Nige why they were so grumpy was told to mind their own business. Maddie wouldn't say anything at all so we left them to it. After a few days the disagreement seemed to be over and they both apologised for their rudeness and bought everyone a round of drinks."

"So did anyone find out what had happened?"

"Nope. There were rumours that it was an argument about instruction techniques. After that there was a sort of . . . coolness between them, which was a bit out of character for both of them."

"What about now, Frank?"

"Like I said, all systems normal these days, or so it seems. They meet each other quite frequently and go together to visit Nigel's Mum in her care home although I think I heard somewhere she's had a stroke and she's in hospital now."

"She sounds quite a character, Nigel's sister."

"You'd like her if you met her. Nigel told me she's still unattached. I'm surprised no bloke has snapped her up. Last time I saw her she still had her looks and her charm."

"How old is she?"

The flight engineer furrowed his brow. "Let's see . . . she'll be two years younger than me and Nige. That'll make her forty-five or thereabouts."

"Past her prime, maybe."

"Can't imagine that, Rodders."

CHAPTER 14

Among the fifty-three passengers waiting for their cases to appear on the baggage carousel at Adams Airport, Little Rock, was Professor Madeleine Maunsell. Standing beside her was Mike Snow, who had met her when she came through Arrivals. More specifically, she had introduced herself to him after spotting him holding up a placard on which her name was printed. Mike was the thirty year old pilot of the Jet Ranger helicopter sent by Southern National Generation to pick her up and ferry her to the industrial area of Strumble, a nondescript town lying twenty odd miles to the northeast of Little Rock. To Madeleine the helicopter seemed a terrible extravagance when a car could make the journey in half an hour. She was obviously not yet fully accustomed to American ways.

It had taken Madeleine the best part of two days to travel to the heart of Arkansas. The previous day a Pan Am 707 had carried her in first class comfort from London to New York, where Southern National had paid for her overnight stay in the luxurious Paramount Hotel in Times Square, accommodation noticeably superior to that provided for her during previous visits to the city for scientific conventions.

Today's travel arrangements were not quite so expeditious. A Pan Am 727 took her to Washington, where she changed to a Braniff DC9 which had brought her via a brief intermediate stop at Nashville to the Arkansas state capital.

Mike carried her suitcase to the helipad and soon the machine was airborne and tilting its rotor towards Strumble, following Highway 67. Included among the documents in the professor's briefcase on the seat behind her was a copy of the map Maxwell Murdoch had previously sent her via Oxford University's newfangled telephone facsimile device, entitled 'SNG Generating Plant, Strumble, AR'. The plan view showed the layout of the various components of the coal-fired power station, of which the dominant feature was the six large cooling towers. A dotted outline extending from the main plant encompassed a blank area labelled only 'Extension'. In a phone conversation with Madeleine prior to her departure from England Mr Murdoch had mentioned his plans for the 'extension' without giving away too much detail. No doubt she would find out more when she met him.

Although a thin overcast filtered out direct November sunlight the visibility beneath the cloud was good and soon after the sprawl of Little Rock had drifted behind the Jet Ranger Madeleine spotted the steam clouds lazily billowing from the cooling towers poking up from the horizon ahead of them. As they descended towards the landing pad at destination Madeleine commented on the mile-long coal train trundling towards the power station on tracks converging from their right side.

"Me silly jet," is what Madeleine thought the pilot said in response.

"Pardon?"

"Miss Illy Jet," said Mike, enunciating more clearly for the benefit of the English lady who spoke a different version of the language of America. "It's short for 'Mississippi Illinois Railroad'. The coal train is the 'jet'."

"So does the 'jet' name derive from the cargo?"

"Excuse me?"

"Jet . . . as in black."

Over the intercom Madeleine heard Mike chuckle.

"Hey, I never thought of that! I assumed 'jet' was a—whaddya call it?—ironical name 'cos they're so slow."

At the head of the train crawling along the tracks could be seen three big diesel locomotives. Madeleine vaguely wondered whether any British freight trains needed the power of three engines to drag them along. Not for the first time she marvelled at the increased scale of all things American compared to their European counterparts.

"That load probably originated in Texas," continued Mike, manoeuvring the helicopter as they dropped lower to stay clear of the cooling towers and the electricity pylons radiating from the power station, "but a lot comes in from the north, from Wyoming." He chuckled again. "The joke about the Miss Illy is that their trains run slower than the rivers it's named for."

As the Jet Ranger settled on its skids and Mike lowered his collective to kill the rotor lift Madeleine noted three people waiting to greet them, hair and clothes rippling in the helicopter's downwash. The central figure was obviously her host, taller and more corpulent than the other two, a man and a woman standing either side.

Madeleine stepped out of the aircraft and walked forward to the trio, holding out her free hand, which Mr Murdoch grabbed and shook vigorously. Wearing a beige suit, he was sporting a moustache and goatee beard matching his mane of white hair, reminding the professor of the photos Verity had dug up when they were researching his background.

"Hey, Madeleine, great to meet with you at last. Welcome to Strumble. How was the flight?"

"Okay, thanks."

"Swell . . . let me introduce you to the guys who run things around here." He pointed to the suited middle-aged man on his right. "Dave Montgomery, my personal assistant. He knows all my secrets . . . well, nearly all of them!" The hand then swung leftward towards the woman, who looked younger, also smartly dressed. "Nedra Springfield, General Manager. She's the boss here at Strumble."

Mike Snow had completed his shut-down checks and approached the group, carrying Madeleine's suitcase.

"Am I to take this to your office, sir?" he asked.

Mr Murdoch turned to Madeleine. "Anything you need from the case right now?" he asked.

"No, thank you."

"Okay," he said to the pilot, "get my chauffeur to take it straight to the hotel, would ya?"

Ensconced round a table in Nedra's office ten minutes later the foursome got down to business.

"You happy with the itinerary, Madeleine?" asked Mr Murdoch. "Anything you want to change?"

The arrangement was for the chemistry professor to stay at Strumble for two nights, then three nights in Wilmington, then home via Norfolk, Virginia. At Strumble she would be briefed on the power plant and the mysterious "extension" and at Wilmington she would visit Southern National's headquarters and discuss additions to the research facility.

"Right," said Mr Murdoch. "First things first. What we talk about here is strictly confidential. All got that?"

Three heads nodded concurrence.

"You want minutes, sir?" asked Dave.

"Yes, but no one gets to access them apart from us four. Lock 'em up when we're done."

"Got it."

"Okay, the big picture. The task is to remove as much carbon dioxide as we can as soon as it's generated by the combustors before it gets into the atmosphere. We react it with methane which we get from natural deposits and turn them into hydrogen and carbon monoxide, which can then be used as fuel or for other purposes. The big problem is you have to heat these gases to thirteen hundred degrees to get them to react, even when you're using a catalyst."

Madeleine was about to correct him, to tell him the required temperatures were much lower. But then she remembered he was quoting Fahrenheit figures in the American way.

"If we generate the temperature by burning fossil fuels then the whole exercise is pointless," continued Mr Murdoch. "We'd be pumping out more CO2 than we were removing."

The CEO of Southern National paused, his eyes scanning his small audience.

"So . . . we heat things up with nuclear power."

There was little reaction to the words. Madeleine had already guessed what was going to be proposed from the hints Mr Murdoch had previously dropped and no doubt Dave and Nedra had been pre-briefed.

"Here's the deal," continued Mr Murdoch. "We build a small nuclear reactor here as an extension to the main generating plant. My guys at Wilmington think we could use the sort of reactor Westinghouse are putting in submarines. Maybe we could even buy one from them but we might have to clear it with the Department of Defense. If they're not happy we'll build one ourselves, a baby version of the reactors currently in use at power stations."

Another pause. "Any questions so far?"

"What catalyst will you use?" asked Madeleine.

Max Murdoch grinned. "What do you suggest, Professor?"

"We've tried various things," replied Madeleine cagily. After agreeing to accept Mr Murdoch's invitation to meet him in the States she had discussed confidentiality with John Davis, who had reminded her that part of the Science Research Council's funding contract specified that no research findings could be released into the public domain without the Council's approval. There were two reasons for pursuing this policy. Firstly, there might be military applications which could take advantage of new discoveries or advances in technology and secondly the Government rightly expected a return on investment if research led to successful commercial developments. In the case of reformation of carbon dioxide by methane, no catalyst had as yet outperformed the tungsten carbide-palladium hybrid her post grads had invented. The big problem was that on an industrial scale the quantity of catalyst required would be prohibitively expensive.

John Davis had given Madeleine discretion as to how much she should reveal. They knew that if too much data was withheld the American might abandon the whole project, which would signify its death as no other definite sponsors had been found to date and the Review Board had already confirmed that no new Treasury funding would be available after mid-January. Several meetings between interested parties, some less than friendly, were required to hammer out the decision that the seventeenth day of the month would see the flow of cash cut off.

"Give me a clue," smiled Mr Murdoch now.

"On an industrial scale it would probably have to be nickel," said Madeleine, knowing that this offering was hardly a revelation and that the American would already know about it. "On cost grounds there would be no alternative."

Mr Murdoch nodded. "Yeah, that's what my d-block guys say."

The professor furrowed her brow. "D-block?" Vague notions of prisoners conducting research for Southern National filtered into her mind.

"Transition metals, honey," grinned the American. "Hell, you're supposed to know about these things, you being a professor and all. Didn't you win prizes for that stuff? That's what I was told."

Madeleine smiled. Another Americanism. They certainly had a way with words, condensing convoluted phrases into pithy abbreviations. So the group of researchers investigating the properties of elements whose highest energy electrons occupied third level d-orbitals became "d-block guys".

"What about photon flux?" asked the professor. "Will electromagnetic radiation be included in the process?"

"I was going to check that with you," responded the CEO. "The Wilmington guys tell me that although the reactor will supply all the energy we need they think it's worthwhile looking into whether we can reduce the energy requirement using electromag and pressure variations so we can minimise the reactor size. I've told them the target is to find the best combination of inputs compatible with minimum cost and complexity."

"Seems sensible," said Madeleine. "We found that photons in the UV energy range boosted yield slightly so perhaps more work could be done on that." She looked down at her notes. "Would you go for steady state or alternating periodic flow?"

"Wilmington says steady state would probably give the best results overall. Is that what you Oxford people found?"

"Pretty much," nodded Madeleine, "depending on circumstances." She tapped her pen on her notes. "What volume of CO_2 are you intending to convert? How much do you emit here?"

Mr Murdoch raised his eyebrows and turned to the General Manager. "You got the figures, Nedra?"

Nedra flicked through the papers in front of her on the table. "Well, we're burning just under three million tons of coal each year. What's the conversion factor? I'm sure somebody told me once."

"Depends on the carbon content," said Madeleine. "The atomic weight of carbon is twelve and the molecular weight of CO_2 is forty-four. So if you're talking about ninety percent carbon content, burning a ton of coal will generate ninety percent of forty-four divided by twelve . . . say just over three tons of CO_2."

"It's a lot, isn't it?" said Mr Murdoch. "So here at Strumble we're belching out nearly ten million tons of the stuff every year. When you think how many power stations there are in the world . . ."

"Yes, and don't forget transport, too," added Madeleine. "All those internal combustion engines . . ."

"Yeah, infernal combustion engines, some people say," laughed Mr Murdoch. "But oil's not as dirty as coal, is it?"

"No, because you're also getting energy from the oxidation of the hydrogen component, but most of the weight of oil is carbon so the ratio is still roughly three to one."

"And nobody cares about it."

"Which is why I'm having trouble funding the reformation project."

"But we care!" said Mr Murdoch, raising his voice and banging his fist on the table for emphasis. "We care because the Lord told us to care. He told us to clean up the planet he created."

There was no response to the CEO's outburst. Madeleine formed the impression that neither the General Manager of the power plant nor Mr Murdoch's personnel assistant shared their boss's divine inspiration, both looking ahead with polite expressionless faces. The chemistry professor found her thoughts straying to the Righteous Warriors of America, the organisation whose mission was to warn the human race of the consequences of its sins, especially communism and other ungodly follies. There was plenty of fire and brimstone enlivening the organisation's gospel. Judging by what she had read, Maxwell Murdoch seemed to be among the more moderate of the members. She cleared her throat.

"We were talking about conversion ratios, Max. From the plant's consumption of coal we should be able to work out the rate of CO_2 production." She lifted the flap of her brief case. "I've got a slide rule here."

Nedra turned a page in her notes and gave an embarrassed smile. "Sorry, folks, I should have realised. We've got that figure here. Wilmington says we emit approximately three hundred cubic feet of CO_2 per minute when we're working at normal capacity."

"So, is the intention to try to convert as much as possible?" asked the chemistry professor.

Mr Murdoch looked at his General Manager. "Nedra?"

"Wilmington are floating the idea that we run all the exhaust gases through the catalyst chamber, adding methane in varying quantities to see what happens and finding the best setting. Maybe we could build in a recycling system to increase the exposure to the catalyst."

"How big would the catalyst chamber be?" asked Madeleine. "How much nickel would you introduce?"

Mr Murdoch grinned. "That's for you to tell us, honey!"

The exchange modulated into a general discussion of technical details, many of which would require Madeleine to consult the research personnel at Wilmington. At one point she asked the others whether the local populace were happy with Southern National setting up a nuclear reactor in their backyard. There was a momentary silence.

"We haven't told them yet," said Mr Murdoch finally.

"Oh," said Madeleine surprised.

Dave Montgomery spoke for the first time. "We don't feel the need to make a statement until the plans have been finalised," he said. "Then it will be for Mr Murdoch to decide the wording of that statement."

"We've got nothing to hide," said the CEO. "There are nuclear power stations with much bigger reactors all over the place. So we'll probably stress the boost to the local economy while the new installation is being built and the pride that will accrue from being a leader in this activity. And of course if it's the will of God that we have a nuclear reactor here, then so be it."

"Right," said Madeleine.

"Plus the mayor is a very good friend of mine. My dollars helped put him in office so we can rely on him to speak out loud and clear when the time comes."

"Excuse me, sir," said Dave. "You asked me to remind you about patents."

"Oh yes, good point, thank you." The CEO turned to the professor. "The legal factors on any patents we file depend on you, Madeleine. We have to determine in what capacity you're working with us. Have you decided whether you're joining us at Wilmington?"

Madeleine shook her head. "I'm sorry, Max, but I still don't know. For me the answer is—probably—if you still want me—"

"I want you!"

"Thank you. But my post grads I'm not so sure about. One or two might join me if I come over."

"Swell. That would be great. But you're the main prize. You'd be running the CO2 research unit at Wilmington. You could talk to our legal guys about your percentage of the ownership of any patents and the revenue deriving from them. On top of your salary, of course."

"Suppose I don't come to Wilmington but you still want me to do research for you back home."

"Either we write a one-off cheque paying you as a consultant or we adjust your revenue stream."

Madeleine smiled. "I won't keep you hanging on too long," she said. "You've been very patient and I promise to let you know within a few weeks."

"Great."

The CEO dismissed his assistant and the General Manager, telling them he had a couple of personal items to discuss with their guest.

"What do you think of our set-up here?" he asked when the others had left.

"Impressive."

"You'll like Wilmington, too. You'd be a great asset if you do decide to come on board. Plus our PR guys would love a beautiful, brainy English lady scientist in the team. I can just see your picture on the front page of our house magazine."

"Uh-huh."

"Okay. I'll get Dave to get the chauffeur to take you to the hotel and you can relax for a while."

"Thank you."

"Swell. I've booked a table for dinner this evening at the Pimiento. It's the best eatery in Strumble . . ." He lowered his voice conspiratorially, ". . . which isn't saying much. They'll be five of us. You, me, my dear wife Clarissa and Dave and Nedra. We'll pick you up at seven thirty."

"Sounds good."

"Oh, I almost forgot . . . you'll be meeting with Gordon McKenzie at Wilmington. He sends you his love."

It took less than a second for Madeleine's memory to retrieve the name and scan the highlights of their erstwhile relationship. "That'll be nice," she said neutrally. "I haven't seen him in years. What does he do at Wilmington?"

"Head of catalyst research."

"Ah, yes."

"I poached him from Harvard," chuckled Max. "That's where you met him, wasn't it?"

"Yes."

"He's a great admirer of yours."

"The feeling is mutual."
"Swell."

CHAPTER 15

The November sky was crisp and clear. The Piper Cherokee was cruising north-eastwards in the bright sunlight over Henley, indicated speed 120 miles per hour. The three ex-ATA pilots on board preferred to think in knots, as they had been trained to do a quarter century earlier. The true airspeed column on Parminder Collins' navigation log was accordingly annotated '105'.

Parminder and Prudence Dismore-Whyte occupied the front two seats, Parminder on the left side. Behind her sat Madeleine Maunsell, with the vacant fourth seat on her right piled with their bags. Madeleine had flown the outbound leg from Booker to Compton Abbas, the little grass airfield sitting on top of a hill four miles south of the Dorset town of Shaftesbury, so it was Parminder's turn to do the return leg. All three were wearing headsets but Madeleine was only half listening to the intercom exchanges between the front seat pilots and their occasional radio calls. She leant over to watch the bend in the River Thames slide past beneath them on the right side.

"Wycombe," transmitted Parminder, "Golf Alpha Victor India Charlie, over Henley two thousand feet for joining."

"Roger, India Charlie" came the voice in their earphones. *"Call downwind for runway two five, right hand pattern, QFE one zero zero eight, watch out for gliders and tugs."*

On the intercom Parminder announced that she would turn left onto a northerly heading to take them west of the airfield for a join on the downwind leg.

"Sounds like a good plan," said Dizzy.

During the return flight Madeleine had been content to relax in the back seat, not paying much attention to what the other two were up to although her instinctive pilot training kept her eyes scanning outside for other aerial traffic that might conflict with their flight path. Under Visual Flight Rules it was up to them to see and be seen. The fine weather had brought out a few other like-minded pilots although, it being a Saturday, there was no military traffic to worry about.

As the chequered green and brown English countryside had floated past underneath the Cherokee Madeleine had been reviewing the latest developments in the Methane Reformation project. Not to mention episodes unrelated to the chemistry of organic gases.

Gordon McKenzie. Well . . . that was certainly not expected.

After their stay in Strumble Madeleine and the Murdochs had flown to Wilmington in Southern National's brand new Gulfstream bizjet and she had marvelled at the glamorous turn her life had taken, such a contrast to the drab mundanities of English academia. No doubt the ulterior motive for Max's lavish generosity was to persuade her that her future lay in the good old USA.

At Wilmington Madeleine found herself impressed with Southern National's research facilities, especially the catalyst lab, where she thought the deference of the staff was delightful but a little overwhelming. Most of the experimental work revolved round investigating methods for cleaning up power plant emissions by catalytic processes. At the impromptu lecture she gave on her methane project her audience listened in rapt admiration, with applause and cheers at the end of it. That never happened at Oxford!

Gordon McKenzie was as she remembered him and the details of his life had not changed much, she discovered after having cautiously accepted his invitation to dinner at his sea front house for her second night in Wilmington. He was still unmarried because he didn't want family obligations. His hobby was racing motorbikes, he reminded her, and a grieving wife and heart-broken kids should he mess up would not be a nice thing to think about. Yes, he liked the company of women and no, he was not currently in a serious relationship.

Madeleine couldn't remember which one of them had steered the conversation towards their romantic fling at Harvard during a science convention several years previously. She had surprised herself by not immediately declining his invitation to stay the night "for old times sake", which Gordon understandably interpreted as an encouragement to take things further.

As they started to kiss after post prandial cocktails delicious bedroom memories stirred for Madeleine. Gordon on top of her. Gordon underneath her. Repeat performances of same, with variations. All at once Toni's comment sprang to mind. What was it her lodger had wanted during her trip to America? Someone to shag her rotten, wasn't it? Was that where Madeleine herself was now headed?

No, it wasn't! Madeleine had suddenly decided before their intimacy passed the point of no return. Still breathless from energetic kissing she pushed him away.

"I'm not ready for this yet, Gordon," she said, voicing her genuine emotion. "I'm sorry for leading you on."

Gordon had not complained. On the contrary, he had stroked her cheek gently and told her that if she ever found herself ready, he would be willing and able to supply her needs.

Madeleine had wobbled for long seconds, wondering if she was doing the right thing, turning down the chance for sexual excitement after so long without.

Gordon had obviously sensed her confusion. "Don't do anything you may regret tomorrow," he had counselled her, helping her back onto safer ground.

"Thank you for understanding," Madeleine had responded. "I'd better get a taxi back to the Murdochs' place before I change my mind again."

Why had she turned Gordon down? That was the question which ran through her mind during the flight home to England. She didn't doubt his assertion that he would not be cheating on a partner if they had gone to bed.

There was the John Davis factor, of course. Or lack of John Davis. It was plain that John wanted to take things further in their relationship and she had told him there was a good chance she would eventually succumb to his advances but asked him not to rush her. But just as she had decided that yes, she was ready to give in, John's bosses had maliciously sent him away to Rio de Janeiro on a four month posting before she had a chance to let him seduce her.

So, was that why she had rejected Gordon? A sense of loyalty to a man who as yet knew nothing of her plans to allow their relationship to develop into something physical? And who was now happily living his life thousands of miles away. Who knew what romantic enticements would make themselves available to John in Brazil?

Was there a reason for her reluctance to embrace male physical intimacy that she was not admitting to herself? Perhaps she was subconsciously suppressing fears that she might not be capable of enjoying the act of sex. She'd had precious little practice in recent years! Perhaps as a forty-five year old woman she should give up the crazy notion of seeking carnal pleasures.

Did absence of love come into the equation? She had not shared a bed with someone she loved since . . . well, since the ATA days during the war. Perhaps it was mere old-fashioned British primness that had kept her pure for most of the intervening years. Good girls do not sleep with men they do not love. Or who they are not married to.

Back home, unusually for her, Madeleine found it hard to decide where her future lay. Relocation to Wilmington would bring material benefits. And better research facilities. And the company of Gordon McKenzie in the lab and—possibly—in his bed. Even if she wasn't in love with him or wearing a wedding ring.

But she would miss her English friends and her brother. And her work colleagues. And the All Souls choir. And the bridge club. And flights with Dizzy and Pam in little aeroplanes. For all its drabness, there was a comfortable cosiness about English academic life which couldn't be defined and which contrasted favourably with the somewhat garish and superficial flavour of the American way. Also in the credit column was the potential offer from ICI of running a new project, looking at improving the physical characteristics of aromatic polyamide fibres by new methods of polymerisation. If it went ahead, the research would be funded by the chemical manufacturing giant, relieving her of the burden of running cap-in-hand to the Science Research Council whenever funds ran short.

And if she stayed in England, a returning John Davis might re-enter her life. And enter her body . . .

Lascivious ponderings aside, the position of the post grads was not much clearer, apart from James, who would not leave England while girlfriend Stephanie was working in London. He told Madeleine he had applied for a post in a research project at Cambridge University looking at inventing new fabric dyes and had been informed the chances of acceptance were very good.

Xui Ying would almost certainly go back to Hong Kong when the methane reformation project was cancelled, she had told Madeleine. Her father had used his contacts to find her a position at the University, mainly tutoring undergraduates but with the chance of research into food preservative technology.

Joe had said he would be happy to go to Wilmington with the professor or even without her. Before she left to come home, Maxwell Murdoch had confirmed to Madeleine that any or all of her protégés could relocate to Wilmington if they wanted to, even if she herself declined. Joe also told the professor that he wouldn't mind joining her on the ICI polymerisation research if that option was available.

Luke's decision depended on Gillian, his girlfriend. If she would go with him to Wilmington then that's what he would do. If she couldn't be persuaded then he would have to look for something new. He didn't want to pressurise Gill, he had told Madeleine, in case she later blamed him if it didn't work out for them. Gill had told Luke she would decide before year's end. Luke had told the professor the chances of her agreeing to Wilmington were about fifty-fifty in his view.

"India Charlie, downwind," heard Madeleine in her headphones as Parminder transmitted their position in the Wycombe circuit.

"India Charlie, call finals," came the reply. *"One ahead on base."*

Instinctively all six eyes looked ahead to confirm separation from the preceding traffic, a Cessna 150 banking right to line up with the runway.

Another benefit of Wilmington, Madeleine knew, would be cheaper flying. Dizzy was delighted when the chemistry professor had informed her she was going to renew her Private Pilot's Licence. Madeleine decided to split her training between the Cherokee for general handling and navigation exercises and the Chipmunk—more expensive—for aerobatics, although proficiency in these manoeuvres was not a mandatory requirement for the licence. Research had shown that hourly costs in the States were only half of the British rates. She could get an American PPL and then she and Gordon McKenzie could go off on romantic trips together . . .

"India Charlie, finals," transmitted Parminder.

"India Charlie, cleared to land, wind calm."

The six eyes now looked ahead, the three brains assessing approach path for descent gradient and centreline tracking. Parminder pulled the lever between the front seats fully up to lower the flaps to the landing position. Three posteriors felt the slight retardation and momentary lift increase. Parminder's right hand briefly let go of the throttle and reached up to wind the trim handle in the roof a turn or two to compensate for the change in pitch trim.

In contrast to the sometimes clumsy handling of control wheel and throttle by some of Dizzy's inexperienced students, her old ATA friend guided the Cherokee towards the runway as if it was on rails, with only tiny adjustments of engine power and flight control inputs needed.

They floated over the newly constructed section of the M40 motorway just short of the runway threshold and Parminder smoothly closed the throttle and lifted the nose, holding the little aircraft a few inches off the surface until it was ready for mainwheel touchdown with a squeak of tyres on tarmac. Then the nosewheel dropped onto the runway and the Cherokee was a creature of the ground again.

"Beautiful," said Dizzy as Parminder applied the wheel brakes and ruddered the aircraft off the runway.

"Agreed," offered Madeleine from the back seat.

"Sloppy speed control on finals," muttered Parminder.

"You mean the horrendous two knot drop before you corrected?" queried Dizzy, tongue in cheek. "Yes, that was so pathetic. I thought I was going to have to take control!"

"I was getting airsick in the back," added Madeleine, matching the instructor's tone. "The plane was all over the place!"

The banter continued as Parminder taxied the Cherokee to its parking stand. But then Dizzy pointed out of the window towards the hangars.

"Hello," she said, puzzled. "What's going on over there?"

CHAPTER   16

"It's a Spitfire," said Madeleine.

"It's the one Chris Coleman is bringing here," said Dizzy. "But I thought that was scheduled for next week."

"Chris Coleman?" queried Madeleine. "The chap we knew in the war?"

"Yes, he found it derelict somewhere. He's just had it rebuilt and now it's flying again."

On the apron there was quite a crowd milling around the fighter. Dizzy told Parminder to park the Cherokee on a remote stand to avoid the potential hazard of an inattentive person walking into their propeller.

The three ATA friends shut down their aircraft and strolled over to the crowd to see what was happening. Despite the intervening years Madeleine easily recognised their wartime colleague. His curly hair was thinner and his figure rounder but the impish face had not changed. Chris was standing by the engine cowling, one hand resting on the leading edge of the Spitfire's slender wing. Pointing at him were several cameras, including a bulky movie camera bearing the logo "BBC Television News". Judging from the notes they were jotting in their notebooks, several of those in attendance were reporters.

Madeleine and her friends edged closer and were able to pick up the gist of the interview.

" . . . a few miles west of Nevers."

"On a farm, did you say?"

"That's right. The chap flying it force landed after the engine got hit by a Jerry 190. The farmer and his friends pushed the aircraft into a barn and covered it with a tarpaulin and hid it behind other farm machinery because they didn't want the Jerries to find it."

"What happened to the pilot?"

"The Resistance tried to get him out of the country but he got captured and sent to a POW camp. Sadly he died before the camp was liberated."

"Was he someone you knew?"

"I knew him slightly. His squadron was based at Biggin."

"So did the farmer tell the British their Spitfire was still in the barn?"

Chris screwed up his eyes. "It seems he didn't bother. Don't forget, when the war ended there were aircraft wrecks all over the place and a damaged Spitfire in a barn probably didn't seem worth bothering about. So it sat there gathering dust for twenty years."

"So how did it get discovered again?"

"Chap on holiday with his wife, touring the area. A local mentioned the Spit and they found it in the barn. The tourist offered to buy it from the farm owners and arranged to have it brought back."

"How did they transport it?"

"It was derigged and packed in crates over there and brought back by road. They took the crates to Duxford. When I heard about it I bought the aircraft from the new owner and set about restoring it."

"How long did it take?"

"Just over three years. It was in pretty good shape, apart from the engine. That took a lot of sorting out."

"So it's still got its original engine?"

Chris gently patted the Merlin's cowling. "About seventy to eighty percent, I should say. We had to cannibalise another couple of defunct engines to get this one going again."

"How much did the restoration cost?"

The pilot grinned. "A lot."

"How much is the plane worth now?"

"A lot more."

The questioning continued and Chris fielded all enquiries with courtesy, even those edging into trivia and his personal life. Answering one reporter, his glance caught Prudence standing at the edge of the crowd. His smile broadened.

"Dizzy! Nice to see you. They told me you were flying today."

Then Chris saw Prudence's two companions and his jaw dropped.

"Good God! Pam Collins! Maddie Maunsell! What the hell are you doing here?"

"Same as you," smiled Parminder. "Except our plane's smaller."

"We took a Cherokee to Compton Abbas," added Madeleine.

"Cherokee?" grinned Chris. "Surely you're not flying them? They've got control wheels instead of sticks and their tailwheels are at the wrong end of the fuselage! Sacrilege!" Noting the puzzled looks on the faces of the assembled reporters and onlookers, he swept his hand towards the ATA pilots.

"Ladies and Gentlemen," he announced, "let me introduce you to the Three Witches of White Waltham."

Dizzy opened her mouth to protest but Chris beat her to it.

"When I say 'witches', I mean three of the best pilots. Aeroplanes——not broomsticks—were their chosen means of transport."

The entourage now turned their attention to Dizzy and her friends and a new round of questioning began.

"What planes did you fly?"

"Did you ever crash?"

"What did the male pilots think?"

"Were your boyfriends pilots?"

"Did you ever get lost when you were flying?"

Chris Coleman sensed that his ex-colleagues were tiring of their interrogation as it descended into the banal and clichéd and announced that the press conference was over. The four pilots posed with forced smiles for the cameras and then Chris physically shepherded the ladies away, thanking the representatives of the fourth estate for their interest as they moved aside to let the objects of their attention pass through.

Out of earshot of the reporters Chris lowered his voice. "They're a pest but it's good publicity for the Spit. It'll bring punters in when we display it at airshows. Listen, are you three free for a coffee and a chat? I've got to put the bird away and there are a few other things to attend to but . . ."

"Me too," said Dizzy. "A bit of paperwork for today's trip but then I'm free."

Madeleine looked at her watch. "No rush for me."

"I'm okay till five," said Parminder. "Then back for GDDs."

"What's that?" asked Chris.

"General Domestic Duties," laughed Parminder.

Twenty minutes later the four were ensconced in the Nissan hut serving as the airfield's eatery. Under Chris's supervision his Spitfire had been pushed into a hangar by mechanics, sharing the space with the Cherokees of the Wycombe Flying Club.

The building in which the four ATA pilots were now seated styled itself the Wycombe Air Park Restaurant but to those who regularly refreshed themselves there the preferred epithet was "canteen". It was already dark outside and there were only two other customers sharing cakes and a pot of tea at another table.

For his companions, Chris summarised his life since the end of the war. He'd managed to find gainful employment in aviation for a while, he told them, including a spell on the Berlin Airlift.

"What equipment?" asked Dizzy.

"The dear old York," smiled Chris. "Ugly brute, but it could shift ten tons of stuff. Flew nicely, though, like the Lanc."

"Same airframe, wasn't it?" queried Madeleine.

"Pretty much. Lanc's wings and tail. Merlins too, of course."

"So you were a civvy pilot on the Airlift?"

"Yeah. Quite a few civvy airlines got in on the act. The military needed all the spare capacity they could get. That's how Freddie Laker got started."

"I remember him at ATA," said Parminder. "Bit of a wheeler dealer, I seem to recall."

"Got his own airline now, hasn't he?" asked Dizzy. "I'm sure I read about it in 'Flight' magazine. Ex-BOAC Britannias, as I recall."

Chris nodded. "Started last year."

"My brother used to fly those," said Madeleine.

"Nigel Nixon," said Chris. "I liked him. What's he up to now?"

"Captain on 707s."

"Nice."

"What did you do after the Airlift, Chris?" asked Dizzy.

"The company I flew for went bust and there were no other jobs going so I went into journalism." Chris jerked his thumb towards the door. "So I'm one of them, really. Though my stuff is more serious . . . well, that's what I hope."

"How did you get started?" asked Parminder.

"Actually, it was the Airlift. I was asked to write a column about it for one of the dailies. They liked it, so they said. So when Freightways went tits up I went back to the paper and asked if I could work for them as their aviation correspondent and they took me on."

"So is that what you do now?"

"Mainly. Plus freelance stuff for aviation and motoring magazines. More people are interested in cars than planes so the ratio is gradually increasing in favour of cars."

"What's the pay like?" asked Dizzy.

"Not bad. Enough to look after a wife and two kids and a dog."

"And buy a Spitfire!"

Chris laughed. "Sadly, no. But my dear old Mum fell off the perch a few years ago. She was a widow and she left us a fair pile of cash, so . . ."

"How much flying do you do?" asked Madeleine.

"Got a Chipmunk at Elstree," came the reply. "Probably do a hundred hours or so a year in that."

"Maddie's been flying the Chippie here," said Dizzy. "I finally managed to persuade her to renew her PPL."

"Great," said Chris. "Come and fly our aircraft any time you want." He turned to the Indian. "What about you, Pam?"

"I've been working on her," said Dizzy before Parminder could reply.

"I'm thinking about it," said Parminder. "You lot seem to be having a lot of fun."

A call came from the woman serving behind the counter. "Closing in ten minutes, folks."

The door opened and a middle aged man walked in, looked around and smiled when he spotted the pilots sitting at the table. It was the Chief Flying Instructor of the Wycombe Flying Club.

"You'll have to be quick, Mitch, I'm closing in a few mins."

"It's okay, Elsie, I'm not after a cup of your excellent tea. It's Dizzy I want to see. Actually, it's Madeleine."

Bob Mitchell came over to the ATA group.

"Hello, boys and girls."

"So I haven't filled in the tech log incorrectly then?" smiled Dizzy.

"Not that I'm aware of, Diz. No . . . there's a phone call for Madeleine. You can take it in my office if you want."

Back at the flying club the chemistry professor picked up the receiver in Mr Mitchell's office.

"Hello?"

"Mrs Maunsell?"

"Miss Maunsell here."

"Okay. This is Laura Wissett. I'm a friend of Jackie Holmbeck, Nigel's partner. I'm very sorry to disturb you but I'm afraid there's some bad news."

"Go on."

"Jackie's been in a car accident."

"Oh, no." Madeleine slowly lowered herself into the Chief Flying Instructor's chair. "Is it bad?"

"She's dead, I'm afraid. I'm so sorry."

"Oh, my God."

"She was crossing the road and got hit by a car. They took her to hospital but she was already dead."

Although reeling from shock, Madeleine's brain started to assess the situation. "Does Nigel know?"

"No, he's away on a trip according to Emma, that's Jackie's daughter. We couldn't find the airline number but we found your number in the address book and your lodger said you'd be at the flying club so that's why I phoned. Are you able to contact Nigel?"

"I can probably find out where he is from the BOAC Operations staff. I think I've got their number at home. Is Emma okay? Does she know when Nigel's coming back?"

"She thinks Monday . . . she's not sure. She's in shock."

"What about Emma's Dad? Does he know?"

"Yes, he's going to try to get down tomorrow. He lives in Barnsley."

"Right. What's going to happen to Emma tonight? Is there anything I can do to help?"

"I'm going to stay here at Jackie's with my daughter, who's a friend of Emma's. We'll be looking after her. Beyond that we haven't made plans yet. Perhaps Emma can stay with us in our house for a while."

Madeleine thought quickly, her mind clearer now. "Thank you for what you've done. I'll try to find out where Nigel is and whether he wants to come back tomorrow if it's possible."

"That would be very helpful. Thanks for that."

"Okay. And if Emma wants to stay at home I'll come down to Epsom myself and stay with her as long as she wants me too."

"Thank you. Sorry to have to pass on such bad news."

"You say you were a friend?"

"Our daughters are in the same class at school. That's how I got to know her. A lovely person."

"That's what Nigel was always saying."

CHAPTER 17

Among the dreaming spires of Oxford were more nondescript buildings, one of which was the town house headquarters of Emcro Film and Television Productions. As with many similar enterprises, Emcro's output varied from mundane documentaries to full-blooded dramas. A dispassionate observer would likewise rate their efforts as variable in quality. Occasionally a turkey would defeat quality control and arouse the opprobrium of the critics. On the other hand, the previous year's production of "Beale Street", a documentary exploring the history of Memphis Blues, had won several prestigious awards.

In a ground floor office, whose door bore the legend "Bertrand Stewart, Senior Production Manager", sat three people. Mr Stewart himself was at his desk, facing Junior Accountant Lionel Willkay and Assistant Producer Antionette Clark.

"The question is," said Mr Stewart, leaning back in his chair and tapping his chin with his biro, "why doesn't anyone else think it's important?"

"But Professor Maunsell and Maxwell Murdoch are big names," said Toni. "Their opinions should count for something."

"Well, the prof sounds like she knows what she's talking about but I'm not so sure about the Yank. Religious nutcase, judging by what you've told me."

"He's a very successful businessman," said Toni, "and he's got a research facility stuffed with eminent science academics. That's what Prof Maunsell told me, anyway."

"Okay, but what about the vested interest angle?" said Mr Stewart. "If these science bods are on his payroll they're not going to rock the boat by challenging what he wants, even if what he wants is no use to anyone."

Toni shrugged, unable to think of an answer.

"She's a friend of yours, isn't she?" continued the senior manager.

"She's my landlady."

"You're sure she's not trying to use us just to get cheap publicity for her pet project?"

"Absolutely not, Mr Stewart. It was my idea. She's very enthusiastic when she talks about it and I thought it would be good material for a documentary."

"Just remind me what she's doing with this carbon . . . whatever it is."

The assistant producer scratched her head. "Carbon dioxide. You get it when you burn oil or coal. They want to find a way of reducing the quantity emitted by power stations and such like by making it react with methane."

"Because it's poisonous, is that it?"

Toni shook her head. "No, it warms up the atmosphere."

"Is that so bad?"

Again Toni shrugged. "Prof Maunsell says it could be in the future. It could mess up the world's climate."

"How far in the future?"

"She says it depends how quickly it builds up."

"So . . . are we talking years, decades, centuries? . . ."

"I don't think it's imminent, from what I can gather. I don't understand the physics of it myself."

"And the people funding the project are shutting off the taps."

"Yes."

"But the mad American wants to do this stuff over there?"

"Yes."

"And he's a member of this weird political group, right?"

"Yes, the Righteous Warriors of America."

"What do we know about them?"

"They say they're doing God's will by fighting Communism. They think they've been specially chosen."

"Maybe that's not so extreme, Toni. None of us are fans of the Reds."

"True enough, but they include ridding the world of Socialism in their remit. Any form of State handouts is anathema to them. Survival of the fittest and all that. Their agenda is based on excerpts from the Bible. They're keen on capital punishment, upholding the ban on abortion, chemical castration for homosexuals, that sort of thing."

"Lots of people would have no objections to these policies."

"They think all religions apart from Christianity should be suppressed. Some people say they're anti-Semitic and racist. They openly say that negroes are inferior to whites."

"Not so good."

"Although Prof Maunsell says Maxwell Murdoch seems to be an affable character who probably wouldn't support the extremists."

Mr Stewart nodded. After a few moments' thought he turned to the accountant.

"What did I say about funding for Toni's idea, Lionel?"

The accountant leafed through some papers. "You put it on the 'strong maybe' list."

"Hmm." The manager pointed his pen at the assistant producer. "You said you wanted to do it as a two-parter, Toni. Is that correct?"

"Yes, with Mr Murdoch as the link. The first part would be the carbon dioxide story and part two the Righteous Warriors."

The pair explored the project in more detail for a few more minutes, all three occasionally jotting down notes. Toni confirmed that Bill Bray, one of Emcro's experienced producers, would consider taking charge if approval was forthcoming. At Mr Stewart's request she also gave him a rundown on Madeleine Maunsell's past and present life, which seemed to perk up the manager's interest. He sent Lionel off to rustle up coffee and biscuits then turned to Toni again.

"I think we can run with this," he said. "But I'm thinking of a different angle."

"What do you mean?"

"Running it more as a human interest story . . . we've got two colourful characters . . . the wealthy religious entrepreneur and the brainy lady who used to fly Spitfires, and they're both keen on this carbon . . . what did you call it again?"

"Carbon dioxide, sir."

"Yeah. Trouble is, that in itself is not very exciting, is it? You've told me you can't see it or smell it or taste it. That's why I'm saying go more for the human interest side."

"That would work, I think."

"I don't think there's enough material for a two-parter, Toni. Maybe we can save a bob or two running it as a single show."

"Okay."

"So we start off with some background, then we say what these two are up to now. Finish off with the carbon whatsit. Tell the viewers we don't know if it's dangerous—keep them worried. It'll make more of an impression if we do that."

"Okay."

"No hope of a romantic angle, is there?"

"Don't think so, Mr Stewart. Maxwell Murdoch seems to be happily married to his third wife and Madeleine . . . she hasn't got anyone special as far as I know."

"She's not a lesbian, is she?"

"No."

"Shame."

"Pardon?"

The manager laughed. "No, I didn't mean . . . hell, you know what I meant, Toni. The entrepreneur and the lady prof . . . would've added some spice."

"Madeleine's friendly with a chap from the Science Research Council—they've dated a few times—but I don't think they're romantically involved."

Mr Stewart sighed. "Okay . . . no sex, then. But I think there's enough colour to make it work . . . so someone would need to go to the States to grill this Murdoch character?"

"A team of four could do it . . . interviewer, cameraman and producer and assistant."

"Who did you have in mind?"

"No one in particular for the camera but perhaps Lizzie Bedford to ask the questions. Bill Bray would have the final decision if he takes the job."

"Lizzie . . . she did a good job on 'Flowers', didn't she?"

Emcro's documentary on the San Francisco hippie movement, "Flowers in Your Hair", had recently aired on national television and had gone down well with the viewers.

Lionel reappeared with three coffees on a tray. As he offered round a plate of biscuits the senior manager spoke to him.

"How did the budget go on 'Flowers', Lionel?"

The younger man wrinkled his brow. "As I recall we came in under budget, Mr Stewart."

"That doesn't happen too often!"

"No, sir, perhaps one in every three productions."

Mr Stewart turned to Toni. "You were first assistant on 'Flowers', weren't you?"

"Yes . . . I'd be happy to do it on the new project."

"Have you got a title yet?"

Toni frowned. "Haven't come up with anything snappy so far." She looked down at her notes. "'Righteous Climate Warriors' was one idea. Or 'The Professor and the Mogul'."

"They don't exactly slip off the tongue."

"I'm sure we'll think of something better, sir."

"Something with a bit more . . . zing . . . how about 'Carbon Time Bomb'?"

"Would people say we're scaremongering, though?"

"Good point. Okay . . . let's put that to one side. Are you thinking monochrome or colour?"

"Would the budget stretch to colour?"

"Well, as always, that depends on length. If we stick to one hour max we could probably run to colour. What do you think, Lionel?"

"If we keep the interview team down to four people then four thousand would probably cover it if there were no unforeseen expenditures. Can't be so definite when there's overseas travel involved. Too many variables. It would be film, I take it, not tape."

"Yeah, it'll be a while before we're colour taping," agreed Mr Stewart. "Plus the equipment's far to bulky to fly it over to the States and it would cost too much to hire it over there."

After another ten minutes or so the basic formula for the new as-yet-unnamed production had been thrashed out and Mr Stewart began to bring the meeting to a close. But as Toni was about to follow the accountant out of the door the senior manager called her back.

"How's it going on 'Lucy and Amy', Toni? Is it looking good?"

"Yes, pretty well. Early days, of course. We swapped it round to 'Amy and Lucy'. The marketing people liked the sound better. The script's just about finalised and we've done the initial casting. Shooting should begin within a week or two." The Emcro production was the drama set in Northumberland during the depression of the 1930s, following the contrasting fortunes of two families in the coalfield.

"You're the first assistant, aren't you?"

"Yes. Donny Mcleod is producing it."

Mr Stewart nodded. "Just had a thought, Toni. The Righteous Warriors . . . didn't we make a reference to them in the hippie thing?"

"Just in passing, sir. We didn't know anything about them until after we got back from filming. Maddie told me about them."

"Maddie?"

"Professor Maunsell. Madeleine."

"Of course . . . another thought. Have you worked out which questions you're going to hit Murdoch with?"

"Provisionally, yes."

"I'd like to see them before we launch."

"Of course, sir."

"Great." The senior manager moved to open the door for Toni but then stopped.

"How about you run the new effort . . . the Righteous Nutcases."

"As producer, you mean?"

"Uh-huh."

"Love to. Would Bill object? He's provisionally been pencilled in."

"I'll square it with Bill. He's got a few other things to keep him occupied. He won't mind . . . he thinks highly of you."

"Thank you."

"We'll take you off the coalfield thing so you're not overloaded. Plus, we'll pay you an extra emolument, of course. Think you can handle it?"

"Definitely."

"It's yours then."

CHAPTER 18

Madeleine and Toni had split the newspaper between them. Emma was reading a magazine for teenagers. All three sat at the breakfast table, the two adults in the toast phase and the child still munching cereal. The landlady was already dressed for work but the others were still in their dressing gowns. From a portable radio on the shelf came the sombre tones of a BBC news reader but no one was really paying attention.

Emma turned a page or two then glanced up at the shelf. "Can I put Radio 1 on, please, Madeleine?"

"I don't mind, if Toni doesn't," smiled the chemistry professor.

"Yeah, we don't want to be L seven, girls. Spin that dial, Emms!"

Madeleine swallowed the mouthful of toast and marmalade she was chewing. "And pray tell . . . what is L seven."

"Shall I tell her?" said the TV producer, looking at Emma, "or you?"

The girl smiled, unable to reply with her mouth full of cornflakes. She gave a "you do it" gesture and stood up to change the station on the radio.

Toni held up her two hands, each with thumb and forefinger splayed at right angles, one wrist rotated one hundred and eighty degrees so that the left thumb touched the tip of the right forefinger and vice versa.

Madeleine frowned, still none the wiser. "No . . . not with you."

"Look, this hand makes an 'L', this one a 'seven'."

"Yes . . ."

Toni laughed. "You're supposed to be the brainy one, Madds!"

"What shape is she making?" prompted Emma.

"A rectangle?"

The other two grinned and Toni rolled her eyes.

"Suppose a rectangle has all sides equal . . ." continued Emma.

"A square?" offered Madeleine, still confused. Then the penny dropped. "Oh, I see . . . let's not be square."

"She's got it! Hooray!" said Toni.

"Actually, we don't say 'square' any more," said Emma. "It's sort of an old-fashioned way of saying 'old-fashioned'."

"So only squares say 'square' now," deduced Madeleine.

"That's right. It would be uncool."

"So something that's up to date is cool, is that it?"

"Yes . . . or groovy . . . or happening."

"I'll make a note. My post grads will be most impressed when I drop 'groovy' into the conversation."

"That's a nice tune they're playing now," said Toni. "Who is it?"

"The Moody Blues," answered Emma. "Nights in White Satin."

"I like it."

"Are you doing homework today?" asked Madeleine.

The girl shook her head. "No, I've done most of the half term stuff. I'm catching the bus into Oxford to meet some friends."

"I can give you a lift if you're ready at half eight."

"Thanks, Maddie, but I'm not meeting them till eleven so the bus will be okay."

"People from the church?"

"Yes, from the Youth Club."

Madeleine switched the questioning. "What time are you going in to work, Toni?"

"Probably for lunch. I've got stuff I can do here this morning."

After the upheaval of her mother's untimely death, Emma Holmbeck had settled into a routine of sorts. When Nigel was home Jackie's daughter stayed with him in Epsom. When he was away on trips Madeleine took herself down to her brother's house to take his place. As a variation, sometimes after school had finished on Friday Emma caught the train to Oxford to spend the weekend with Madeleine, returning on Monday. The headmistress at her school gave her permission to report late as long as she caught up with the missed work. Recently—an encouraging sign of the healing process––the girl had said she was ready to spend nights alone in Epsom, with a list of friends she could call on in case of difficulties.

With Nigel flying during the February half term week Emma had agreed to come to Woodstock to stay. Madeleine was pleased that the girl had already made friends at the Sunday Youth Club run by All Souls' Church. All in all Emma seemed to be coping with her bereavement, as was Nigel, although BOAC had readily granted him compassionate leave immediately after his partner's death, agreeing that he should not be at the controls of a 707 until his frame of mind was restored.

Likewise, the University had been accommodating to Madeleine when she needed to take time off to look after Emma in Epsom, and the Remunerations Board had confirmed that there would be no deductions from her pay. Taking advantage of his natural leadership qualities, the professor had appointed Luke Stirling as her Methane Reformation Project deputy in her absence, after impressing upon him the responsibilities that went with the post, including punctuality and tactfulness when dealing with people who might challenge his new authority, such as Safety Officer Bill Syerman. Luke had assured her that his new car was completely reliable and his respect for Mr Syerman would never be in question.

All in all, Madeleine's life had entered calmer waters since the New Year after the calamity of Jackie Holmbeck's death and the uncertainty of the future of the MR project. After much heart searching she had decided to decline Maxwell Murdoch's offer to join Southern National Generation's research facility in Wilmington, the need to offer support to her brother and Emma in their bereavement finally tipping the balance. She renewed her offer to place the Oxford Organic Chemistry Gas lab at Mr Murdoch's disposal if they could agree an agenda and a mutually acceptable finance package.

Eventually a deal was struck between Southern National and the Science Research Council in which SN would pay the MR running costs in areas where the project coincided with the generating company's requirements and Madeleine's consultant fee would feed into the SRC's coffers, which in turn would resume funding the four post grad students in her MR team and meeting costs not covered by SN. There was a small hiccup when ownership of patents and percentages of potential royalties came into the equation but finally an acceptable contract was signed by all parties.

By coincidence, Toni's first Emcro task for 1968 was supervising the interview with Maxwell Murdoch as part of the programme she was producing which delved into the activities of the Righteous Warriors of America. After returning from Wilmington with her team she concurred with Madeleine that if you overlooked the religious zealotry the CEO was an impressive character, if somewhat eccentric. She also let it slip that she had found her way into Gordon McKenzie's bed and he had indeed "shagged her rotten". Madeleine's secret thought was that perhaps it was just as well that she herself would not now have to face the dilemma of whether she should make herself similarly available.

The Righteous Warriors programme was currently in the editing phase and Toni as producer had the final say when editorial decisions were necessary. The transcript of the latest cut now lay on the breakfast table and Madeleine's lodger's stated intention was to work on it after she was washed and dressed. As the professor picked up her car keys to set off for work, Toni thanked her yet again for originally bringing the Righteous Warriors to her attention and reminded her she would need to check her own inputs before the final cut went to air.

```
EMCRO PRODUCTIONS "THE PROFESSOR AND THE RIGHTEOUS WARRIOR"
                    (provisional title)

Version 5C

Producer: Antoinette Clark (AC)
Editor: Callum Lloyd (CL)
Interviewer: Elizabeth Bedford (EB)
Camera: Hugh Crosland (HC)
Assistant: Michael Watson (MW)

Contributor: Maxwell Murdoch (MM)
Contributor: Madeleine Maunsell (PROF)

EXT. AERIAL SHOT OF COOLING TOWERS, FERRYBRIDGE POWER STATION.
INTRODUCTION.

EB (voice over): Every year, power stations all over the world
burn millions of tons of coal. The heat of combustion turns
water to steam, which drives the turbines which generate the
electricity which lights and heats our houses and cooks our
food and powers our factories. The gases released into the
atmosphere by the power stations when the coal is burnt are
water vapour and carbon dioxide. Both are harmless substances.
But some people are concerned that increasing levels of carbon
dioxide may bring problems in the future. We spoke to
Professor Madeleine Maunsell of Oxford University.

INT. PROF MAUNSELL'S OFFICE, OXFORD. EB INTERVIEWING PROF
MAUNSELL.

EB: Can you tell us, Professor, what your concerns are?

PROF: When the sun's radiation hits the earth's surface, some
of it is reflected back into space. This stops the earth from
overheating. But if the reflected radiation hits molecules of
carbon dioxide it is absorbed, making them vibrate. That
energy is transferred to the surrounding atmospheric
molecules, making them warmer. Each year more carbon dioxide
is injected into the atmosphere by mankind's activities so the
atmosphere will become progressively warmer.

EB: Is that harmful?
```

PROF: At the moment, no. There are roughly 320 parts per million of CO_2 in the atmosphere--
EB: What does that mean?

PROF: In every million molecules, roughly 320 are molecules of CO_2.

EB: That doesn't sound like much.

PROF: No, it doesn't, but if the number increases it could have detrimental results.

EB: Such as . . .

PROF: It could affect the climate--it could cause droughts and flooding and increase the occurrence of extreme weather events, like hurricanes. Heat waves could start forest fires.

EB: How much would the CO_2 have to increase before the climate was affected like that?

PROF: Experts say that 400 parts per million would be the danger point.

EB: When will the CO_2 levels get to that figure, Professor?

PROF: At our current rate of production, eighty to one hundred years.

EB: No need to panic, then!

PROF: No, not yet!

EB: What should we do about it? Can we ignore it?

PROF: It's a controversial subject. Conventional wisdom is that alternative sources of energy will be developed before we get to the critical stage.

EB: What are the alternatives?

PROF: Hydro-electric, solar, wind, tidal--they're all renewable sources of power and they don't generate CO_2. Nuclear power is non-renewable and has the problem of disposal of radioactive waste but it doesn't generate CO_2, which is obviously helpful.

EB: We're already using some of these alternatives, aren't we? Like nuclear and hydro-electric?

PROF: Yes, but we need to introduce more renewables so the CO2 level can be stabilised. Then future generations won't have to worry about climate change.

EB: And your team's working on a method for doing that?

PROF: What we're doing is working out ways of removing some of the CO2 so it doesn't build up before the renewables come in. The remainder will be absorbed by the Earth's forests.

EB: How are you doing it?

PROF: We mix the CO2 with methane and heat the mixture to make hydrogen and carbon monoxide.

EB: So do these gases get released into the atmosphere instead of the CO2?

PROF: No, they're used in other applications . . . to make other chemical products or to use as fuel.

EB: Are there any disadvantages?

PROF: Yes . . . if you burn fossil fuels for the heating process then you're generating CO2, which defeats the object of the exercise.

EB: So what's the answer?

PROF: Using alternative energy sources to heat the gas mixture.

EB: You're working with an American company looking at using nuclear power for the heating process. Is that correct?

PROF: Yes.

EB: But didn't you just say that nuclear has its own drawbacks?

PROF: Yes, the same as nuclear power stations. Disposal of radioactive waste products and preventing radiation leaks.

EB: Do you think the nuclear option will work?

PROF: In theory, yes. We're doing research to confirm viability.

EXT. SOUTHERN NATIONAL HEADQUARTERS, WILMINGTON, NORTH CAROLINA, USA.

EB (voice over): Our team went to Wilmington, North Carolina to interview Maxwell Murdoch, Chief Operating Officer of the Southern National Generation Company.

INT. MAXWELL MURDOCH'S OFFICE. EB INTERVIEWING MAXWELL MURDOCH.

EB: Thank you for agreeing to talk to us, Mr Murdoch.

MM: It's my pleasure, ma'am. My friends call me Max.

EB: Can we confirm that Southern National is working on the same area of research as Professor Maunsell's team?

MM: That's correct.

EB: So you're also convinced about the need to stop CO2 building up in the atmosphere?

MM: Yes . . . we're pumping out billions of tons of this stuff into the air each year and we have to reduce that.

EB: Many experts say the danger level won't be reached for another eighty to a hundred years and by then we'll have different energy sources. So why go to all this trouble now? What about the cost and effort involved?

MM: It's the will of God that we look after the planet he gave us. We have to do God's bidding.

EB: Is this a doctrine of the Christian Church?

MM: Okay, the Bible doesn't mention carbon dioxide directly . . . that's true. But God created the world for our benefit so we can infer a responsibility for looking after it, which means preserving the atmosphere as it is and not polluting it. The same goes for the oceans and the continents.

EB: But most of mankind's industrial activities cause pollution of some sort, don't they?

PROF: We gotta get the balance right, Lizzie. The business of America is business, to quote President Coolidge. We all agree on that. But while we're doing it we gotta keep the planet as clean as we can.

EB: Southern National is one of America's biggest generating companies, isn't it?

MM: I'm proud to say that is so.

EB: So you produce a lot of CO_2?

MM: Yes . . . we're aware of that and that's why we want to cut it down by using this Reformation process that we're pioneering.

EB: Professor Maunsell agrees with you that nuclear power for heating the Reformation gases is a viable option, is that right?

MM: Yes, her team was looking at ways of reacting the methane and the CO_2 together but they couldn't come up with a source of heat that didn't produce CO_2 itself. So that's where we came in with the nuclear option.

EB: You're a member of the Righteous Warriors of America, aren't you, Max?

MM: Yes.

EB: Are they involved in this process?

MM: They agree with my team that we have to protect our planet.

EB (to camera): Later on we'll be talking to Mr Murdoch about the Righteous Warriors but in the meantime here's a presentation on the chemistry and physics of—

"Do you want a cup of tea, Toni?"

The producer looked up from her work to see Emma holding the kettle.

"Lovely."

The girl pointed to the sheaf of papers under Toni's hand. "Is that the programme with Madeleine in?"

"Yup. She'll be a TV star soon."

"Groovy."

Toni flicked through a few more pages of the transcript and came to the second interview session with Maxwell Murdoch.

EB: Some people are concerned that some members of the Righteous Warriors hold extremist views. What would you say to them?

MM: If you're talking about extremists, think of the Moaists in China. Look at the atheist regimes of Stalin and Hitler. If the forces of good had been more determined in fighting the forces of evil maybe those bad guys would never have gotten as far as they did. We have to be stronger than the enemy or they will defeat us.

EB: It's been reported that your colleague Ronald Bateman has called for tactical nuclear weapons to be used against the North Vietnamese forces. Would you support this policy?

Toni picked up her red biro and added an annotation in the margin. *Is "reported" sufficiently strong to guarantee we're safe from libel accusation if R.B. denies he said this? Recheck by Legal?* She took a sip of tea and resumed her task.

MM: No, I wouldn't, unless under extreme provocation.

EB: Such as . . .

MM: If the Reds started sending down mass reinforcements from China.

EB: That would be grounds for the US bringing in nuclear weapons?

MM: We would warn them first. If you don't stop what you're doing, we'll nuke you.

EB: On another topic . . . some members of the Warriors have been accused of racist statements. What's your view of that?

MM: It's a controversial area. What cannot be denied is that some types of humans are more advanced than some other types.

EB: Can you give me an example?

MM: Okay, take technology. The white races worked out the laws of science. They invented the steam engine and the gasoline engine and the airplane. They invented factories. They invented rockets that can take men into space. Or take politics . . . how many non-white races have democratically elected governments?

EB: So are non-white races intellectually inferior?

MM: I think they just haven't evolved as quick as the rest of us--not yet anyway. And there are plenty of exceptions, such as Martin Luther King--a very gifted man.

EB: Some people regard these views as offensive. What would you say to them?

MM: It's a case of stating what we know to be true. Some facts are unpalatable but that doesn't stop them being true.

EB: The teaching of the Christian Church is that all men are created equal. Are you saying you don't agree with that, Max?

MM: They have equal rights. That's doesn't mean they're equal in ability because we know that's certainly not true. There are plenty of black folk who are cleverer than some white folk.

EB: Are women equal to men, would you say?

MM: Again, that's a facile question, Lizzie. There are so many—

The phone rang in the hall. Toni got to it before Emma, who was halfway down the stairs.

"Hi, Toni, it's Bertie Stewart."

"Good morning to you. How are you?"

"I'm fine. Listen Toni, we've got a bit of a problem."

"Uh-huh."

"Actually it's worse than that. We're in a pickle and Donny suggested I give you a call."

"Go on."

"It's 'Amy and Lucy' . . . you were going to be AP on that weren't you?"

"Yes. You took me off it so I could produce the Warriors."

"Right. Anyway, you probably remember the part of Hermione?"

"The schoolgirl friend of the other girls?"

"That's right. They were supposed to be shooting in Aldbury today but the girl playing Hermione has gone sick and we can't find a replacement."

Although the drama was set in Northumberland, Emcro preferred to shoot it closer to their Oxford headquarters to minimise the cost and complication of transport. For the external shots they had chosen a row of terraced houses in Aldbury, a village near to Berkhamstead in Hertfordshire to stand in for the cottages inhabited by the families of the miners. For the interior scenes, sets had been constructed inside a warehouse in Berkhamstead hired by the production company.

There was the extra complication of school lessons for the child actors. To meet legal requirements Emcro had to hire peripatetic teachers to tutor the children on set when shooting schedules kept them out of school, which added to the difficulty of sourcing replacements when required. Some parents of young actors would only allow them to perform during school holidays.

"What about continuity, Mr Stewart?" asked Toni now. "If we bring on a sub—"

"Mary's going to be unavailable for a while apparently. She's got to go to hospital—nothing too serious but she won't be able to work for us. There's no problem with continuity 'cos she's only done a couple of scenes—we'll reshoot them. But we can't find anyone to take her place. None of the usual casting agencies can help us so Donny suggested giving you a call to see if you could come up with any ideas."

"Just remind me . . . how old is the character?"

"Just a minute, let me check . . . she's eleven."

"Appearance?"

"Not critical."

"I may be able to help, Mr Stewart. There's a friend of my landlady staying with us at the moment. She's on half term. I could ask her if you like."

"Has she done any acting?"

"School drama, I think."

"If she could do it it would get us out of a big hole, Toni."

"When would you need her?"

"ASAP. The whole crew are at Aldbury. There's some other stuff they can shoot first but we really need her today if you can fix it. Sunset is just after five and they want to shoot in strong daylight."

"I'll call you back in a few minutes."

CHAPTER 19

Xiu Ying Song was glad she had brought gloves and a scarf. A stiff northerly wind overpowered the feeble efforts of the late February sun at raising the temperature and swept discarded rubbish along the platforms of Oxford station. Her train was running late and Xiu Ying decided to take refuge in the grubby waiting room, which would at least shield her from the arctic blast.

There were three others similarly taking shelter, two middle aged ladies and . . . Joe Curtis. The two post grads looked at each other in surprise.

"Susan! What are you doing here?"

"Catching the train to London."

"So am I. You never said you were going to London."

"Nor did you."

Along with Luke Stirling and James Ibsley, Xiu Ying and Joe had attended the morning meeting called by Professor Maunsell in Lab 2C. She said she would tell them the latest news from Wilmington and how it affected their own work and then they could take the rest of the day off. It was an occasional Friday treat for the post grads to give them a long weekend.

There had not been major changes in 2C since Southern National had taken over funding of the Methane Reformation project. Initially Wilmington wanted the Oxford team to work exclusively on improving catalyst performance and drop their other areas of research but Madeleine fought successfully for permission for the post grads to pursue their other investigations, pointing out that funding for these came from the British Science Research Council rather than from Southern National. The Americans had decided that Wilmington would discontinue periodic flow experiments and concentrate exclusively on steady state processes although they agreed to let Oxford carry on, a relief to Xiu Ying and James, whose PhD thesis hinged on this work.

With a blare of its two-tone hooter the London train announced its imminent arrival and the waiting passengers watched the maroon diesel locomotive approaching the station. But Joe's attention was distracted by a whistle from the opposite direction.

"Well, look at that. The last of the few."

Xiu Ying followed his gaze. On the northbound line a steam locomotive clanked through the station at the head of a mixed goods train, Joe's gaze following it until the London train cut off his vision.

"Is it special?" asked Xiu Ying.

"It's a rarity," came the reply. "The steamers will all be gone by the end of this year."

"Are you a trainspotter, Joe?"

"Used to be, when I was a kid. That's a Stanier Black Five."

"If you say so! It looked very dirty."

"It'll be scrap metal soon . . . sad."

A few minutes later their own train was accelerating southwards and Joe asked his colleague why she was headed for London. It seemed that Xiu Ying's father had arrived from Hong Kong earlier that day. He was part of a delegation formulating the colony's legal system for implementation after the transfer of sovereignty to the Chinese government in 1997.

"But that's thirty years away!" said Joe.

"I think the plan is to finalise the arrangements and bring them in before the handover so it's not a sudden change. It's a bit complicated 'cos they have to take into account the demands of the Chinese."

"Hong Kong isn't it going to be part of China then?" asked Joe.

"That's what the Chinese want, but not the Brits. Daddy says it's going to be a Special Administration or something. The big argument will be about the form of government, how much independence the Chinese will allow."

"Didn't you tell us there were riots there last year?"

"Yes, but things are a bit quieter now, thank God."

"I thought you Buddhists didn't have a God!"

"We don't, Joe! I've picked up English ways of speaking."

"So how long will your Dad be in London? How long are you staying?"

The delegation was expected to last a week, said Xiu Ying. Her father was staying in a hotel in Chelsea and had booked two rooms––one for himself for the duration of the delegates' assembly and another for his daughter for the weekend.

"So a bit posher than your digs, then Suzie!"

Unusually for a post grad student Xiu Ying had never left the accommodation in Headington she had taken up residence in at the start of her undergraduate course. She got on very well with her landlady and the convenience of having her room cleaned for her and her laundry done outweighed the loss of privacy as far as she was concerned.

"Are you still in your flat in Botley?" asked Xiu Ying now.

"Yup."

"I hope it's tidier than it used to be when we last saw it! Everything caked in dust and the kitchen sink full of dirty pots. And the toilet . . . ugh!"

"We're a bit more civilised now, Suzie. We actually clean pots after we've used them rather than just before and we pay a cleaner to come in once a month. The toilet bowl gleams."

"Civilisation is the War against Entropy," laughed Xiu Ying, quoting the embroidered text hanging in a frame on a wall in Professor Maunsell's office. "There was plenty of entropy in your flat, Joe!" In the world of science the term "entropy" meant "disorder". At an early stage in their undergraduate careers it was impressed on chemistry students that the Universe naturally tended to greater disorder and that only expenditure of effort and energy could reverse the trend.

"Ever since the wall hanging appeared," continued Xiu Ying, "Bill Syerman's always coming out with that quote when he's telling us off for untidiness."

Joe nodded. "He's a bit obsessive, isn't he? I'm sure someone told me that in Bill's house his wife had to make sure that the saucepan handles were never overhanging the edge of the cooker in case they got knocked over."

"I'm glad the Prof was able to keep the team together," said Xiu Ying, changing the subject.

"Me too," concurred Joe. "Although it would have been great to have gone to the States. I'd like to go there one day. Maybe after I've done my PhD."

"What about . . ." started Xiu Ying but then looked down, her question unfinished.

"Maria?" said Joe. "Is that what you were going to ask?"

The girl nodded silently.

Joe smiled. "Haven't seen her for yonks. We're not going out any more."

"Oh."

Joe smiled. "What about you, Suzie?"

Xiu Ying shook her head and looked out of the window.

"Tickets please."

The youngsters looked up to see a middle-aged uniformed inspector in a peaked cap holding a clipper. They handed over their tickets and Joe noticed the script on the signet ring on the inspector's finger.

"GWR. God's Wonderful Railway."

The inspector bowed his head a little and lowered his voice conspiratorially. "You said it, sir. We're not supposed to wear items from the old days but some of us can't help ourselves."

"Are you sorry to see the end of steam?"

"Personally, no," replied the inspector. "Mucky, smelly things. Mixed feelings for the engine men though. They were proud of their mounts—true—but their job's a lot easier and a lot cleaner with the diesels."

"We saw a Black Five at Oxford. It looked a bit tatty."

"Foreign equipment," smiled the railway man. "Belonged to the 'Ell of a Mess before nationalisation."

"The what?"

"LMS—London Midland and Scottish, as it used to be. You would never have found a Stanier loco on our road, even though he started his career with us before he defected. But there's only a few steam locos left now and they wander all over the place."

"I saw 'Mallard' at King's Cross just before it was withdrawn," said Joe, referring to the locomotive that held the world speed record for steam traction. "Filthy, it was. The only clean bit was where the crew had wiped the cab side so they could see the number."

The inspector harrumphed. "We kept our engines clean even when standards were slipping on the other companies, like the Late and Never Early Railway."

Joe laughed, amused at the inspector's interpretation of the acronym for the London and North Eastern Railway, the owners of the A4 locomotive class of which "Mallard" was a member. "It's funny," Joe said, "when I was a spotter I used to buy the magazines to find out what was going on and it was surprising to see we were still building steam engines when the rest of Europe went for electric and diesel."

"The country was broke after the war, don't forget. We couldn't afford the new technology," explained the inspector. "And we had loads of cheap good quality coal so it made sense to carry on making the steamers. Shame, though, some of them they're scrapping now are only ten years old. They were built to give thirty years of service."

"What a waste," said Joe.

"You said it!" came the response. "Anyway, nice talking to you, sir. Enjoy your journey." The inspector handed back their tickets and moved along the carriage.

"Sorry," smiled Joe, looking at his travelling companion. "That must have been a bit boring for you."

"I didn't mind. I was thinking about what we were talking about."

"Boyfriends and girlfriends, you mean?"

Xiu Ying blushed. "No . . . the Professor."

"What about her?"

"Well, you think she would be happy that all her hard work paid off. She saved the MR project."

"You don't think she's happy?"

The girl frowned. "It's difficult to put into words. When she smiles it's like a sad smile. Sometimes it's like she looks a bit tired."

"Can't say I've noticed, myself," said Joe. "But it must have been hard for her, dealing with Southern National and the SRC, working out a deal that everyone's happy with. Perhaps she's feeling the effects of that."

"Yes," said Xiu Ying, "that could be it. And her brother's partner was killed in a car crash, wasn't she. If her brother's sad then maybe she's sad too."

Joe nodded and thought for a moment. "Tell you what, Suzie . . . how about we organise a surprise party for her. To say thanks and cheer her up a bit?"

"That's a lovely idea, Joe," beamed Xiu Ying. "I'm sure the others would help us to organise it."

"That's settled then."

The girl reached into her pocket and extracted a bar of chocolate.

"Like a bit?"

Joe nodded. "Thanks."

They nibbled at the confectionery for a minute or two and then Joe restarted their chat.

"It could be a bonus that we're working with Southern National," said Joe. "Maybe after I've got my PhD they'd take me on. There's no doubt the States is the place to be in our line of work. Everything's better there . . . more resources, better pay."

"I can see the temptation," said Xiu Ying.

"What about you, Suzie. Do you think you'd be interested?"

The girl wrinkled her nose. "Maybe. But I also feel I should go back to Honkers to be with my family. They've been so supportive of me since I first came to Oxford."

"Incredible, isn't it? Who would have thought when we both started as undergrads back in '63 we'd still be working together five years later. What a great team, eh?"

"Yes, I like being with you . . . in this team."

Their eyes met for a moment and then deflected towards the window to watch the Berkshire countryside flashing past their carriage at ninety miles an hour. For a minute or two an uneasy silence prevailed. Then Xiu Ying looked at her colleague again.

"Why are you going to London?"

"To stay with my folks in St Margarets."

"Where's that?"

"Near Richmond, on the River Thames."

"Is it nice?"

"Yeah . . . wouldn't mind living there myself one day."

"What does your Dad do?"

"He sells insurance. Works in an office between St Margarets and Twickenham. Tomorrow we're going to the Imperial War Museum."

"In London?"

"Yes, it's in Southwark, south of the river, just to the east of Waterloo. They're putting on a special exhibition about the Desert Rats."

"What's the Desert Rats?"

"The British Army Division who fought against Rommel in North Africa during the war. My Dad was a sergeant in the Rats."

"Why were they called 'rats'? That's not very nice."

"The Germans said the British were trapped like rats during the siege of Tobruk. So our chaps thought it would be good counter-propaganda to refer to themselves that way."

"Is that where your Dad fought the Germans?"

"No. My Dad was at the battle of El Alamein in Egypt. He was wounded and got shipped home. They gave him a medal."

"So it'll be nostalgic for him to see things about the Desert Rats?"

"Yes, he's meeting a few of his wartime chums there, too, so it'll be a reunion as well."

"And you're going too?"

"Yes. It'll be interesting, I reckon."

"Why didn't you drive to London, Joe? Wouldn't that have been more convenient?"

"My car's kaput, Sue. Leaking radiator and I can't afford to get it fixed, not at the moment anyway."

"Pity."

"Yeah."

The train cruised through another mile or two of windswept English farmland. This time it was Joe who restarted the conversation.

"Have you got your driving licence yet, Sue?"

"Not yet. Still taking lessons. Got my test booked for next month."

"Will you buy a car?"

The girl grinned. "It'll be one of the suggestions I make to my Dad!"

"You're gonna ask him to buy you a car?"

"Isn't that what Dads are for!"

"So will you be with him all weekend?"

"Not tomorrow. He's meeting some of the others for a preliminary session of some sort."

"What will you do while he's in this meeting?"

The girl shrugged. "Look around the shops, maybe."

Joe looked earnestly at her. "I don't want to spend all afternoon listening to Dad and his mates going on about the war."

"No," said Xiu Ying, blushing a little.

"Do you want to meet somewhere later in the afternoon?"

Xiu Ying returned his gaze. "Is this a date, Joe? As in . . . date?"

"Do you want it to be?"

"Yes, I do."

"Then yes . . . it's a date . . . as in date."

CHAPTER 20

The three men exited the front door of the Majestic Lodge ski hotel and turned left along Broomstick Road. Their exhaled breath condensed into billowing clouds in the sub-freezing temperature, their boots crunching into the fresh snow on the sidewalk. All three immediately donned sunglasses against the bright winter sunshine. The anonymity they afforded was a bonus, of course. With the Aspen ski season in full swing the three had to occasionally weave past people clomping the other way towards the gondola station in their cumbersome ski boots.

After a hundred yards or so the trio turned left again into the driveway of one of the Alpine style chalet bungalows lining the road. One of the men, a heavy set square-jawed individual with cropped grey hair, removed his glove to get a key out of his pocket. The other two followed him indoors. The men had enjoyed a convivial lunch at the Majestic but now they needed privacy for further business.

The owner of the bungalow and the cropped hair was retired General Ralph Hartmann, ex-United States Army. His guests were Louis Maguire, Deputy Finance Director at the Luchini-Franks Investments Group and Maxwell Murdoch, owner and Chief Executive Officer of the Southern National Generation company. All three were senior members of the Righteous Warriors of America.

The trio had arrived separately that morning in their business jets, which would wait at Aspen-Pitkin airport with their crews until the following morning, when they would whisk their owners respectively back to Houston, Texas, New York and Wilmington, North Carolina. The men's fortunes came respectively from manufacture of military hardware, high finance and power station construction. They had chosen Aspen to minimise the chance of recognition by members of the public and General Hartmann's ski chalet would guarantee that their conversations would not be overheard.

The ex-military officer sat his guests down and served each of them and himself a glass of rye on the rocks. Before long clouds of cigar smoke wreathed the men as they made themselves comfortable in the leather armchairs.

"Okay, guys," said the general, "we're not here for the skiing. This is where we get down to the nitty-gritty." He turned to Max. "Louis and I have already done a bit of research and we're kinda headed towards a provisional plan."

The original motivation for Operation FAB was the apparent lack of determination from the government of President Johnson to eliminate the Communists in North Vietnam once and for all. The acronym stood for "Fire and Brimstone" and the general and his guests all knew the passages in the Bible in which the wrath of God took this apocalyptical form. The reluctance of the President to resort to tactical nuclear weapons in the war-torn country brought much teeth-gnashing from the more resolute Righteous Warriors, who could not understand why this God-given advantage was not being deployed to destroy the atheists.

The general briefed Max succinctly, with Louis occasionally adding a comment. The gist of the briefing was that a group within the Warriors were exploring methods of covertly attacking Hanoi with some sort of nuclear device, independently of any military or political organisation. As he spoke, the general and the finance director watched their associate closely to gauge his reaction. When Ralph finished speaking Louis nodded discreetly at the general, a pre-arranged signal that indicated that further elaboration was in order.

"That's the bare bones of it," said Ralph now. "What do you make of it, Max?"

"Are you saying the Government want no part of this?"

"We haven't even made overtures. As soon as the politicos get hold of something secret the leaks start. Sadly that's also true in the senior echelons of the military. The bleeding heart liberals would have a field day. At the moment there are only six people involved, seven if you want to be part of it."

"Why did you approach me, Ralph?"

"Before I answer, can you confirm you're in our out, Max. If you're with us we'll talk details. If not we stop here and you tell no one else about this conversation. It never happened."

The CEO nodded slowly. "You need to tell me a bit more, Ralph. It sounds okay in principle but where do I fit in?"

"As I just said there are two logistical problems—obtaining a weapon and delivering it. For obvious reasons we can't approach the Pentagon. There are other sources but—apart from the British—

they're not on our side. Our investigations indicate it's difficult to corrupt the Brits so that's probably a dead end."

"There's Israel," interrupted Louis, "but we think they wouldn't be interested 'cos they don't officially have nuclear weapons and they couldn't risk being involved in something that would give the game away if it got found out. And Vietnam ain't their problem."

"So," continued Ralph, "if we can't get hold of a weapon it would have to be some other kind of device, like a reactor. Like the sort of reactor you're bringing in to clean up your power stations."

Max frowned. "But how can you use a reactor as a weapon?"

"You're using pressurised water reactors, yes?"

"Uh-huh."

"With graphite moderators?"

"Or water. We're looking at both configurations."

"And water for cooling and heat transfer."

"Uh-huh."

"So what would happen if the pressure vessel was breached?"

"Nasty . . . the water would boil. You'd have to drop the control rods to stop the core overheating."

"Suppose the control rods weren't inserted?"

"Even nastier. The heat would turn more of the water into steam, which would react with the zirconium fuel cladding . . . you'd get a build up of hydrogen."

"And if the hydrogen was ignited by the reactor heat?"

"A big bang."

"So you'd get an explosion of radioactive material?"

The CEO smiled. "Impressive, Ralph. Not exactly a Hiroshima but you'd upset a few people, for sure."

"Would it work?"

"Probably. How would you breach the pressure vessel?"

"Blow out panel activated by a timer."

"Okay . . . where would you deliver it? How do you deliver it?"

The general took a drag on his cigar. "Like I said, we deliver it to downtown Hanoi. The 'how' bit ain't easy. I mentioned four methods, didn't I? Let's elaborate a bit.

"By missile . . . impossible for the bad guys to intercept but getting hold of one powerful enough to lift the reactor would be virtually impossible, as would concealing the launch pad. Plus the guidance systems on these things ain't accurate enough. By airplane . . . no chance . . . our guys would destroy anything on the radar they weren't happy with and the NLF have got their own radars—if our guys missed it the enemy have got Soviet SAMs to knock out any intruders." Ralph was referring to the National Liberation Front of North Vietnam and surface-to-air missiles. "Never mind the problem of retrieving the pilot if he got through.

"Which leaves land and sea, or river to be more precise. Okay, let's say we stick the reactor on a truck and send it to Hanoi from the south. What are the chances of it making it to the target without interception, even if it's disguised as an NLF vehicle? And the roads ain't exactly six-lane freeways. Farm tracks would be a more accurate description. It would take a long, long time for the truck to get to the target. And again we'd be faced with the retrieval of the delivery team."

"So it would have to be a boat, then," deduced Max.

The general ground out his cigar in an ashtray. "There would be other problems but we reckon they would be easier to solve. The reactor would be carried on a vessel entering North Vietnam via the Red River estuary. It's around 170 miles from the coast to Hanoi along the river so we're talking twenty hours or so. The crew get to Hanoi, moor the boat and then escape on an inflatable dinghy, back the way they came. Our guys pick 'em up in the Gulf of Tonkin."

"What sort of boat?," asked Max.

"Again, we've got three options. Some sort of small naval vessel carrying NLF markings or a civilian supply vessel or a fishing boat, which would be more anonymous. A naval boat would be difficult to source—it would have to be a type the Commies use. Plus it would be expected to be in radio contact with the NLF, which adds an extra layer of complication, language-wise. Supply vessel—easier to get hold of but likely to be challenged as it headed inland—the NLF are as paranoid as our boys about saboteurs. So . . . we use a fishing boat. It discreetly joins a fleet at night in the Gulf and heads for the river when the others do, lagging behind a little. We're pretty sure Vietnamese fishing vessels don't carry radio so a challenge is unlikely. It'll be picked up on their coastal radar of course but they'll assume it's one of the fishing fleet. In the river it'll drop a little behind the others and drop anchor in Hanoi harbour when it gets there. Our team will set the timer on the blowout panel and quietly launch their inflatable and head downstream. It might still be daytime by then but a tiny boat headed towards the sea is unlikely to raise interest, even if the locals had somehow heard about a big unexplained explosion in Hanoi."

"Long way for fishing boats to go," said Max. "A hundred and seventy miles. Why wouldn't they land their catches on the coast and send them to Hanoi by train?"

"Yeah, they normally offload at Hai Phong but our guys have bombed hell out of the railways so it's not unusual for the fishing boats to take their catches up the Red River."

Max nodded. "Sounds okay . . . I like it."

"Great. You can supply a reactor, then?"

"Yeah, I reckon. A graphite moderator would be best, to make sure the heat keeps coming if it loses all the water."

"I don't know much about this nuclear stuff," said Louis. "What does the moderator actually do?"

"My Wilmington guys explained it to me," answered Max. "The uranium atoms in the fuel pellets shoot off neutrons which hit other uranium atoms which shoot off more neutrons so you get a chain reaction and . . . if you don't control it . . . boom!"

"So the moderator controls the reaction?"

Max laughed. "No . . . the opposite! The neutrons are too fast to hit other uranium nuclei which means you don't get a nuclear reaction. So you've gotta slow 'em down. That's what the moderator does. You can use graphite or water."

"So without the moderator . . . no reaction?"

"Correct. No reaction, no heat. Just a cold, useless lump of uranium fuel shooting off high speed useless neutrons."

"So if the moderator is water and we blow the pressure vessel the water disappears as steam and the reaction stops?"

"That's right. It's a safety feature of water moderated reactors. Lose the water and the reaction stops. That's why I'm suggesting graphite for what you've got in mind. Even if you lost all the water you'd still get a nice big radioactive bang."

Ralph smiled. "That's great, Max. That's just what we wanted to hear."

"A thought," said Max. "What about evidence? It wouldn't be much fun if the NLF could show proof that the Yankees had made their capital city uninhabitable. The Soviets might decide to retaliate with their nukes. We'd have to make sure there were no serial numbers stamped on components, that sort of thing."

The general shook his head. "Not a problem. The US government genuinely wouldn't know anything, would they? Our guys could suggest that they put out a bulletin saying the Hanoi explosion was obviously the Commies trying to make nuclear weapons to use against the South Vietnamese Army—outrageous, we'd say. Doubly outrageous if they're trying to blame Uncle Sam for the misfortune they brought on themselves."

"Actually, there wouldn't be a helluva lot of evidence to find," said Max. "Tiny bits of highly radioactive debris."

"And we could lay a false trail anyway—stamp some Chinese lettering on the components."

"Who did you plan to recruit for the delivery team?" asked Max. "South Vietnamese guys or American?"

"We've got someone working on that," said Louis. "We've found three Chinese guys who escaped from a jail in Peking and fled to the States. They were political prisoners who got tortured by Mao's thugs so their motivation would be revenge against the Communists—any Communists. Plus a big bundle of greenbacks if they need more persuading."

"If they say yes we'll take them out in a fishing boat a few times so they can learn the ropes," said Ralph. "We'll buy a fishing boat in Hong Kong and modify it to suit our needs. That's where the mission starts if we go ahead."

"So how will the reactor get to Hong Kong?"

"On a freighter as general cargo—billed as engineering components or something like that."

Max nodded. "Probably best not to build it at Wilmington. We don't want prying eyes. Somewhere on the West Coast would be better for subsequent shipping to Hong Kong. Somewhere quiet."

"We were thinking along the same lines," said Loius.

"I'd have to have a cover story for the guys building it," said Max.

"Tell them it's secret work for the CIA."

Louis chuckled. "So secret even the CIA don't know about it."

The three plotters smiled at each other in mutual admiration.

"Gentlemen," said Ralph. "We have a plan."

CHAPTER 21

The ward sister and duty doctor stood on one side of the bed and Nigel Nixon and Madeleine Maunsell on the other. They were all looking at the body of Dorothy Nixon, who had just breathed her last. The half-siblings had been summoned to St John's Hospital in Maidenhead after Nigel's mother's condition had quickly deteriorated after her most recent stroke. For privacy during their grieving a screen had been erected round the bed.

"We'll leave you alone for a while," said the doctor now.

"Thank you," said Nigel. "Would she have been in pain?"

"Doubt it," said the doctor. "She never fully regained consciousness after her first stroke, did she? When was that?"

"Last October," said Madeleine.

"Ah, yes."

"What's the procedure now?" asked Nigel. "Will there be a post mortem?"

"I doubt that's necessary. There was no uncertainty over the cause of death. How old was she?"

"Seventy-nine," said the ward sister and Madeleine simultaneously.

"Sister Williams will come back in ten minutes," said the doctor. "She'll advise you on what happens next."

"Our condolences," said the nurse with a sympathetic smile as she and the doctor left, the screen curtains dropping back into position after their departure.

Nigel bent forward and kissed his mother's forehead. "Bye, Mum."

"She's at peace," said his half-sister. "Her quality of life was pretty grim before the first stroke, wasn't it."

The pilot nodded. "Yes, it's a relief, I suppose. What would things have been like if she'd regained consciousness? I shudder to think."

"A shame she was so unhappy."

Nigel looked at his sister. "It's a dreadful thing to say, Maddie, but trying to look after her was more of a chore than a pleasure. It was only filial duty which kept me going to Fearnville. I must be worst son in the world."

"No, you're not. She made things difficult for you. And for herself."

"And for you, too."

"I wasn't a blood relative so she wasn't so concerned about me."

The pilot turned his eyes towards his mother's face. "As you said, she looks more peaceful now than when she was alive."

For a minute or two they discussed the funeral arrangements that had been previously formulated and Nigel confirmed that he would phone Henry, their father, to notify him of his first wife's death and to invite him to stay with him and attend the funeral.

"I'll have to go through her effects," said Nigel. "My house is full of the stuff we didn't dispose of when we sold her house."

"What sort of things?"

"Clothes, documents, jewellery, knick-knacks."

"I can help with that, if you like."

The pilot smiled. "Thanks, sis. Might take you up on that."

"The clothes could go to a charity shop."

"Yes, and I could sell or auction the jewellery, I suppose."

Nigel briefly picked up his mother's hand and shook his head sadly. "Sorry, Mum. A bit disrespectful, isn't it, talking about getting rid of your chattels when you've only just left us."

There was a pause.

"Will there be anyone else but us and Dad at the funeral?" asked Madeleine.

"There's no other family, as you know."

"Perhaps some of the Fearnville staff might go."

"They might not be too enthusiastic, given how she treated them."

"She was nicer when she was younger though, wasn't she? You've told me that before."

"She's always been stubborn. One of my earliest memories is her arguing with the vicar over something or other. My step-father intervened on the vicar's side and Mum went into a rant, screaming at them both. It was quite frightening."

"Dad used to call her 'aqua regis'," said Madeleine. "Have I told you that before?"

"Yes, but I can't remember the derivation."

"It was because . . . no, maybe I shouldn't say. It's disrespectful, with Dorothy lying there."

"She can't hear you."

Madeleine smiled sadly. "It was when I was a schoolgirl. I was doing chemistry homework and Dad asked me about it. I told him 'aqua regis' was a very strong mixture of concentrated nitric acid and hydrochloric acid. It was highly corrosive and would just about dissolve anything, even gold. So after that Dorothy became 'aqua regis'."

"I think my step-father worked out a way of putting up with her," said Nigel. "Most of the time he would let Mum get her way for the sake of domestic harmony. But now and then he would take a stand and ignore Mum's tirades. He was like a rock with raging floodwaters foaming round him. I admired his fortitude."

"So sometimes she lost the argument?"

"She had various counter-attacking tactics. Sulking was the initial response. If that didn't work the tears would come. We lost count of how many times Dad broke her heart. Me, too, for that matter. Emotional blackmail, he called it."

"He stuck with her, though."

"He was an honourable man. He took the wedding vows seriously even though he had obviously fallen out of love."

Madeleine shook her head. "Sad."

"And so had I," said her brother.

"What do you mean?"

"I stopped loving her." The pilot stared at the dead body of his mother. "Truth be told, I don't think I ever really loved her, even when I was young. See . . . I told you I was a bad son."

* * * * *

Later that day the pilot and the chemistry professor were taking afternoon tea at The Content Angler Hotel, which sat at the western end of the bridge carrying the A4 road over the Thames in Maidenhead. They had explored all the factors they could think of relevant to Dorothy's interment and the conversation drifted into other areas.

"How's the film star?" said Madeleine after a lull.

"Eh?"

"Emma."

Nigel laughed. "I'm surrounded by celebrities now, aren't I?"

Madeleine frowned. "Which other celebrities did you have in mind?"

"You! You daft thing!"

"Pardon?"

"'The Spitfire Ladies'," said Nigel, quoting the headline over a newspaper article.

"Oh, that."

The report describing the arrival of Chris Coleman's Spitfire at Wycombe Air Park and the fortuitous coincidental arrival of three of the wartime ATA female pilots one day the previous November had initially occupied half an inside page of the local *Chiltern Herald* newspaper. Evidently the report had found its way to the national news desks, resulting a few weeks later in requests for interviews with Madeleine, Parminder Collins and Prudence Dismore-Whyte for a feature in the Sunday Times Magazine.

"Didn't you say Toni was going to turn it into a TV documentary?" said Nigel now.

"I think she's pitched it to Emcro, and they're going to think about it."

"So you'll be a TV star again!"

"I don't think the programme about the methane project and the Rightous Warriors was the most watched programme recently."

"Emcro seem to be happy with what Emma's doing for them."

"How is she coping with the attention?" asked Madeleine. "I saw the report about the programme on the BBC news. They showed clips from the rehearsals."

"Yes, I saw it too. She's started getting fan mail!" said Nigel. "She's handling it pretty well, I should say. No airs or graces. Her school friends think it's great. She signs autographs for them. If it takes off let's hope she keeps her feet on the ground."

"Toni says the crews like working with her."

Emma Holmbeck had turned out to be a natural, so much so that producer Donny Mcleod had asked the scriptwriter to beef up her part. An unexpected bonus was that the three child actors had become firm friends, with no hint of professional rivalry to poison the atmosphere on set. The camera crews and assistant directors and grips and stage hands were amused to note that the three girls had taken to adopting their Northumberland accents for general conversation on and off set. And at home, Nigel told his half-sister.

"She's even got me doing it," the pilot said now. "So now the latest craze on our flight decks is talking like Geordie miners."

"I'm glad she's enjoying it," said Madeleine. "No doubt it helps to take her mind off Jackie."

"Yes . . . although sometimes we have weepy days."

"Anytime she wants to visit me she's more than welcome."

"She likes you, Maddie. She's said so more than once."

"I like her."

"She asked me why you weren't married."

There was a pause.

"What did you say?"

"That you nearly did get married but your fiancé died during the war."

"Uh-huh."

"She said you were the nicest person she knew . . . after her friends, of course."

"And you."

"She was kind enough to say that." Nigel took a bite of Battenburg and washed it down with a sip of tea.

"What happened to that chap from the . . . what do you call it? The people who were financing your project."

"John Davis. Science Research Council."

"Didn't you go out with him a few times?"

"Yes, once or twice. And we played bridge together."

"So . . . what happened?"

"Nothing. He went away—they posted him to Rio. He's supposed to be coming back soon."

"Will you see him again?"

The professor shrugged. "Maybe."

"It evidently wasn't the romance of the century, then."

"No."

"Do you keep in touch while he's away?"

"No."

"Not like you, Maddie, clamming up like that."

"Nothing to say, Nigel."

"Fair enough."

"On the same subject—what about you? Have you got a new girlfriend?"

"Nope. Been out with a girl a couple of times but I don't think it's going anywhere. I told her I'm still recovering after Jackie so she might move on soon. I don't know."

The pilot summoned the waitress and asked for the bill.

"How's the methane thing going, sis?"

"Okay, thanks. We're sort of settled. I have to go Stateside now and then but that's no great burden."

"He's a bit weird, isn't he, your American sponsor? That's the impression I got from Toni's TV programme."

"Eccentric, certainly. Peel away all the religious stuff and the dubious views on race and you find a decent chap. You should meet him one day. He gets over here now and then to see how we're spending his money."

The waitress returned and Nigel took out his wallet, waving aside his sister's insistence that they split the bill.

The two of them walked to the car park, where Nigel's Vitesse was parked next to Madeleine's Morris Minor.

"You'll be able to take the lid off that thing now spring's here," said Madeleine, pointing to the Triumph's soft top.

"Minimum air temp sixty-five for that," grinned Nigel. "For me, anyway."

His half-sister grinned back. "Don't tell me . . ."

"Correct! SOP, dear thing!"

"Talking of aeroplane stuff," said Madeleine, "I forgot to tell you. I've got my PPL back." She was referring to her Private Pilot's Licence.

A broad grin settled on her brother's face. "Fantastic, Madds. I always thought it was aviation's loss when you gave up serious flying. We must do a trip together. You can show me how it's done. What type are you flying? The Piper spam can?"

Madeleine nodded. "And I'm checked out on the Chippie too so I can do aerobatics."

"I'd love a go in the Chipmunk. It's ages since I've been upside down in a flying machine."

"Pammy's doing it too."

"Pammy Collins? Our ATA chum?"

"She couldn't stand back and watch me requalify while she got left out, she said, so she's doing her PPL too, although she's sticking to the Cherokee for the moment. We still do trips together with Dizzy . . . what are you laughing at?"

"Something Frank Armitage said a while ago. An ironic comment about the likely lack of conversation when three women were flying together."

"Cheeky sod!"

"That's Frank!"

CHAPTER 22

There was a full complement in Lab 2C except for James Ibsley, who was off sick with a cold. The three remaining PhD students were perched on their lab stools as usual. Safety Officer Bill Syerman had brought in two chairs for the remaining attendees—Professor Madeleine Maunsell and sponsor Maxwell Murdoch, who had finished speaking just as Verity Abbott brought in four teas and a coffee and a selection of biscuits.

The post grads were thoughtful, assessing the impact the American's words would have on their lives. Madeleine had already been briefed so the new situation was no surprise to her. Max had asked her not to spill the beans until he had had a chance to speak to the post grads directly himself. He thought it his duty to reassure them personally, he had told her. Max also relayed a more personal message from Gordon McKenzie asking Madeleine to pass his love on to Toni but to keep some for herself.

"Just to be clear," said Luke Stirling now. "Although you're pulling the plug on methane reformation research in Wilmington you'll carry on funding us on our PhDs?"

"That's right," said the CEO of Southern National Generation. "As I said, I'm quitting the project myself because I've realised it's a waste of time and money. No one thinks the CO_2 in the atmosphere is gonna cause a problem for while and some people say if the levels go up the world's forests will soak up the extra. But the new deal is that I'm gonna start closing down my coal power stations and converting them to nuclear 'cos I think eventually nuclear will make electricity cheaper than coal can. So that'll cut down the CO_2 I'm pumping out. Once I start doing it others will follow. So it's bye-bye to methane reformation. Sorry, Madeleine!"

The chemistry professor smiled wistfully and shrugged.

"So you're saying nuclear power generation will cost less than coal?" asked Joe Curtis, raising his hand, which had been resting on Xiu Ying's on the work bench. "How long will that take to happen?"

"The more nuclear we build the cheaper it gets 'cos we get better at building them, Joe. All I've gotta do is spread the word of the Lord. He wants clean air and we're gonna give it to him. Like I said, we'll switch from coal to nuclear. So we don't have to mess with methane."

"What about other countries?" asked Xiu Ling. "They'll still be burning coal, won't they?"

"And the Soviets won't be bothered about the word of the Lord," observed Luke. Madeleine quickly glanced at her protégé, ready to scold him if it looked like his words might offend their sponsor, but there was no smirk of sarcasm on his face this time.

Max laughed. "Well, they don't pump out as much as we do," he said. "We can cut them a little slack to start with. Like I said, if we build enough nukes the power they produce will eventually be cheaper than coal so unless the Reds convert they'll lose out economically."

"So you can convert them to nuclear even if you can't convert them to God," joked Joe.

"You said it, fellah!" grinned Max. "But maybe we can change that, too. You know, there are millions of Russians who secretly worship the Lord. Probably the majority of the population. They would love to see Christ the Saviour rebuilt."

"What's that?" asked Luke.

"Cathedral in Moscow. Stalin knocked it down in the Thirties. They were gonna build a government building there but gave up when the Germans invaded. Khrushchev turned it into a swimming pool, which it still is. But who knows . . . maybe one day they'll get their cathedral back."

"So we're going to carry on as before, are we?" asked Joe, bringing them back to the essence of the discussion.

"Unless something unexpected crops up, then yes."

"Why would you carry on funding us," continued Joe, "if you've decided to scrap methane reformation research?"

"Because I said I would. Plus I like what you're doing with improving catalyst performance so SN would get some benefit out of that."

"What'll happen at Wilmington?" asked Madeleine. "Are you cancelling the small reactor experiment?"

"Not for the moment," replied Max. "It's useful work, even if we don't use it to fry methane and CO2. If it works okay we can maybe offer our design for other purposes, like ships and subs. The Pentagon might be interested. We could snatch some of the

Westinghouse contracts. A bit of healthy competition won't do them any harm."

The conversation moved into more technical areas and when the questions explored topics beyond the sponsor's expertise, Verity was summoned to commit them to paper for the Wilmington team to look at.

"Well, if you'll call a cab for me, I'll make my way back to my hotel," said Max eventually. "Are we still on for dinner tonight, Madeleine? Dave Montgomery—my PA—will be with us of course."

"Yes, I'd like that, thank you."

"What about you kids?" asked Max. "You coming along too? My tab."

"Could I bring my girlfriend?" asked Luke. "She's free tonight. She would pay, of course."

"No, sir! No deal! She sits at my table, I pay. That's the way it goes. Joe and Suzie—what about you?"

"We're going to the pictures," said Xiu Ying, "but thanks for the invitation."

"Whaddya gonna see?"

"Bonnie and Clyde."

"Good movie. You'll like it. I know the director—Art Penn—he knows how to make good pictures."

The meeting concluded, Verity was asked to phone for a taxi, to be told that there might be a wait because demand was unusually high.

"They say fifteen to twenty minutes," said Verity.

"I can live with that," said Max.

"Would you like to see the other labs while you wait?" asked Madeleine. "We've got some interesting stuff going on here."

"Sure, honey. Lead on."

"Did I read somewhere you were setting up a new Christian society in Britain?" asked the professor as they strolled along the corridor.

"Yeah. I'm doing an interview with the BBC about it tomorrow. A few people approached me to start a chapter of the Righteous Warriors here in England."

"What sort of people?"

"Can't give you names, Madeleine, not yet anyhow. Confidential. Some business people, some politicos, couple of bishops, couple of movie actors. When the time is right we'll make an announcement."

"Will it be the same as the American set-up?"

"Pretty much."

"What will it be called?"

"We haven't decided yet. I'd kinda like to use the name we have now."

"Talking of TV programmes, my friend Toni enjoyed producing the interview with you in Wilmington," said Madeleine. "I'm sure she'd be happy to do it again if required."

"I'll keep it in mind. I liked her, even if she was a bit too hippy for me." The American chuckled. "She hit Gordon McKenzie like a tornado. Wore him out, I reckon!"

"I can imagine," murmured Madeleine.

In Lab 1D the professor and her visitor found a male student in deep discussion with an older man of Asiatic appearance.

"Everything okay?" asked Madeleine with a smile. She made the introductions.

"Max, this is Larry Rutherford, who's doing his PhD, and Dr Phan Hoc, who's his supervisor. They're looking at various methods of synthesising cyclopentane."

"What's that used for?" asked the American.

"It can be turned into other useful products."

"We're getting strange results according to Fido," said Dr Phan, speaking accented English. "If it's giving us correct info it means we're getting better yields from cracking cyclohexane than we would expect and we don't know why this should be."

"What's a Fido?" asked Max.

Madeleine laughed. "It's a flame ionisation detector." She pointed to a Walt Disney character clumsily drawn in felt tip on the equipment's metal casing.

"Hey, that's Pluto!" grinned Max. "Not Fido!"

"If you say so!"

"It's analysing the gases, is it?"

Dr Phan nodded. "Yes, we use it a lot in gas chromatography."

"Say . . . tell me if I'm out of line here, but are you Japanese?"

"No. I am from Vietnam."

The American raised his eyebrows. "Whereabouts?"

"Hung Yen."

"Is that anywhere near Saigon?"

"No, it's about thirty kilometres southeast of Hanoi."

"North Vietnam?"

"Yes."

After a pause, Max cleared his throat. "How long have you been in England?"

Dr Phan smiled. "Over ten years. I did my degree here at Oxford."

"That's swell. Will you go back after the war is over? Have you got family there?"

The scientist briefly ran through his personal history. Despite the harsh Japanese occupation during the war and the subsequent fighting between the French colonial forces and the communists after the defeat of Japan life was stable enough for him to complete his education. His tutors were impressed with his grasp of physics and chemistry and sent in an application for their student to attend Oxford University to study for a degree. When he was accepted a fund was set by his family up to pay for the course.

"How did that work?" asked Max, interrupting the narrative. "Wouldn't they have wanted to send you to college somewhere in Vietnam?"

Dr Phan nodded. "There was a problem with this. Some people said my Dad was a traitor, sending me to a Western university but he insisted I go to Oxford like my tutors suggested. He was a war hero so they respected his wishes."

"A war hero?"

"He killed many Japanese soldiers during the occupation. And some French soldiers too when they tried to take over."

"Impressive. So do you ever get to see your family or is it impossible now because of the war?"

"I haven't been to Vietnam since I left to study here. But I'd like to go back to visit them as soon as I can. My mother is not very well and I would like to see her. It might be possible for me to travel via Moscow. I've been corresponding with the Soviet Embassy in London and they say it might be possible."

Intrigued, the American probed further into the Vietnamese scientist's background. It seemed his younger brother was a teacher in their home town of Hung Yen. The town had been bombed a few times when South Vietnamese or American forces thought the Viet Cong had set up bases there but the town council had managed to keep the school going. The last Dr Phan heard, the classes were being conducted in a warehouse while the school was being repaired after bomb damage.

"How do you keep in touch?"

The scientist smiled. "Via Moscow. I like to send money home to support my parents and my brother."

The American was incredulous. "You send them English pounds?"

"No. I go to the bank here and buy roubles. I send the roubles to a friend in Moscow. He is a friend of my Dad's who works there and he sends the letters and roubles to Vietnam. I think they go on Aeroflot flights and military transport planes."

"And your family can use roubles in Vietnam?"

"No, they have to change them into dong, the local currency. The arrangement isn't strictly legal so a large percentage of the money goes in . . . commission."

"So what does your Dad do?" asked Max.

"He's got a shop . . . three shops in fact. He sells fruit and vegetables."

"So do your folks live near your brother?".

"No, the shops are in Hanoi, two in the centre of town, one on the riverfront."

"The riverfront?"

"Yes. It's a good place for a shop 'cos it's very busy these days. A lot of fish comes in by boat 'cos the railway from the coast is often out of use . . . it gets bombed a lot. So people come to buy fish and then drop into my Dad's shop to buy vegetables." The scientist laughed. "Last letter I got from Dad, he told me to stop sending roubles 'cos he's making enough money in his shops."

"That's good," said Max, looking down, his voice strangely muted.

"Ah, there you are," came a call from the doorway. It was Verity Abbott. "I've been looking for you everywhere. Your taxi's here, Max."

CHAPTER 23

Overall the party had gone well, thought Madeleine, cruising along the short embryonic stretch of M40 motorway between Stokenchurch and High Wycombe in her Morris Minor. The April air temperature would nudge sixty-five degrees later in the day according to the forecasters, warm enough for her brother to "take the lid off" his Vitesse if he was driving it that day.

Apparently the bash at the Hart and Serpent pub in Oxford was Joe Curtis's idea. The original intention was to organise a surprise party for the post grads' mentor to thank her for all the effort she had put in to secure continued funding for their PhDs. Then somebody found out that the proposed date was close to the professor's birthday and the post grads decided to let Madeleine in on their plans so she could invite friends and relatives if she wanted. So the celebrants included Bill Syerman and Verity Abbott and Nigel Nixon and Emma Holmbeck and three members of the All Souls church choir, including Toni Clark.

Neither Prudence Dismore-Whyte or Parminder Collins were able to attend but they had arranged a separate do at Wycombe Air Park, where Madeleine was now headed.

During the party at the Hart and Serpent a phone call had come through from Wilmington, North Carolina and Maxwell Murdoch and Gordon McKenzie had both come on the line to wish Madeleine Happy Birthday. For some reason she found tears in her eyes and she found herself apologising to the assembled company for her lack of self control. Toni had thrust a glass of wine into her hand and told her to "let it all out, sister".

Later in the evening she had been sitting chatting to her brother and Emma, quizzing the girl on the TV programme she was taking part in. Nigel got up to talk to another guest and Madeleine turned to her young friend.

"Are you enjoying the acting, Emma?"

The girl nodded. "Mostly, yes. They're nice people I'm working with. One or two of the actors are a bit . . . bossy. They keep telling us what we should be doing. The director is a bit of a perfectionist—makes us do retake after retake until he's happy."

"So it's hard work?"

"A lot of it is just hanging around, Auntie Maddie. Even on set we're only rehearsing or filming a small part of the time. In between scenes they're changing the lights and camera angles and it seems to take ages."

Madeleine laughed. "Sounds a bit like airline flying. Nigel says his job is ninety-nine percent boredom and one percent terror. I'm sure he's exaggerating."

"Some of the actors say it was even worse before videotape was invented," said Emma. "At least nowadays they can see straight away what's been recorded. In the old days they would have to wait for the rushes to be developed."

"Will it be colour?"

"Black and white. For colour they would have to use film, which would slow things down even more. I suppose one day they'll use colour videotape."

"But overall, it's enjoyable?"

"Yes. We get treated much better than the extras. I feel really sorry for them."

"Why?"

"They don't get paid much and they get pushed around from pillar to post. On meal breaks they have to wait until the actors and all the crew have been served before they're allowed to choose their own food. One or two of them are a bit nasty to us."

"Perhaps they're jealous of you."

"Yes, that's what some people say."

"Will you carry on acting after this job?"

"I'd seriously think about it if people said I was good at it."

"What are your favourite school subjects?"

"Art . . . and Maths . . . and French . . . and Physics. Nigel finds out from the headmistress what lessons I'm missing at school when I'm acting and he's fixed up tutors for me so I don't miss any important work."

"If you like maths and physics you could be a scientist when you leave school."

"Like you, you mean?"

"Yes."

"Nigel says you're an expert on pollution in the atmosphere. You must be very clever to understand that."

"I'm not an expert. I do research into carbon dioxide emissions from power stations to see if we can protect the atmosphere."

"Is it a bad thing, then?"

"It could be in the future."

"Why? What does it do?"

"Do you know about atoms and molecules?"

"We've just started learning about them."

"Right. The sun's rays hit the earth and some of the rays are reflected back towards space as infra-red. But some of the reflected infra-red hits carbon dioxide molecules and makes them spin and vibrate. This extra energy is transferred to the other gases in the atmosphere and heats them up, which could change the weather, say some experts. More floods, more droughts, more hurricanes and so on."

"Is there a lot of CO2 in the atmosphere, then?"

"About one molecule in every three thousand."

"That's not much! Why don't the other molecules spin and vibrate—the oxygen and the nitrogen?"

"There are only two atoms in those molecules, Emma. CO2 molecules have three atoms. It takes photons of different wavelength to affect the other gases, not infra-red. Have you learnt about photons yet?"

The girl shook her head. "No."

"I expect you will in time."

"So nobody's worried about the CO2 at the moment?"

"That's just about it, Emma. That's why our project is winding down. But in the future they may need to use the research we're doing now."

"Maybe I could be a pilot, like you," said the girl in a non sequitur.

"It's more difficult for girls to get into flying, but maybe that'll change in the future. Has Nigel taken you on any of his trips?"

"Not yet. He says he will when I'm older."

"It'll be fun."

Emma tilted her head. "If Nigel gives me permission, would you take me up in a little plane, Auntie Maddie?"

"Of course! I'd love to."

"I'll pick a moment when he's in a good mood."

"Are things okay at home now?" asked Madeleine.

"You mean, after Mum's death?"

"Yes."

"Yes, thanks, Auntie Maddie. Sometimes we both get sad but we can cope with it."

"Is Nigel still going out with . . . I've forgotten her name."

"Sophie? I'm not sure. I haven't seen her for a while so I don't know if they're still going out. It's strange, sometimes Nigel sits in a chair at home and goes into a sort of trance."

"Trance?"

"Yes, like he's turning things over in his mind."

"About your Mum, do you think? . . . or Sophie . . . or something else?"

"Don't know."

"Do you ask him about it?"

"No . . . I wouldn't know what to say."

"How long do the moods last?"

"Not long. He'll take a deep breath and sort of shake his head and then he's back to normal." Emma looked into Madeleine's eyes. "You shouldn't be sad, Auntie Maddie."

The professor quickly smiled. "Was I looking sad?"

"Not at the moment. But sometimes . . . you're like Cordelia . . . you look as if things are not going right."

"Sometimes I feel a little . . . I don't know . . . weary, maybe."

"You're better off than Cordelia 'cos you've got people who love you, like all the people here at your party. Most of all, I love you and Nigel loves you."

Madeleine blinked rapidly and took a tissue out of her handbag to blow her nose. She cleared her throat.

"I love you too, Emma."

*   *   *   *   *

Well, I'm not sad or weary today, thought Madeleine now as she turned off the motorway. It was a lovely day and she was going to meet two of her oldest friends.

Parking the Morris alongside the clubhouse the professor noticed a couple of gliders circling over Marlow a mile to the south at about two thousand feet. Approaching the airfield itself the Piper Cub tug aircraft was sinking towards the grass area south of the tarmac runway, its trailing tow cable flicking in the aircraft's slipstream. A Wycombe Flying Club Cherokee was also approaching to land. Madeleine instinctively watched the two arrivals. The Cub flared just above the grass, touched in a perfect three-pointer and rolled quickly to a stop. The Cherokee arrested its descent ten feet up, losing speed, and then started to drop. Madeleine waited for the crunch when it gave up the struggle to keep flying and hit the runway. But a sudden roar of engine caught the fall and held the aircraft in its intended medium. Recovering from its abuse, the machine groggily climbed away for another circuit. Madeleine wondered whether the saviour was an instructor taking control or a student realising his or her error and correcting accordingly.

On the apron were a few more Cherokees and Cessnas, the Wycombe Flying Cub Chipmunk and a vintage Auster. By the hangars stood Chris Coleman's Spitfire and a replica First World War Fokker Triplane.

Very agreeable, thought Madeleine as she locked her car, enjoying the sights and sounds of a busy airfield on a lovely spring day.

"Happy birthday to you . . . happy birthday to you . . ."

Parminder and Dizzy were walking towards her with grins on their faces.

"Thunder, lightning and rain," greeted Madeleine.

"Hurly-burly," responded her friends in unison.

After a few minutes swapping gossip and news the three ATA friends went to the restaurant for morning coffee and Parminder and Dizzy handed over their cards and presents.

"Ah, the birthday girl." It was Chris Coleman, who had just come in carrying a small book. "They said you'd be here yakking to Pammy and Dizzy."

"Damn cheek!" protested Parminder. "We do not yak—we exchange views."

"If you say so! So, girls, are we on for the quiz?"

"Yes," said Dizzy.

"Ready when you are," said Parminder.

"What quiz?" asked Madeleine.

"Right," said Chris, "let me grab a coffee and we'll start."

There was some discussion about which of the ATA pilots had dreamt up the idea of testing their memory of the technical details of one of their wartime mounts, namely the Spitfire Mark V, an example of which was basking outside in the spring sunshine.

Seated back at the table, Chris held up the book he had brought in. The others could see that the title read "Spitfire Mark V Pilot's Notes".

"Right," said Chris. "I'm the question master and umpire. In the case of disputes my decision is final."

"Hang on," said Madeleine, addressing her friends. "If you two knew about this you've probably done some revision."

"Heaven forfend!" smiled Parminder.

"That would be cheating," added Dizzy.

"Okay, what's the prize?" asked Madeleine.

"Magnum of Bollinger chilling as we speak," said Chris. "The winner is expected to share it with the losers . . . and the quizmaster." His glance swept round the three ladies. "Let battle commence. First to answer correctly wins the point. Question 1 . . . max take-off RPM and boost?"

"Three thousand, plus sixteen inches," said Parminder.

"Two eight fifty, twelve inches," said Dizzy.

"Three thou, plus twelve," said Madeleine.

"The birthday girl has it," said Chris. "Question 2 . . . what is the normal air pressure in the pneumatic system?"

"Three hundred," chanted three voices simultaneously.

"Good . . . a point each," said Chris.

Madeleine pleasantly surprised herself with the accuracy of the gen that her memory had retained but it didn't take long for her to realise that her friends were letting her win the competition. Eventually she held up her hands. "Okay, you lot, that's enough. Let's call it a draw and throw that Bolly down our throats."

"That was only part one of the quiz," said Chris. "Now we'll go outside and check your practical knowledge."

Chris had arranged for a photographer to take pictures of the ladies in front of the Spitfire. There was only one flying suit and one pair of boots so they took it in turns to don the wartime garb and pose for the camera. First was Dizzy, then Parminder and finally Madeleine.

"Right, let's have a piccie of you in the cockpit," said Chris.

"Love to," said Madeleine, climbing on to the wing and stepping inside. "Haven't done this for . . . what . . . over twenty years."

"Here . . . hold this," said Chris, standing on the wing outside the cockpit. It was a flying helmet. "Now smile for the camera."

And then the professor saw the trolley accumulator being wheeled up to the aircraft by two of the maintenance crew. Her mouth opened in surprise as they plugged the accumulator cable into the aircraft's external power socket. She turned to look at the Spitfire's owner.

"Chris, what's happening? Are you going to start the engine? Are you going to fly it?"

"No, Maddie, you are."

"What!"

"Happy birthday!"

The chemistry professor began to protest but her wartime colleague held up his hand to silence her.

"You've just proved you still know the important stuff. And Dizzy told me you've got recent tailwheel time on the Chipmunk. So now you're going to show us you can still handle a Spit."

Madeleine looked down to see Dizzy and Parminder grinning at her, thumbs up. And standing next to them . . . Nigel! Her brother had appeared from somewhere, joining in the fun.

"You horrors!" she grinned back. "I'll kill you! You set the whole thing up!"

"Guilty!" laughed Dizzy.

CHAPTER 24

Chris continued the pre-flight briefing while the ground crew fussed round the aircraft. A growing crowd of onlookers watched the proceedings with interest.

"There are a few differences since the last time you drove one of these, Madds. You'll see the old push-button four-channel VHF is gone but the new 360-channel set is located in the same position. It's a Narco, just like the Cherokee. You've got a VOR as well if you need it. The press to transmit button is on the throttle, like it used to be."

"Okay."

"Airspeed in knots—some of the earlier Spits had miles."

"I noticed that."

"I've deactivated the boost control override to preserve the engine but there's still plenty of power. Feed in the throttle slowly on take off and remember it's right rudder to keep straight—the opposite of the Chippie."

"Will I need to do a full power run-up?" asked Madeleine, fastening her parachute harness and seat straps.

"No—we did one earlier. Maybe a mag check at two thousand RPM but keep an eye on radiator temp. You should be okay as long as you don't loiter—the air's not too warm today."

"Rad setting when airborne," said Madeleine. "Red triangle, as before?"

"Correct, but check the temp and adjust if necessary."

The professor's right hand moved to the top of the stick and her thumb caressed the guard over the gun firing button. She grinned at the Spitfire's owner. "Okay to shoot any one down who gets in my way?"

"You'll have to go dak-a-dak-a-dak yourself, Maddie. Sadly the guns are empty."

"Shame."

"Quite. I've kept the button there for authenticity for when I do film work. Ditto the VHF—I've made a cardboard replica of the original which covers the new kit."

The discussion moved on to navigation and airspace restrictions. Chris pulled out a chart from the stowage left of the pilot's seat.

"You know the local area well from your Cherokee flying, don't you," said Chris. "Stay below the TMA till you're north of the Chilterns. I usually head towards the Vale of Aylesbury for general messing around and if you stick below eight thousand you'll be legal. Base of Amber One is eight-five out there on ten thirteen." The Spitfire owner was referring to airspace restrictions and altimeter setting procedures.

"No shortage of landmarks round here," said Madeleine. "It would be difficult to get lost . . . famous last words!"

"Plus you've got the VOR. It's tuned to Garston but you can change it if you want, of course."

"Okay."

"You've got full tanks, Maddie, so if you fly for an hour, say, you'll have plenty of reserve."

"Okay."

"Give the tower a call when you're on the way back and we'll launch the camera ship."

"Camera ship?"

"Nigel and I will get airborne in the club's Comanche and head out to meet you. If you formate on us we'll take some photos."

"Right, but tell me if I'm getting too close—it's twenty years since I've done any formation."

"We trust you!"

"Good . . . what's my callsign, by the way?"

"We've told the tower to listen out for 'Spitfire Maddie.'"

"Easy to remember."

Chris held out a conventional headset to swap for the wartime helmet. "This'll be more comfortable for you." He plugged in the headset, checked Madeleine's straps and stood back.

"You're on your own, Madds."

"Thanks so much, Chris. I'll try not to bend it!"

The Spitfire's owner hinged the door up to the closed position and Madeleine secured it from inside the cockpit. Her eyes swept the controls and indicators one more time and then she looked outside to see Chris clearing spectators and unnecessary equipment out of the way.

Madeleine gave the engine a few squirts of primer and then checked the primer was locked. She set the throttle slightly open and checked outside to make sure the areas round the big three-bladed prop and behind the tail were clear.

She switched the magnetos on and called out, "Contact!" Outside, Chris was giving her thumbs up.

Madeleine held the stick back with her left hand and with her right pressed the starter and booster coil buttons. The starter turned the Merlin through a few blades and the engine coughed and then caught, one cylinder . . . two . . . soon all twelve cylinders were firing and a whiff of exhaust smoke took the pilot back twenty years.

Oil pressure, fuel pressure, brake pressure good. Gyros erecting. Check the individual magnetos. All good. Madeleine signalled "Disconnect equipment and chocks away" and watched the ground crewman move the trolley accumulator clear.

Madeleine pressed the transmit button. "Wycombe, Spitfire Maddie taxy for local area."

"Good afternoon, Spitfire Maddie, taxy for runway two five, QNH one zero one seven, do you want the grass or the hard?"

"Grass runway, please."

"Roger, advise ready for take-off."

The pilot acknowledged and set her altimeter. A glance right and left. At the wing tip Chris gave a wave and Nigel blew her a kiss. Release brakes—an inch forward on the throttle and the Merlin crackled its compliance. The Spitfire rolled forward and Madeleine brought her throttle back to idle. Even at minimum power the mighty Merlin could pull the aircraft across the apron at a comfortable pace.

Taxying out, weaving the nose to check all clear ahead, with frequent glances at the radiator temperature, the pilot was again transported to White Waltham during the war years. Those beautiful elliptical wings, the rumbling of the idling Merlin a tiny hint of the thousand horses it could unleash, the smell of the cockpit. Was it really over two decades ago?

Five minutes later Madeleine was at the holding point, checks complete, hood forward and closed. Rad temp still rising but below limits. Time to go!

"Spitfire Maddie, ready for take-off."

"Spitfire Maddie, cleared take-off, two five grass, wind one nine zero, four."

Madeleine released the brakes and slowly inched the throttle forward with her left hand, the growl of the Merlin swelling into a roar. A generous bootful of right rudder kept the aircraft's nose pointing straight. A glance at the airspeed indicator confirmed speed building. With the stick neutral the Spitfire quickly lifted its tail and after two hundred yards or so eased itself into the air.

Happy that her aircraft was climbing and accelerating, Madeleine swapped hands on the stick and selected undercarriage up. Remembering Chris's entreaty to nurse the engine she throttled back a little and reduced prop RPM, adjusting attitude to stabilise the speed at one sixty knots. A bank to the right swung the nose northwards until the fat concrete pillar of the Stokenchurch telecomms tower was visible ahead and left as an orientation check. Conscious of controlled airspace above, Madeleine brought the throttle further back to reduce her rate of climb.

Then the Chilterns were behind her and the pilot opened up her Merlin again to soar to six thousand feet, where she levelled off and pulled into steep turns left and right to check her position and confirm there were no other aircraft nearby, the 'g' forces pushing her down into the parachute pack cushion acting as her seat.

The next twenty minutes were sheer delight. Madeleine threw the aircraft around in a sequence of loops, barrel rolls, chandelles, stall turns, and slow rolls. Inexplicably she found herself laughing out loud and singing snatches of the old Max Bygraves song "Out of Town". Absurd! A grown woman edging towards her half century fooling round like a kid with a new toy!

"Spitfire Maddie, are you still on frequency?" came the voice in her headset earphones.

"Affirmative."

"Roger, we have a request."

"Go ahead."

"RAF Halton have requested you do a low level pass."

"Of course."

"Roger, change to one two two point one and give them a call. They're expecting you."

"Roger."

Madeleine did an orbit to check her position. As she said to Chris, how could anyone get lost round here? In the excellent visibility she could see the new Didcot power station cooling towers, the line of the Chilterns with a lazy column of smoke drifting upwards from the cement works at Chinnor and the three railway lines converging at Princes Risborough. To the east lay Aylesbury, to the west Oxford and closer were the disused airfields at Oakley and Westcott.

Madeleine stopped her turn on a gyro heading of one zero zero and throttled back to start descent. A check of the instruments and gauges and fuel remaining confirmed all in order. She changed the transceiver frequency and called Halton.

"Good afternoon, Spitfire Maddie, advise when you're five miles out and we'll clear you in for a run in."

"Roger," replied Madeleine.

A different voice came on the air. *"Spitfire Maddie, this is Benson, do you read?"*

Madeleine remembered that 122.1 was the common frequency for aircraft fitted with civilian radios to contact military bases. Benson was an RAF base a few miles east of Didcot.

"Loud and clear, Benson," responded the pilot.

"Roger, can you fit us in after you've beaten up Halton?"

"It will be my pleasure!"

* * * * *

Madeleine roared along runway zero one at Benson as low as she dared with the airspeed indicator needle quivering at 270 knots, just over 310 mph. Briefly the old Air Force joke flashed through her mind:

"Poor old so-and-so pranged yesterday."

"What was he doing?"

"About three hundred knots!"

Passing the airfield boundary she pulled up vertically until the speed dropped below two hundred then eased the nose forward while rolling on right bank.

"Thank you, Spitfire Maddie," said the Benson controller in the pilot's earphones. *"We liked that. When you're clear change back to Wycombe."*

On the Wycombe frequency she heard Chris's voice. *"Spitfire Maddie, this is Lima Oscar, if you read us, say your position."*

Madeleine straightened up on heading zero four five, altitude two thousand. "Reading you, Lima Oscar, I'm three miles west of Stokenchurch. Say your intentions."

"Okay, Maddie, we're just south of Thame at two point five. We're heading one eight zero, looking for you."

The chemistry professor looked left. A black speck was visible just above the horizon.

"Tally ho!" she transmitted. "Got a visual on you, Lima Oscar."

"We see you too, Madds. We're indicating one forty knots, two point five, turning left heading zero five zero."

"Roger, Lima Oscar," replied Madeleine, "where would you like me, port or starboard echelon?"

The Comanche's speed was a mere dawdle for the Spitfire and Madeleine closed rapidly, throttling back to formate behind its tail to the right, lifting her nose a little to increase the wing's angle of attack to compensate for the reduced airspeed.

"That's a good position, Spitfire Maddie. Hold it while we take some pics."

After easing over to the other side of the Comanche for further photos Chris announced he was turning left to reverse course, Madeleine holding station throughout.

"Good stuff, Madds," heard the professor after a few more minutes. *"We've got what we want. See you back at Booker. We'll follow you in but we'll hold clear if you want to do a run and break."*

"Roger, Lima Oscar, I'm breaking off."

Madeleine swung east towards the Hughenden valley running from Princes Risborough to High Wycombe and called up the field.

"Wycombe, Spitfire Maddie, rejoin."

"Welcome back, Spitfire Maddie, join right base or long finals for runway two five. You're number one for a run in."

For the third time that day Madeleine roared along a runway a few feet off the ground at breakneck speed, the Merlin snarling in front of her. Controlled airspace above Booker precluded a vertical climb so the pilot eased the throttle back and lifted her nose slightly to bring the Spitfire up to circuit height while killing her speed. A gentle right turn brought her on to the downwind leg and she lowered the undercarriage and ran through her landing checkist.

"Spitfire Maddie, downwind."

"Spitfire Maddie, cleared to land two five grass, wind light and variable."

A curving descending approach lined her up with the grass runway. Madeleine lowered her flaps, the hiss of pneumatics touching another chord of nostalgia. She let the drag ease her back to 80 knots, the Merlin crackling a smidgen above idle power.

Over the new M40, sinking towards the runway threshold, throttle fully back, hold her off, back on the stick . . . back . . . back . . . she touches! Three pointer! A little rudder to keep her straight and a touch of brake to slow her.

"Nicely done, Spitfire Maddie, you're cleared to taxy to the apron."

CHAPTER 25

Well, that was a lovely birthday present, thought the chemistry professor as she headed home in her Morris Minor. Thank you, Chris Coleman! And thank you Dizzy and Parminder and Nigel, who together with Chris had organised it all. She would have to find some way of repaying them.

Another group of enthusiasts had welcomed the Spitfire on its arrival back at Booker, including the ex-ATA pilots. When Madeleine pulled the cut-out to stop the engine and slid her hood back a mighty cheer arose, audible even to her Merlin-deafened ears. Amongst those pointing cameras and cine-cameras at her she recognised the *Chiltern Herald* reporter who had previously interviewed her.

It took half an hour to answer all the questions thrown at her and satisfy the photographers who wanted pictures of her and her ATA friends in various combinations. Finally the spectators drifted away and Chris began to organise the Spitfire's post-flight handling, telling the ladies he would join them in the restaurant for Bollinger and cucumber sandwiches.

If Emma could see me now she wouldn't think I was sad, thought Madeleine as she steered her Minor along the A40 through the little village of Tetsworth. A couple of hours previously she had been cavorting in the clear blue sky above this little corner of Oxfordshire in that beautiful flying machine. What a perfect spring day. The Spitfire. Her friends. The champagne. She had declined a third glass so her brain wouldn't be too befuddled on the drive home.

No, life was good, all things considered. Maxwell Murdoch had promised that Oxford's methane reformation project would be kept going until Madeleine's post grads had got their PhDs, after which it would almost certainly die. So be it. Perhaps the professor had indeed "backed the wrong horse", as her old rival Alfie Grove at the Science Research Council had suggested. If the rest of the world judged atmospheric CO2 levels to be unworthy of attention perhaps it was Madeleine who was out of step rather than them.

The students were already planning their futures. Luke Stirling was now engaged to girlfriend Gill Bradshaw and it looked like he would join the Bristol based pharmaceutical company where Gill worked once his doctorate was successfully completed. Luke's version of events was that she had proposed to him and he had accepted as long as he didn't have to join the Labour party. Gill's account, as relayed to Madeleine by Xiu Ying, was that Luke had popped the question and she had accepted as long as he didn't try to prevent her becoming a party member.

Xiu Ying and Joe Curtis seemed to be headed for a future together in Hong Kong. Joe had visited his girlfriend's parents and announced to his colleagues that he had "passed Suzie's Dad's selection procedure".

James Ibsley's girlfriend, Stephanie, had proved her devotion by turning down a leading role in a dance show in Sydney, Australia so she could stay with him in the UK. Suppose Broadway summoned, he had queried. James related to his colleagues that her response was that she'd immediately dump him. He hoped she was joking.

Madeleine's future looked promising, academically at least. She had taken up ICI's offer of managing their new aromatic polyamide development project. As the methane reformation work ran down the ICI task would build up momentum. To confirm his support for the remaining MR work Maxwell Murdoch had even gone so far as to present Madeleine with a legal document obligating Southern National Generation to funding as required (up to defined limits) until the post grads had finished their practical work, or until 1159 p.m. Eastern Standard Time, December 31, 1968, whichever came first.

Strange man, Maxwell Murdoch, thought Madeleine. Religious zealot and larger than life character, following the American tradition of ruthlessly amassing a fortune and then dispensing some of it in philanthropic largesse, although it was likely that MR funding was a mere trickle compared to the flow of Murdoch dollars into the Righteous Warriors coffers. Strange man, getting stranger, judging by a report Madeleine had read in her paper recently.

It seemed that a faction in the Warriors had hatched a plan to plant a nuclear bomb in the centre of Hanoi, North Vietnam, to force the Communists to surrender and so end the civil war tearing the country apart. The plotters had supposedly approached Max Murdoch and asked him to supply a suitable nuclear device but instead of allying himself to the cause he had reported the plan to the CIA and the FBI. The plotters had refuted the allegation and were now threatening to take legal action against Max, citing defamation of character and accusing him of attempting to falsely stigmatise them with the intention of displacing them from the ruling council and taking more control himself. The Warriors were now dividing into the Murdoch camp and the Hartmann camp, hurling insults at each other. Both factions declared that God was on their side.

Which group was telling the truth? wondered Madeleine. Max was certainly a bit eccentric, but to invent such a far-fetched story . . . that would really be weird. A determined man, yes, but she'd never thought of him as a power-crazed schemer. Anyway, she would be able to ask him all about it soon because she would be meeting him on his native territory.

The trigger for the New York trip was her brother's roster. Heathrow to JFK, two night stopover and return. Nigel had offered to take her along as supernumerary crew and she had readily agreed. The chemistry professor had mentioned the visit in a phone call to Max and he had said he would see her in New York as he had a business meeting there. By coincidence he would be travelling to England on the same day as Madeleine. He told her he would cancel his reservation on Pan Am and rebook on Nigel's flight.

A further bonus for Madeleine was that according to Nigel the rostered flight engineer was Frank Armitage, another colleague and friend from ATA days. She hadn't seen him for a while and it would be a good chance for a catch up. A faint bell in her memory told her that he made reproduction antique clocks as a hobby. A new clock for her dining room had been on her wish list for a while. If the price was right maybe Frank could supply one.

Perhaps the only missing element in Madeleine's life was romance. Sometimes she would find herself wondering whether she should have given in to Gordon McKenzie during her first Wilmington visit. Of course that was not really romance . . . rather raw sex. Which did she really want? The honest answer . . . both!

Surely she wasn't too old for physical love? But there weren't any available men in her orbit. Those who had made advances in recent times were either married or . . . yuk! There was a glimmer of hope, however. John Davis had sent a letter saying he was back from his Rio posting and would she like to meet him for bridge or a trip to the theatre?

Yes, she would! And she would certainly give in to him if he made advances. Judging from his enthusiasm on their last date before he went off to Brazil there would be no reluctance on his part. She had held him at bay—shades of Gordon—and he, ever the gentlemen, had cooled his ardour. If for some reason he was slow to start the ball rolling this time round, she would take the lead! *In the arena of love these days, anything goes!* So the younger generation thought, anyway. So why shouldn't the older types join in the fun!

Driving through Woodstock Madeleine turned her thoughts back to New York. Nigel had booked her a room in the Santa Clara Plaza, the same hotel as him and his crew. When she said she would pay he said no, she wouldn't. Especially as he had managed to swing airline discount on the nominal rate. Or so he said, anyway.

A rough plan was forming. They would be departing Heathrow mid-morning on the Monday, arriving early afternoon local time. After a nap and a shower the crew would meet up in the hotel bar for pre-prandial drinks. Some BOAC captains insisted that all the members of their crew should go out together for meals and entertainment. Some even demanded equal payments from them when settling the subsequent bill, wilfully ignoring the fact that some crew members had deliberately chosen cheaper options in order to save as much of their allowances as possible for other purposes.

Nigel was not one of these captains. Once they were off duty, his crew could go wherever they wanted with whoever they wanted. His only stipulations were that their behaviour should not bring the airline into disrepute and that they observed the abstinence from alcohol rules prior to the return duty. When it came to paying restaurant bills, Nigel's policy was to ask everyone to chip in what they thought they owed, including a tip. Only very rarely did the kitty fall short and the captain was happy to unobtrusively throw in a bit more of his own cash to preserve the genial mood and avoid embarrassment.

For their free day, Nigel had suggested that he and Madeleine and anyone else who wanted to join them could buy tickets for the country music festival scheduled at the Delacorte Theater in Central Park. Madeleine happily agreed, although she pointed out she would need some time for wandering around the Fifth Avenue shops. And the Metropolitan Museum of Art. For the second evening Nigel thought they should see if there were tickets available for a good Broadway show. He was happy for Maxwell Murdoch to join them for any or all of these diversions.

The return flight left JFK at 1910 local time on Wednesday evening and was scheduled to land at Heathrow soon after the night jet curfew finished at 0600 on the Thursday.

As Madeleine approached her house she noted a familiar car parked behind Toni's in the road outside. She turned into her driveway, searching her memory for the identity of the owner. Ford Anglia, two tone blue . . . John Davis! What a coincidence!

She let herself in and checked the silver platter in the hallway for second post. Nothing today. Her lodger and her visitor were not in the lounge or the dining room or the kitchen. Madeleine dropped her handbag on the kitchen table and opened the door to the garden. Not there either.

She thought she heard a noise from upstairs. She went slowly up, wondering whether to call out, "Anyone at home?", before an instinct told her to keep her silence. The door to Toni's room was ajar and there was certainly some sort of activity going on in there, realised Madeleine. She pushed the door open.

John was on his back on Toni's bed, naked. Toni was astride him, also naked, groaning and grinding her pelvis on her partner's. No imagination was needed to deduce the geometry of their more intimate anatomical parts.

Madeleine's jaw dropped. "Oh . . ."

In mid thrust Toni turned towards her. "Maddie . . . hi." She grinned. "Get your kit off and join us if you want. We're having a great time."

John also turned his face towards the door and his look transformed from eyes half closed ecstasy to embarrassed surprised. "Oh . . ." he echoed.

"Sorry, Madds," continued Toni, momentarily resting herself on her lover's lower abdomen. "I didn't think you'd be home so soon."

"No . . ." said her landlady, still trying to regain her composure.

Toni looked down with a broad smile. "John, darling, you've gone all limp on me. What can I do to stiffen you up again?"

"I'll leave you to it," said Madeleine, turning away. "I'm just . . . going to the shops. I'll be back in about half an hour."

"Lovely, Madds. We'll be done by then . . . probably. Buy some cakes and we'll all have afternoon tea."

*   *   *   *   *

Normality of a sort had been restored an hour later. Toni and John were dressed and seated at the kitchen table, sharing tea and Vienna slices with Madeleine. John looked a little sheepish but Toni was chirpy.

"It's all my fault," she said. "I dragged him upstairs for a shag after he had told me he wasn't actually going out with you."

John smiled weakly. "I didn't exactly resist, did I? But what Toni says is true. None of it was planned. I was passing through and thought it would be a nice surprise to see you, if you were at home."

"I was certainly surprised," said Madeleine wryly.

"I actually wrote out a note to drop through your letterbox in case you weren't in," continued John, "suggesting meeting somewhere. Toni invited me in and we got chatting over a glass of wine. She asked me if we were together and I told her, 'no'."

"Well, no harm done," said Madeleine. "And no reason why we shouldn't meet socially . . . without any obligations."

The visitor smiled ruefully. "I suppose you think I'm a bit of a cad."

Madeleine returned the smile. "Actually, no, John. I know you're a decent sort. When Toni is out on the prowl not many men could resist."

"That's me," grinned Toni. "When the urge hits me I've got to find someone who's available to service it. Poor old John didn't stand a chance. As soon as he told me he was free I practically tore his clothes off. I was desperate for a shag."

"For what it's worth I thoroughly enjoyed it," said John. "You certainly know how to get a man going."

Toni tilted her head playfully. "Are we on for a return match?"

"Not here!" said John. "That would be disrespectful to Maddie."

"Look," said the landlady. "If you must, then you must. But perhaps wait until I'm not here."

"You could always join in. What do you think, John? Shall we give our Madeleine a good seeing to?"

"No!" said the other two together.

"Shame!"

CHAPTER 26

The sign on the door said "B.O.A.C. Crew Report" but everyone knew it as "Room 221", the room's designation before it migrated to the Central Area Oceanic Terminal from the old North Side buildings.

On this busy Monday morning in early May there were various aircrew in the room coming and going, carrying out pre-flight preparations or post-flight admin tasks or just standing around swapping gossip with chums. It had taken a few weeks for the mood in Room 221 and the rest of the airline to lighten after the incident the previous month in which a BOAC Boeing 707 had burnt to a skeleton after an emergency landing at Heathrow. An engine had caught fire just after take-off and the crew hastily brought the burning aircraft back. After landing the fire had spread uncontrollably. During the evacuation, five passengers and a stewardess, Jane Harrison, had died. Jane had stayed at her post until overwhelmed by smoke, giving up her life trying to save her passengers.

This May morning fine weather helped to lift the spirits of the BOAC aircrew in Room 221. At one table sat Captain Nigel Nixon, First Officer Rick Mountford, First Officer Jeremy Bosworth, Senior Engineer Officer Frank Armitage and Professor Madeleine Maunsell. Nigel had introduced their supernumerary crew member to the two copilots, who listened respectfully when Frank added a little of Madeleine's history to the introduction. The flight's cabin crew occupied a nearby table and Nigel had already greeted them and checked relevant information with the Chief Steward.

As always, a considerable chunk of the pre-flight work revolved around documentation detailing expected fuel requirement, flight time, navigation, payload details and estimated take-off weight. Later on the three pilots would make their way to the Briefing Room to check winds, en route weather, destination and alternate airport weather and Notices to Airmen, "Notams" in aircrew parlance. At the same time Frank would proceed to the departure stand to do his engineering checks on the aircraft.

"So, they've given us Foxtrot Golf," said Nigel now. "It's already here on Stand forty-one, fully serviceable as far as we've been told." He tapped the sheaf of papers on the table. "We'll be on the heavy side, but the headwinds aren't too bad, minus forty to forty-five overall. At the estimated take-off weight take-off performance shouldn't be a problem if we're using two eight right." Runway 28R was the northernmost Heathrow runway, pointing east-west in parallel with the old A4 Bath Road just outside the airport and the new M4 motorway a little further out.

Madeleine understood most of what was going on. Apart from her own aviation background she had accompanied her brother on the occasional trip in the past, both on the Britannia turboprop and latterly the Boeing jet.

"Hello, boys, what mischief are you getting up to?"

They looked up to see a middle-aged man in BOAC captain's uniform, cap at a rakish angle. The moustache on his grinning face was even more luxurious than Frank's.

"Hello, Mike," greeted Nigel. "Good trip?"

"Super. Fell in love with my 'A' girl . . . again!"

"You're incorrigible, you old rogue!"

"You're taking Fox George, aren't you?" Captain Michael Beechwood was one of many older pilots who sometimes lapsed into the pre-1956 phonetic alphabet.

"Uh-huh. You brought it in, didn't you?"

The other pilot nodded. "She's a good ship. Couple of minor snags. Nothing to stop her taking you lot to JFK if you treat her nicely." He tilted his head and smiled at Madeleine. "I don't think I've had the pleasure . . ." he began.

Again Nigel made the introductions and Mike looked at Madeleine quizzically. "I do know you," he said. "Or I know of you. Now I come to think of it, I've seen your photograph."

"She was in the papers recently, flying a Spitfire," said Nigel. "Is that what you mean?"

"No, I didn't see that. No, I'm talking about an earlier photo, when you were with the ATA. You were Terry Walker's girlfriend, weren't you?"

"We were engaged," smiled the professor.

"We were both on 139 Squadron—Mosquitos—Terry and I," explained Mike. "He didn't make it back from the Dortmund raid. I was on that raid myself."

"They told me he got shot down by a night fighter," said Madeleine.

"No, someone must have got their wires crossed. That's not what happened . . . Sorry, Nigel, I'm interrupting your briefing."

"We've got a few minutes to spare, Mike. You carry on."

The other captain pulled up a chair and sat himself down.

"It was nasty. We found the target okay but we got a real pasting from the flak guns. We were pathfinders, dropping markers for the main force. Terry was one of the first in. After he laid his markers he announced he would make a dummy attack to draw the flak. Well, actually he did it three times. Then the CO told him to break off and head for home and someone else would take his place. Terry said he'd do one more run and then bugger off. That's when they got him."

There was a respectful silence around the table.

"He was a good chap," said Mike. "One of the best. Good instructor, too. He checked me out in the Mossie. Got a posthumous DFC after Dortmund, I believe."

"Yes," said Madeleine. The Distinguished Flying Cross was awarded for "exceptional valour, courage and devotion to duty whilst flying in active operations against the enemy".

"I'm curious," continued the professor. "Where did you see my photo?"

Mike looked at her. "Terry had a picture of you he used to take with him on missions. He'd stick you on the instrument panel by the DI with a bit of plasticene." The direction indicator was a gyroscopically stabilised instrument telling the pilot the aircraft's compass heading. "He said you would always point him in the right direction."

"I don't remember him telling me about the photo," said Madeleine.

"Anyway," continued Mike, "on the outbound leg someone asked Terry on the VHF if his Madeleine was keeping us on track. He replied that he'd forgotten to take the photo with him and they would have to rely on their navigators. So then there was a bit of banter before the CO told us off for breaking radio silence. When we got back we found the photo among his effects."

Another pause. Then Madeleine said, "They must have sent the photo to his parents. I never saw it. Thanks for telling me about it, Mike."

"And now we've got the real thing to guide us on our way," smiled Nigel, pointing at his sister.

"Nice to have met you, Madeleine," said Mike, standing up and holding out his hand. "I'll bid you all adieu." In a stage whisper he said, "Nigel doesn't like his briefings interrupted, you know. It's not SOP."

"Bugger off, you clown!" grinned Nigel. As his colleague walked off his expression turned more serious. "Okay, chaps, let's get back to business."

*   *   *   *   *

Foxtrot Golf rolled slowly along the outer taxyway towards the two eight right holding point. In the cockpit three of the four men in the crew were running through the before take-off check list. There were only two aircraft ahead of them in the departure queue so they didn't anticipate any delays. In the copilot's seat sat Jeremy Bosworth, who had been offered the leg as handling pilot by the captain. The crew's other copilot was seated at the navigator's station behind Madeleine, who herself occupied the observer's seat alongside the flight engineer. Rick Mountford had accepted the captain's request that Madeleine take the place he would have otherwise expected to be seated in himself. Of course, he wouldn't have dreamt of declining, but he appreciated being "requested" rather than "ordered". He could think of several captains who would not have bothered to consider his feelings on the matter.

Like the others, Madeleine was wearing a headset but she wasn't really paying attention to the ritual the 707's crew were going through. The photo Mike Beechwood had mentioned in Room 221 was taking her memory back to the war, to Terry, to White Waltham and to the adventures she and her friends went through, some enjoyable, some frightening, some sad, some . . . painful. And here she was, quarter of a century later, sharing the cockpit of an airliner with Nigel and Frank, two of the White Waltham brigade. Who would have thought they would survive the war, let alone find themselves together in a large jet airliner about to whizz across the Atlantic?

The cockpit door opened and the Chief Steward came in. He waited until Nigel could spare his attention. "We're all strapped in here, Captain," he announced. Nigel smiled and gave him a thumbs up.

"What's on the lunch menu, Barry?" asked Frank. "I'm getting peckish."

"Bread and water, unless you're good boys."

"We'll be good!"

"I'll see what we can rustle up."

"I can help in the galley, if you like," said Madeleine.

"Thank you," said Barry, moving back and opening the door to leave. "When you're fed up with these adolescents, come and chat to us. There's a seat for you in First Class. See you later."

Frank Armitage, thought Madeleine now, reminiscing again. Sitting there at his panel, stroking his moustache. His claim to fame

174

was the time he brought his Lancaster home after the pilot got hit by flak, American if she remembered correctly, a B17 driver who had converted to Lancs. Married with a couple of kids, wasn't he? She'd have to ask him about his family later on.

Nigel Nixon, her half-brother, Moths to Lancasters to Boeing 707s—quite a journey. Taught her instrument flying to keep her out of danger. Nigel, who might have ended up as her husband if . . .

"Speedbird five nine two, advise when ready," came the voice in the headphones.

In the copilot's seat Jeremy looked across at the captain and raised a querying eyebrow. In turn Nigel turned towards the flight engineer.

"All set, Frank?"

"Aye skip, before take-off checklist complete."

Nigel nodded to Jeremy, who pressed the transmit button on his control wheel. "Affirmative, we'll be ready on reaching."

"Roger, Speedbird five nine two, after the departing Trident line up runway two eight right."

"Okay, chaps," said Nigel on the intercom. "Essential chat only until further advised, please."

Rotating the tiller wheel by his left knee, the captain turned Foxtrot Golf's nosewheel to swing the aircraft round onto the runway and then he applied the parking brake.

"Your aircraft," he said to Jeremy. "My radio."

"Speedbird five nine two, cleared take-off, wind two five zero, seven."

Nigel acknowledged the clearance, then glanced quickly at his copilot and flight engineer.

"All set, chaps?"

"Yes, sir."

"Aye, skipper." Frank glanced at Madeleine and gave her a wink.

"Okay, let's go."

"Brakes released," said Jeremy, holding the control wheel.

Nigel's right hand slowly pushed the four throttles forwards and the Conways roared in response. He lifted his hand an inch or two and Frank, leaning forward, adjusted the throttles.

"Power set, all good."

Nigel's hand dropped back to the throttles.

The jetliner gradually accelerated, Madeleine noting various callouts from the other three over the intercom.

"Speed building both sides."

"One hundred knots."

"Vee one . . . rotate."

Jeremy pulled his control wheel steadily rearward and Foxtrot Golf lifted its nose in response. Out of her window Madeleine watched the runway drop away.

"Vee two, positive rate."

"Gear up."

Madeleine watched the pilots as they retracted the flaps and let the aircraft accelerate. Jeremy called for autopilot engagement and released the control wheel.

"Ready for the after take-off checks, Jeremy?"

"Yes, sir."

"Here they come."

"Speedbird five nine two, airborne one four, contact radar, one two three decimal seven."

With the 707 safely airborne the atmosphere in the cockpit relaxed a little and the crew released their shoulder harnesses. Frank grinned down at their supernumerary crew member.

"Five minutes gone, Maddie. Only seven hours and four minutes to go."

CHAPTER 27

"Looking after you, are they?"

Madeleine looked up to see Frank's smiling face. She had been half asleep, the magazine she was reading open on her lap. The first class cabin was only half full and the chemistry professor had been able to spread out her accoutrements on the seat next to her.

"Yes, just had a lovely lunch—smoked salmon paté and delicious canapés, washed down with a glass of bubbly."

"Yes, Lorraine sneaked in a few bits and pieces for us at the sharp end. No bubbly, sadly."

"They seem like a nice bunch, your crew."

"It'll be a good trip, Maddie. Always is when Nigel's running the show. He looks after his crews. Some captains treat the cabin crew like dirt, not to mention their copilots and engineers."

"His partner used to be a stewardess, didn't she?"

"Yes . . . Jackie Holmbeck. Lovely girl. We were shocked when she got killed."

"I suppose she would have given Nigel an insight into what life as cabin crew was like. Maybe some pilots don't think about that."

"That would be part of it, Maddie, but it's mainly a character thing. Nigel's just a nice bloke. No chance of him getting payback pie."

"What's that?"

Frank settled himself into the seat across the aisle. "May I join you for a while? . . . one of the kids is looking after the panel. Nigel likes to give us breaks away from the grindstone. Helps to keep us happy. Some captains don't bother."

"Of course, Frank. Always nice to chat with you. What's this 'payback pie?'"

"You'd think that pilots would take care not to antagonise the folk that were bringing things they eat and drink. The bleeding obvious, isn't it? If a pilot annoys the cabin crew they get their revenge by . . . modifying the food and drink they take into the cockpit."

"Modifying?"

The flight engineer listed a few of the treatments, Madeleine's look of surprise gradually turning to horror.

"Surely they wouldn't do that, Frank?"

"So I'm always courteous and considerate to the cabin crew . . ." He smiled up at the "A" girl approaching them, ". . . especially gorgeous creatures like the lovely Lorraine here."

The stewardess smiled at Madeleine. "Is he bothering you, Miss Maunsell?" She winked at Frank. "I'll ask the captain to divert to Keflavik and offload him."

"That wouldn't work, Lorraine. I'm the only bloke round here who knows how this flying machine works."

"Has he come out with the thing about pilots and engineers yet?"

Madeleine smiled. "I've heard it many times over the years—pilots break aeroplanes and engineers fix them."

Lorraine rested her hand on Frank's shoulder. "Would you two like a cup of our superb first class coffee? I'm just refilling the perc."

As the stewardess returned to the forward galley Frank leaned towards Madeleine. "So you enjoyed strapping a Spitfire to your backside again?"

"It was wonderful, Frank. Complete surprise, fixed up by Nigel and the ATA girls. Took me back twenty years plus. Pure nostalgia."

"White Waltham. They were good times, weren't they, mostly."

"Mostly, yes."

They swapped war stories for a few minutes, sipping the fresh coffee Lorraine had brought them and then compared more recent history.

"You've got two boys, haven't you?" asked Madeleine.

"Yep, teenagers now. Absolute terrors."

"Are they following their Dad into aviation?"

"My eldest wants to be a racing driver and his brother a deep sea diver . . . wants to make films about marine wild life. I've no idea where that notion came from."

"What are you doing when we get to New York?" asked Madeleine. "Are you coming to the concert in Central Park?"

"Yeah, reckon so. I think most of us are, though Barry, the Chief, is off to see some relatives somewhere in New Jersey, I think. You want to hit the shops, didn't Nigel say?"

"Yes, and maybe the art museum. Have you seen it?"

"Uh-huh. Though as a bit of a philistine that sort of stuff isn't really my thing. Is Nigel going to the museum with you?"

The professor shook her head. "No. He said he's got a lot of paperwork to do. He's got a meeting with our solicitor on Friday morning to go over his Mum's will and other documents. He reckons when we get back on Thursday he'll spend most of the day sleeping so he wants to do the paperwork over the next couple of days."

The flight engineer nodded. "And you're meeting this mad American bloke, aren't you?"

"Yes. And he's taking us to Broadway to see 'This Time With Feeling', picking up on Nigel's idea. Max has sorted out the tickets apparently. Is all the crew going, do you know?"

"I think so, apart from Barry."

"Sounds like fun."

"I read about this Max bloke in the paper. Something to do with blowing up Hanoi with a nuke."

Madeleine supplied the developments that she knew of. It seemed that the warring factions in the Righteous Warriors had come to some sort of truce. A statement had been issued explaining that the whole fracas was actually the result of a misunderstanding. The Hanoi plot had originated as a joke but some people had taken it seriously.

"Actually, the settlement was purely a device for avoiding law suits," confided Madeleine to Frank now. "Max told me the Hanoi thing was genuine but no one wanted an expensive court case so the deal was struck. The only fly in the ointment is that the other lot want him to stand down from the steering committee and he's refusing. Watch this space!"

"He sounds a bit of an oddball to me. God-botherer, isn't he?"

"Yes, his evangelism is a bit flamboyant but I think he's basically a decent man. He's still funding our research work in Oxford even though he won't get much benefit from it himself."

The flight engineer frowned. "What is it you do? Filtering carbon dioxide out of the air, isn't it?"

"Don't be scornful, Frank. One day Planet Earth might need to make use of our work."

"Nigel told me you told him it would take a hundred years to get to the danger level."

"At present rates of emission that's true. But if nations such as China and India start to industrialise it'll happen more quickly."

"What happens if we get to that level?"

"It adds extra enthalpy to the atmosphere."

"What's that?"

"Heat energy. If the atmosphere gets hotter it can hold more water vapour. So more water evaporates from the sea, which means we'd get more precipitation."

"More rain, then."

"Yes, in many areas, but less in others. According to some climate scientists we'd get more extreme weather events generally. Violent storms, floods, droughts, more hurricanes."

Frank nodded. "You're a lone voice, aren't you, Maddie?"

"So was Churchill when he warned us about the Nazis."

"Fair point."

The professor looked earnestly at her friend. "This aircraft uses paraffin for fuel, doesn't it?"

"Er . . . we call it jet fuel. To the Yanks it's kerosene."

Madeleine nodded. "How much will we consume today?"

Frank frowned again. "What was the expected burn? Just over 35 tons, as I recall. I'd need to check my log."

"Right. There are around twelve carbon atoms in each molecule of kerosene. So when you burn it you produce twelve molecules of CO_2 for every molecule of fuel."

"Sounds a lot."

"It is. When you take the atomic weights of carbon and hydrogen and oxygen and do the maths you find a three-to-one ratio. So the 35 tons of fuel we burn today will release more than 100 tons of CO_2 into the atmosphere."

"Wow! Dirty buggers, aren't we?"

"Yes. And at some point in the future it'll have to be dealt with."

"Perhaps the world will run out of oil and coal before then. So there'll be other sources of energy."

Madeleine nodded. "That's what everyone's assuming." She sighed. "Which is why they think I'm wasting my time." She paused. "Maybe they're right," she added quietly.

Frank looked at her. "You're doing other things, though, aren't you, Madds?"

"Yes. There's plenty of other work to keep me occupied." The professor smiled ruefully. "To be honest, Frank, I'm getting a bit tired of the methane reformation project myself. It's like I'm banging my head against a brick wall. Dreadful thing for a researcher to admit . . . almost heresy!"

"Do you ever wish you'd stuck with flying after the ATA?"

"It was difficult, Frank. I loved flying but I loved chemistry too. Plus there were plenty of unemployed pilots of both sexes looking for work. I think I made the right choice."

The flight engineer looked at his watch. "I'd better be getting back soon. Mustn't abuse Nigel's consideration."

"I've enjoyed our chat. Especially reminiscing about the old days."

"Perhaps I can meet your ATA friends one day."

"I'm sure they'd love to meet you again, Frank."

"And you must come to us for dinner soon."

"I'd like that."

"With a partner?"

The professor tilted her head. "You never know."

The flight engineer stroked his moustache. "Did talking about Terry make you sad earlier?"

"A bit. But it's many years ago now."

"Considering how many people we lost the morale held up pretty well at Waltham, didn't it?"

"I'd say so."

"You had a bad spell, as I recall."

It was the professor's turn to frown. "When was that?"

"It was the time you and Nigel were doing night circuits or something."

Madeleine paused, as if undecided how to answer.

Frank filled the silence. "We were expecting both of you to join us for a few bevvies in the bar after the detail but you never showed. And you were both grumpy for several days afterwards. We never found out the reason."

"It was after Terry was killed," said Madeleine slowly. "Nigel said I should be grounded in case my emotions affected my flying."

"And I take it you did not agree."

"I did not. I told him it was not his place to make that judgement. So then he suggested I see the medicos. The discussion got quite heated."

"You should have come to the bar," laughed Frank. "We'd have plied you with booze until you made up."

The professor fixed the flight engineer with her gaze. "I hit him, Frank. I slapped his face. He swore at me and stormed off. Sadly you and the other Waltham people got caught in the aftermath."

"It was so out of character . . . for both of you. We all wondered what had happened."

"Well, now you know."

CHAPTER 28

On some stopovers the captain's accommodation was identical to the rest of the crew. By contrast in one or two Middle Eastern hotels a captain could expect an entire luxurious apartment to rattle around in by himself. In the New York Santa Clara Plaza the captains were allocated Executive Suites, larger than the standard rooms and with more amenities, including a lounge separate from the bedroom and bathroom.

In the lounge in Executive Suite 832 BOAC Captain Nigel Nixon was dispensing ice cubes. Around him sat Senior Engineer Officer Frank Armitage, stewardess Lorraine Compton and chemistry professor Madeleine Maunsell. The night caps which had just been delivered by room service were Scotch for the men, cointreau for Lorraine and a Cinzano with lemonade for Madeleine. Nigel put the ice bucket back on the table and then sank into the tan leather sofa.

The captain raised his glass. "Cheers, everyone."

The others followed suit. "Bloody good evening," said Frank. "Thanks for organising it, Nigel."

"Well Maddie and Max Murdoch had a hand in that." Nigel raised his glass again. "I salute them."

"Not to mention the gig at the Delacorte," continued Frank. "Even the youngsters appreciated the country music stuff."

"Or pretended to," said Nigel. "I'm sure I overheard Jeremy saying something like 'rock'n'roll it ain't'."

Their second evening in New York had gone pretty well according to plan. Apart from Chief Steward Barry Linton, who was staying with relatives in Asbury Park, New Jersey, the entire crew had eaten at a Manhattan steakhouse prior to strolling to the Maison Royale Theater on Broadway to see the musical, "This Time With Feeling". Waiting to meet them there was Maxwell Murdoch, who Madeleine introduced to the others. When asked about how he had managed to book two blocks of seats close together in a packed house at short notice, the entrepreneur had smiled and mentioned that the theatre's General Manager was a friend.

After the show the younger members of the crew took themselves off to a night club in Greenwich, except for Lorraine, whose stated intention was a night cap and then bed. To remind the others of the requirement not to blot the airline's escutcheon Nigel resorted to light-hearted irony, telling them to get paralytically drunk and make sure they reported for work the next day nursing monumental hangovers.

Max Murdoch had also taken his leave, telling the others that he was meeting friends but adding that he was looking forward to seeing them again at JFK the following evening for the flight to England.

"This must be a bit boring for you, Maddie," said Frank after he and the others had been swapping airline stories over their drinks.

"Not at all," said the professor, hoping she sounded sincere.

"Ah . . . knew there was something I'd forgotten to tell you," continued the flight engineer. "I met a fan of yours a few months back."

"A fan?"

"American chap. Stan Hazzard."

Madeleine shook her head. "Doesn't ring any bells."

"B17 driver. I was his engineer when he converted onto Lancs."

"Ah, yes, I vaguely remember. Wasn't it his aircraft you brought back and landed after he got wounded?"

"Yep. That was him."

Lorraine's eyes widened. "You landed a Lancaster?"

"Yep. It didn't survive my arrival though. They had to turn it into saucepans after I wrecked it."

"Frank's too modest to tell you that it had been badly shot up," said Nigel. "He did a brilliant job and got a DFC for it."

"Anyway," said Frank, "Stan told me he asked you for a date when you brought a new Lanc to Wyton." He grinned at the others. "The ice maiden here turned him down."

"I don't remember that," said Madeleine.

"To be fair, you might have been engaged to Terry at the time."

"Ah . . ."

"Right," announced Lorraine, "I'm going to hit the sack. I'll probably see you lot at brekkies."

"Sleep well, Lorraine."

Frank left shortly afterwards, expressing his intention to drag his battered old frame to the dining room for breakfast if he woke up in time.

Madeleine drained off her glass. "I think I'm ready for bed, too. Thanks for a lovely day, Nigel."

"My pleasure."

"A good bunch you've got with you. They all seem to like you."

"I bribed them. Didn't want you thinking your brother was a pompous martinet of a captain."

"I knew it all along!"

"It's great working with Frankie."

"Yes, we had a long chat on the flight over." Madeleine put her glass down and lowered her eyes. "He asked about the big bust up we had at Waltham."

The captain looked at her. "What did you tell him?"

"I had to think up an answer pronto. I told him you wanted to ground me after Terry got killed. Then we argued about it and I hit you."

"Did he accept it?"

"I think so."

"Well, I hope he doesn't bother to look into it in more detail. He'd realise what you told him wasn't true."

Madeleine sighed. "You're right, of course. The . . . incident . . . happened before Terry and I started going out. When Frank questioned me I couldn't think of another explanation on the spur of the moment."

"Our log books would confirm the dates if anyone looked at them."

"Well, let's hope Frank doesn't bother about it any more," said Madeleine. "There's no reason why he should—it was a trivial episode in his life."

"It was a major one in mine."

"And mine."

The two of them held each other's eyes.

"I think we were wrong to chastise ourselves for what happened at Waltham," said Madeleine quietly.

"I've often thought that myself. I shouldn't have been beastly to you when I knew it was partly my fault—"

"I was mortified," interrupted Madeleine. "For me the anger I showed to you was really against myself for what we'd done."

"Do you still regret it, Maddie?"

"No . . . I don't think so. It was forces beyond our control."

Nigel began swirling an ice cube in his whisky glass while Madeleine toyed with her necklace.

"I loved you then," began Nigel, still looking down, "and I love you now . . . not brotherly love but . . ."

Madeleine rose from her chair and sat herself next to her brother. She took his hand in her own.

"I've always loved you the same way, dear Nigel," she murmured. "Even when I was engaged to Terry it was you who held my heart. I loved you when we were teenagers and when you got married and when you started going out with Jackie. I'll always love you. I can't help it."

Nigel put his hands on his sister's shoulders.

"This is how it started that night at Waltham."

"Yes."

"Do you want to—"

"Yes."

The first kiss was tentative but the exploration soon intensified, their pulses quickening. Then their mouths separated as they paused for breath. Nigel released his embrace and his hands strayed to the front of Madeleine's blouse. Slowly he undid the top button and then the one below, then another. Inside Madeleine's bra he could see her breasts rising and falling with her rapid breathing. With yearning eyes Nigel looked up for reassurance.

Madeleine nodded.

"Are you sure?" he whispered.

"I'm sure. I've wanted to do this ever since Waltham . . . ever since Sywell. We'll share your bed here . . . tonight . . . and we'll make love. But it will never happen again."

"I love you, Maddie."

"Then show it."

CHAPTER 29

All five occupants of the cockpit were entranced, staring out the 707's windows. To their left and ahead and above them the iridescent curtains of the Northern Lights shimmered and swirled in the night sky, which was black except for a faint glow of twilight marking the northwestern horizon. The pilots had turned the cockpit lights down to minimum to accentuate the fantastic display outside. There were no lights below to distract them. The jetliner was cruising northeastwards across Labrador and any signs of human habitation were obscured by a thick blanket of cloud.

"Shall we ding the cabin crew, sir?" said First Officer Rick Mountford in the copilot's seat. "I'm sure they'd enjoy the view."

"Good idea," replied the captain, reaching up to the overhead panel to press the button labelled "Cabin Call".

"Max might like to see it as well," said Madeleine, sitting in the observer's seat, leaning back a little so that the view of First Officer Jeremy Bosworth crouching beside her was unobstructed.

"The more the merrier," commented Frank, peering ahead from the flight engineer's station.

"Yes, but not all at once," suggested Nigel, picking up the interphone handset from the central console.

"Yes, Barry," the others heard him say. "Nice fireworks outside. Want to take a look? . . . yes . . . yes, but two at a time, eh? . . . no . . . and ask Max if he wants to see them as well . . . ta very much."

In the end the captain followed the Chief Steward's suggestion that they extend the invitation to view the Northern Lights from Foxtrot India's cockpit to any first class passenger who might be interested. During the entertainment the coast of Labrador passed unseen below them, announced by Jeremy at the navigator's station. Underneath the cloud would be the pitch darkness of the Atlantic Ocean. The last visitor was Maxwell Murdoch.

"Ain't that somethin'," he breathed. "I seen plenty of sights in my time but this sure beats most of 'em."

"It's beginning to fade now," said Nigel. "We often get a good show in this area at this time."

"And they pay you guys for this!" laughed Max. "You should be paying them!"

"Don't suggest it," said Frank dryly. "They're always looking for excuses to avoid pay rises." The flight engineer had already lost interest and was reading a copy of the *Washington Post*, his spotlight a harsh contrast to the surrounding gloom.

"Say, how do you guys know where you're headed?" asked the American.

"We're using Loran," came a voice from behind. Jeremy was again seated at the navigator's table. As Max turned round the copilot pointed to the glowing cathode ray tube.

"Is that an electronic map?" asked Max.

"No . . . its based on hyperbolic position lines generated by radio beacons. Two position lines will give us a fix but we usually go for three for greater accuracy and confirmation."

"So you don't need to follow the stars, then?" chuckled Max.

"We do astro if we need to, if the Loran isn't good. Sun, moon and stars."

In the dim light Jeremy could see the American raise his eyebrows. "Really? I said that as a joke."

"We've got Consol, too," joined in Frank. "It's actually a modded version of what the Germans were using during the war to navigate their U-boats."

"You don't say! How about that!"

After a few more minutes of aeronautical chat Max thanked his hosts and said he'd head back to the cabin.

"I'll come back with you," said Madeleine. "I think the boys need some woman-free space for a while."

"That's right," chuckled Frank. "I've got a couple of new jokes the boys might like to hear. Not suitable for a lady's ears."

"Why does that not surprise me?" commented the professor wryly.

In the cabin the conversation between Madeleine and Max ranged over topics many and varied. The Righteous Warriors storm appeared to have abated, with Max issuing a public statement apologising for impugning the honesty and integrity of General Hartmann and Louis Maguire. In turn they had withdrawn their demand that their accuser resign from the steering committee. The Warriors public relations department had added their own bulletin saying that they were grateful that prayers to the Lord for a resolution had been answered.

"It stinks," said Max now. "It sticks in the craw that I've had to say sorry to those lying bastards but . . . the greatest good for the greatest number and so on. Lots of our members were upset by what was happening. It was suggested that the quarrel was the work of the devil and he needed to be vanquished."

"Is that your opinion?"

The American grimaced. "I've done a lot of thinking about it, Maddie. I haven't lost my faith but I reckon there's plenty of people in the Warriors that don't exactly follow the guidelines Jesus Christ set down. I may resign . . . I don't know. Crazy, isn't it? Here I am going to London to promote the cause and I can't even convince myself."

"I lost my faith many years ago. I couldn't stop wondering why an omnipotent God would ignore human misery."

"That's a tough one," admitted Max. "And I can't answer when people ask me who created God."

"Perhaps Jesus was just advocating a form of socialism. You know . . . the fortunate helping the less fortunate."

"I'm not comfortable with that word, Maddie."

The professor smiled. "Okay, Max. He was a good man trying to persuade others to be good people."

"I like that better."

"But perhaps he wasn't divine. Perhaps he was deluding himself when he claimed to be the Son of God."

"Well, we'll have to part company on that one."

"Okay."

"But you're a member of the Church of England, aren't you?"

"Notionally yes, but only because I sing in a church choir."

"So who inspired all those composers of religious music and hymns if it wasn't God?"

"I can't answer that, Max."

The American smiled. "So, moving on to lighter things, did you have a good time in New York?"

"Loved it, Max. All the crew did."

"I thought the Broadway show was great."

"That was the best evening for me."

"Your brother is a great guy. You could see all the crew like him."

"Half-brother."

"Yeah."

Chief Steward Barry Linton appeared from the galley and moved slowly down the cabin, advising each row of passengers that he would soon be dimming the lights and would anyone like a nightcap beforehand.

"Cinzano and lemonade, please," smiled Madeleine when it was her turn.

"Scotch on the rocks for me, Barry," said Max.

"Certainly, sir."

"Max."

"Certainly, Max."

$$*\quad*\quad*\quad*\quad*$$

The chime interrupted the tranquil inactivity in the cockpit, stirring Jeremy in the observer seat from his doze. Frank looked up from the newspaper crossword at the blue light in the overhead panel labelled "HF2".

"It's Box 2," said Rick, referring to the high frequency communications radio. "Not Oceanic . . . it's Ops."

"I wonder what they want?" asked Nigel, also looking up.

"We didn't leave anyone behind, did we?" joked Frank.

The captain smiled. "You need new material, Frankie."

"Shall I take it, sir?" asked Rick, reaching for his headset.

"Yes, please. Let's put the speaker on so the rest of us can listen headsets off."

Rick donned his headset and pressed his transmit button.

"Speedbird London, Foxtrot India answering Selcal."

Through the crackle of static came the reply. *"Fox India, we've had a Telex from Ops New York. Advise when ready to copy, over."*

Frank pulled a notebook towards him. He nodded to the copilot.

"Go ahead," transmitted Rick.

"Okay, Fox India. We don't know how authentic this message is. We got it a few minutes ago, over."

"Go ahead," repeated Rick.

"Okay. Message reads 'Strong possibility repeat strong possibility an explosive device has been placed in the hold baggage of passenger Maxwell Murdoch on flight BA565 operating JFK to LHR. Source currently unverified. Advise captain's intentions', over."

The three crew members looked at their captain. Jeremy's langour had vanished. Like the others he was alert, expectant.

"Tell them to stand by," said Nigel. "We'll get back to them ASAP."

"Roger," came the reply, *"we'll maintain listening watch. Out."*

"Where's the nearest diversion, Jeremy?" asked Nigel, twisting round in his seat. "We could head back to Goose, maybe. Weather was crap, though, wasn't it, when we overflew it?"

The copilot began to move to the nav station. "I got a good Loran fix about ten minutes ago, sir. I'll work out a DR fix. Two minutes."

"The Goose forecast was bad, wasn't it," said Rick. "Cloud on the deck, heavy precip and gale force winds."

Nigel drummed his fingers on the glareshield. "Okay . . . what do you think, fellahs? Is it a hoax?"

"Max has got enemies, hasn't he?" said Frank. "Powerful enemies in that religious organisation he belongs to. He's not a random passenger. To me that means the chances it's real are higher."

"Could still be a hoax, though," said Nigel. "Someone just stirring up trouble . . . any thoughts, Jeremy . . . Rick?"

"Diversion would be tricky if we can't make it back to Goose," said Jeremy.

"Which is the greater risk," said Rick, "diverting into crap weather or taking a chance on the threat being a hoax?"

"Okay," said Nigel. "Let's get some updated weather reports if we can . . . Goose, Gander, Saint John's, Sondrestrom, Narsarsuaq, Keflavik. Will you start on that, Rick?"

"Okay, sir, I'll see what Gander Volmet's offering. I'll need to use Box 2 for that."

"No problem," said the captain. "We don't need to talk to Ops until we've got a plan. But if it takes too long to cycle through the Volmet sequence we'll ask Oceanic."

Rick reached up to change frequency on the HF radio to the weather broadcast station.

Frank leant forward. "Narsarsuaq, skip? Short runway, poor approach aids . . ."

"I know," said Nigel. "But if we're desperate and that's the closest, that's where we'll go."

"Can't argue with that, skipper. If we divert, we'll have to think about dumping fuel to get the weight down for landing."

"Yeah."

"And once we've dumped it we'll have fewer options if things don't work out."

"I know."

Jeremy had untaped the plotting chart from the nav table and now he brought it forward to show the captain. Nigel switched on the overhead floodlight and all four screwed up their eyes at the unaccustomed blaze of light reflecting from the chart.

"We're here, sir," said Jeremy, pointing to a small triangle he had drawn off the southern tip of Greenland. "As of two minutes ago. Accuracy within ten miles, I'd say. Ties in with a bearing from the Prins Christian NDB." The Non-directional Beacon broadcast its signals from Cape Farewell.

"What ranges are we talking about for diversion?" asked the captain.

The copilot read from a list he had compiled. In the right front seat Rick was writing weather reports, both his ears covered by the headset to cut out the chat from the others.

" Narsarsuaq . . . 130," began Jeremy, " . . . Sondrestrom . . . 460, but that's direct from here—shooting the approach would add a few miles. Er . . . Goose . . . 580, but we'd be bucking an eighty knot headwind."

"I was thinking the same," said Nigel. "Ops gave us this northern track to ride the westerly jetstream. Plus the weather isn't looking too pretty there."

"Gander's further, of course," continued Jeremy, "and we'd have the headwind problem."

"What about Keflavik?" asked the captain. "We'd have this eighty knot tail if we went there."

"720 miles, sir."

"Sondrestrom would be still air from here, would it?"

"Slight tail, I would say, ten knots maybe."

Rick took his headset off and handed over the paper on which he had been scrawling meteorological data.

"Latest actual and forecast for Goose, Sondrestrom, Narsarsuaq and Keflavik," he announced. "Goose is crap. The others aren't brilliant but they're workable. Narsarsuaq is just on limits but the forecast is not so good."

"Just remind me, boys, what approach aids have Sondrestrom got apart from their NDB? It's a USAF base isn't it? They'll have Tacan, won't they?"

"Yes, sir, and PAR, and according to our brief they've got a VHF frequency we can use. The military use UHF for their own comms." The Tactical Air Navigation system would give distance information and the Precision Approach Radar would guide them to the runway.

There was silence for a minute or so as the Boeing put another ten miles behind, the tension mounting as the copilots and flight engineer waited for a decision. "Bugger!" muttered the captain at one point, bringing wry smiles from the others.

CHAPTER 30

"Right," said Nigel finally, "it's Sondrestrom, then. Jeremy—"

"Turn left onto zero zero five initially, sir," said the copilot, anticipating the captain's question. "I'll give you a more precise heading in a minute when I've plotted the track. I'll steer us towards the initial approach fix to start with and hopefully Sondrestrom Radar will pick us up as we get closer."

Nigel turned the heading bug on his instrument panel and the 707 dipped its left wing in response.

"That's allowing for variation, is it, Jeremy?"

"Of course, sir!"

"Sorry, a bit rude of me to doubt your nav—you're probably better at it than me."

"It's about twenty-eight degrees in this area but it'll increase a bit as we get further north."

The captain turned to Rick. "Get a clearance from Gander, would you. Tell them we're diverting for operational reasons. We won't mention the supposed bomb—ATC exchanges get picked up by ham radio operators and passed to the newspapers."

"Okay, sir. If we have to change level because of other traffic, what shall I ask for?"

The captain turned in his seat. "Frank . . . what's the cabin altitude?"

"Seven point three." Although the 707 was cruising at 33,000 feet the pressurisation was maintaining the passenger cabin and baggage holds at an altitude of seven thousand three hundred.

"If we go up—"

"I know what you're thinking, Nige. The cabin alt would also go up a bit and if the bomb had a barometric trigger . . ."

"So down would be better."

"I think so."

"But not too low—we'll lose true airspeed." The captain addressed Rick. "If we can't stay at three three ask 'em for twenty-nine."

The copilot donned his headset again and began transmitting to Gander Oceanic, who controlled the airspace over the western half of the Atlantic.

"How about I keep the cabin at this altitude?" suggested the flight engineer. "The diff would decrease as we descend and if the bomb went off it would do less damage. When we got below seven thou on the descent I could completely depressurise the cabin so the diff is zero."

"Yeah, let's do that. Rick, tell Gander we want twenty-nine. That way we'll reduce the diff without losing too much TAS."

"Okay, sir. Are we definitely elbowing Narsarsuaq?"

"Too dodgy," said the captain. "Marginal weather according to the last report, NDB approach, no radar, short runway."

"We could make an approach to Narsarsuaq and if we couldn't get in then divert to Sondrestrom."

"It would add to the overall flight time, Rick. If the thing is on a timer . . ."

"Good point."

"Talking of which, let's put a bit more coal on the fire. Forget fuel economy—we've got more than we need anyway." Nigel pushed the four throttles forward. "We'll go for point eight five, Frank. That'll save a few minutes." On the instrument panel the gauges confirmed that the Conways were winding up to achieve the new cruise Mach number—eighty-five percent of the speed of sound.

"Fuel checks, Skip?"

"Not necessary. We'll be dumping some later anyway."

"Okay." Frank leant forward to fine tune the throttle settings.

The captain scowled. "If this is a hoax, I'll be bloody angry!"

"I'll be bloody angry if it isn't," muttered the flight engineer.

"Here's the clearance, sir," said Rick.

"On the speaker, please."

"*Speedbird five six five, Gander clears you direct Sierra Foxtrot, flight level two nine zero. Advise when level two nine zero.*"

Rick repeated the clearance to Gander to confirm he had received it correctly as Nigel switched the autopilot to speed lock and brought the throttles back a little to start a gentle descent. The flight engineer turned to his panel and adjusted the pressurisation controls to maintain the cabin altitude.

A few minutes later the 707 was level at 29,000 feet, the four Conways set to hold the Mach 0.85 cruise.

"Next job," said the captain to Rick. "Ask Gander if there's an HF frequency for Sondrestrom so we can talk to them directly."

"I did that, sir. Nothing doing. We'll have to wait until we're in VHF range. I've already asked Gander to notify them we're inbound."

"Good man. And when you've got a mo, call Ops and tell them what's going on."

Jeremy appeared alongside Frank. "Come right three, sir, zero zero eight. That'll take us direct to the Sierra Foxtrot beacon."

"Thank you. Let's get the approach plates out for Sondrestrom and give ourselves a thorough briefing. We'll call up Barry and tell him what the picture is."

"Is it worth talking to Max, Skipper?" suggested Frank. "He might have useful information. He strikes me as a cool character—I don't think he'd panic if we told him what we've been told."

"Yes—good idea. Let's bring Maddie up here too. Her knowledge of gas chemistry might be useful if there's an explosion."

*   *   *   *   *

It was quite crowded in the cockpit. Nigel and Rick occupied the two pilot seats, with Madeleine at the flight engineer's station. Max sat in the observer's seat with Frank standing between him and Madeleine. Behind them Jeremy was at the nav station, seated but leaning forward to join the discussion group.

"So you guys are saying there might be a bomb in my luggage?" asked the American incredulously.

"The message mentioned you by name but the source was unverified."

"Jesus! Can we land somewhere quick in case it goes off?"

"We're headed towards Sondrestrom right now," said Nigel, "landing in about . . . Jeremy?"

"At our current groundspeed, around forty-four minutes from now."

"Shit!" muttered Max, "is there nowhere closer?"

"We're over Greenland," said Rick. "We're flying at max speed. There aren't any other options."

"Shit!" said Max again. "Hey, the flight attendants didn't say anything about this to us."

"We don't want to alarm the passengers," explained the captain. "Most of them are asleep now. Barry—the Chief Steward—will do a PA when we start our descent. He'll say we're landing at Sondrestrom for operational reasons and the passengers will get more info after we land."

Frank looked down at the American. "Is there any way your bags could have been interfered with since they were packed?"

"Let me think," said Max. "My wife packed my suitcase back in Wilmington—only one case as I'm only away for six nights total. She knows what clothes and other stuff I need. I packed it myself this morning in the hotel."

"And was it out of your sight at any time?"

"A bell hop took it down to Reception."

"How long was it out of your sight?"

"A few minutes, I guess. When I got down to Reception it was in the lobby."

"And it was definitely the same bag as you packed?"

"Yes . . . well, I think so. It looked the same. I didn't examine it closely."

"And after checking out you caught a taxi to the airport?"

"Yes."

"And the bell hop took your case to the taxi?"

"Yes."

The captain sighed. "Personally, I think this is a hoax, but we're not taking any chances. We'll land ASAP at Sondrestrom, deplane the passengers and have the baggage checked before we go anywhere else."

"So how long will we be in the Sonder . . . whatever you call it? I need to get to London pronto."

"Sondrestrom. Difficult to say, Max. Obviously we'll do all we can to minimise inconvenience to you and the other passengers."

Jeremy looked up from his plotting chart and leant forward. "Heading change, sir. Come left four degrees, new heading zero zero four."

"Okay, thanks." The captain adjusted the autopilot and twisted in his seat to address the American. "Thanks for your cooperation on this, Max. You can go back to the cabin now. If you think of anything else that might be useful, tell Barry."

"You got it. Say . . . good luck!"

When Max had left the cockpit Frank slipped into the observer seat so that it would be easier for the captain to talk to Madeleine without having to turn round.

"So," began Nigel, "the question is . . . if there's a bomb in a suitcase, what damage would it do if it went off? Any thoughts, Maddie?"

The professor frowned. "There would be many variables. What type of explosive? How much explosive? How fast does the flame front travel? What material is the case made of? Would the explosion be attenuated if the case was surrounded by other bags?" She smiled grimly. "A shame Bill Syerman isn't here."

"Who's he?"

"Safety Officer in our labs. Used to be in bomb disposal in the Army."

"He couldn't help us here, though. We've got no access to the baggage holds."

"Can you reduce the differential pressure in the cabin? That would help to prevent catastrophic structural failure."

"Rock and a hard place, Maddie," said Frank, running through the conflicting technical aspects of the problem.

"Would you prefer me to stay here or go back to my seat?"

"What would you prefer?" countered the captain.

The professor reached forward and squeezed the four-stripe epaulette on her brother's shoulder. "I would prefer to be here with you."

Rick turned round and managed a half smile. "Best looking flight engineer I've ever flown with."

"Are you saying I'm ugly?" said Frank, matching the forced levity.

"Right, boys and girls," said Nigel. "Back to the serious stuff. Frank, will you go back to the panel now? We'll talk about fuel dumping procedure."

"Speedbird five six five," came over the speaker. *"When able, contact Sondrestrom Radar one two four decimal seven. They have your details."*

Rick acknowledged the instruction and called Sondrestrom on the VHF frequency but as they were not yet within reception range there was no response.

The crew ran through the approach briefing. The runway pointed northeast, situated at the end of Kangerlussuaq fjord. The area radar controller would direct them towards the approach and then hand them over to the precision controller, who would guide them all the way to the runway. If for some reason radar was not available there was a localiser beam approach with a Tacan distance measuring beacon to give them ranges. Not so easy as the pilots would have to compute their own glideslope. Worst case scenario—a non-directional beacon approach using a stopwatch for distance calculation, which would not be suitable if the weather was poor. The main factor concentrating their minds was the surrounding terrain. The lower they descended in the pitch blackness the more precise the navigation demanded.

A Selcal chime on the Ops HF frequency interrupted the briefing. The four crew men and their female guest were all now wearing headsets so they all heard the message concerning facilities at Sondrestrom. They learned that the military barracks could accommodate two hundred and sixty persons, roughly double Foxtrot India's one hundred and thirty-eight passenger load. Currently forty-seven rooms were occupied by USAF personnel so there was plenty of space for the passengers and crew of the 707. Some rooms had only beds and communal toilets and showers, others—intended for officers—were equipped with en suite facilities. The commissary department was supposedly well stocked although most provisions were processed rather than fresh food.

"Bravo Alpha five sixty-five this is Sondrestrom Radar," came an American voice in their earphones. *"Are you on frequency?"*

Having endured several minutes of silence and inactivity, with the crew wondering whether their lives and those of their passengers were about to be violently terminated in the exploding fuselage of their Boeing 707, the call from a safe haven lifted their spirits.

"Affirmative, Sondrestrom," responded Rick. Over the intercom he asked, "shall we use the BA callsign, sir, or ask them to use 'Speedbird'?"

"BA's fine," said Nigel. "Don't want to confuse them."

"Bravo Alpha five sixty-five, I copy you reading me, turn left one five degrees for radar identification."

The captain twisted his heading bug as Rick read back the instruction to the radar controller.

"Wouldn't have thought there'd be much traffic round here in the middle of the night," noted Frank dryly. "Identification indeed! Who can he muddle us up with?"

"SOPs, Frank!"

"Bravo Alpha five sixty-five, you're identified one thirty-five miles south south east, stand by for vectors, advise ready for descent. Are you requesting a Precision Approach Radar approach?"

Rick confirmed their requirement for a PAR and asked for the latest weather.

"Five sixty-five, copy latest Sondrestrom weather. Wind three zero zero one two knots, visibility one quarter, blowing snow, cloud scattered below one hundred, overcast two hundred, temperature two eight, dew point two eight, altimeter two nine six five. Runway swept, cleared width one hundred feet, braking action unknown."

"Yuk!" exclaimed the captain. "Quarter of a mile vis. That'll be statute miles of course, four forty yards. That's worse than the forecast."

"We'll be okay with a PAR, skip," offered Frank.

"Scattered cloud blow minima," said Rick. "That's not brilliant." The lowest height permitted on the approach without adequate visual reference for landing according to the BOAC Operations Manual was two hundred feet for a PAR approach.

"Let's just hope the cloud's in the right place," muttered Nigel.

"Tailwind as well. That won't help," said Rick. "The runway's not overly long. If it's slippery . . ."

"We should have got Max to pray for good weather," said Frank. "He's supposed to have a direct line, isn't he?"

There was no response to the flight engineer's comment and Foxtrot India put another few miles of Arctic air behind it. Then Jeremy reached forward offering a piece of paper.

"Here's the DME table, sir," he said, "in case the PAR becomes unavailable." The copilot had calculated the correct check heights for the final approach, based on their range from the Tacan distance measuring beacon.

"Good work, Jeremy," acknowledged the captain, taking the paper from him. As if to confirm the copilot's initiative the DME readout sprang into life on the instrument panel.

"Right chaps, only essential chat from now on. Frank, start the fuel dump checklist. Rick, request descent at DME ninety."

"Bravo Alpha five sixty-five, we're aware of your . . . ah . . . security situation. After landing we require you to stop on the runway and shut down your engines without delay. There will be fire trucks illuminating all four exits and we will position standard KC135 stairs at all exits. Alternatively, you can deploy your escape slides and overwing exits if necessary. Advise please."

"If we use the slides there'll be injuries, skip," said Frank.

"I agree," said the captain. "It'll only take a little longer to deplane using stairs." They knew that "KC135" was the USAF designation for the military Boeing 707s.

Rick transmitted their intentions and asked for descent.

"Roger, Bravo Alpha five sixty-five, cleared altitude one seven thousand, advise approaching for further."

"Descent checklist, complete, Frank?"

"Yes, skip, still dumping though. I'll advise when we're done."

"Okay," said Nigel, "Down we go." He switched the autopilot to speed lock and slowly brought the four throttles fully back. Foxtrot India lowered its nose in compliance and the Conways quietened to idle thrust.

Another check on the weather brought the terse response, *"No change."* A position notification by the controller tallied closely with the crew's own assessment. The descent was uncomplicated, almost routine, with only the putative explosive device and the grim airport weather to trouble their minds. Frank announced fuel dumping complete, all dump valves shut and cabin depressurised.

"Five sixty-five, position one five miles southwest, maintain heading three six zero, maintain altitude three thousand, contact Precision one one nine decimal nine. If no contact come back to me."

Precision picked them up, turned them on to final approach and told them to initiate descent at the rate to maintain the correct approach angle. Nigel, flying manually now, called for undercarriage down and the landing checklist. Outside there was nothing to see as the 707 sank into the blackness, bouncing in the turbulence.

"Five sixty-five, you're on course, on glideslope, confirm three greens." The controller was checking that the Boeing's undercarriage was down and locked. Rick transmitted an affirmative response.

"Five sixty-five, you're a little left, new heading zero nine eight, on glideslope. Do not acknowledge further transmissions. Confirming you're cleared to land from this approach. If you hear no transmission for five seconds go around, climb altitude four thousand and contact Sondrestrom Radar."

"One thousand," called Rick over the intercom, looking at his altimeter.

"Check," responded Nigel, eyes scanning his instrument panel.

"Flaps fifty, landing checklist complete," from Frank.

"Five sixty-five, on course, on glideslope, range three miles . . . Five sixty-five, on course, thirty feet low, adjust descent . . ."

Nigel eased the nose up a fraction and edged the throttles forward.

"Five hundred," called Rick.

"Check."

"Five sixty-five, on course, range one decimal five, on glideslope . . ."

"Landing lights on," called Nigel, looking up for a second. Two spears of light lit up trillions of swirling snowflakes streaming past the cockpit windows, dazzling the crew.

"Off again!"

"Five sixty-five, you're slightly right, turn left two degrees, new heading zero nine six, twenty feet low, adjust descent."

Rick was alternately watching his instruments and looking ahead for visual contact, any light that might penetrate the bumpy black void. The 707 was descending through three hundred feet towards the minimum of two hundred. "One hundred above," he called.

A terse "Check" from the captain.

"Five sixty-five, on course, turn right one degree, new heading zero nine seven, on glideslope . . ."

"Decision . . . no contact," called Rick as the altimeters sank through two hundred.

"Continue," called Nigel in defiance of SOPs, which required a go around and climb away if visual contact had not been achieved at Decision Height.

"Five sixty-five, on course, ten feet high, range zero decimal five."

"Runway lights ahead!" called Rick and Frank simultaneously.

"I got 'em!" called Nigel, looking up.

"We have contact and we're landing," transmitted Rick.

"Five sixty-five, we're ready for you."

CHAPTER 31

The briefing from BOAC Operations describing the set-up at the United States Air Force Sondrestrom base turned out to be reasonably accurate, although it underestimated the quality of food available. Besides the more predictable burgers, hot dogs and ice cream nutritious comestibles on offer in the officers' dining room included locally caught fish and fresh fruit and vegetables flown in regularly from the homeland.

Seated at one of the tables, enjoying a late afternoon snack, were Captain Nigel Nixon, Senior Engineer Officer Frank Armitage, Professor Madeleine Maunsell and CEO Maxwell Murdoch. At adjacent tables could be found the other members of Foxtrot India's crew and some of the first class passengers. Besides the meal, the crew and passengers were also digesting the news they had just watched on the large television screen fixed to the wall, around which they had crowded as the reported events unfolded. A USAF officer had apologised for the grainy quality of the monochrome image but pointed out that since the only signal available was from a satellite it was the best they could expect.

Outside in the afternoon sunlight a row of F4 Phantom fighters could be seen on the apron, and a couple of C130 Hercules transports, all adorned with a blanket of snow, the residual evidence of the previous night's storm. Further away the BOAC Boeing 707 stood on a remote taxyway at the far end of the airfield, where it had eventually been towed when the base's commanding officer deemed it safe to do so. After Nigel had brought it to a halt on the runway, max reverse thrust compensating for potential unreliable braking on icy patches, the expeditious deplaning of the passengers had gone according to plan, after which all personnel and vehicles were quickly removed to a safe distance.

Twelve hours later, following consultation with senior officers on the mainland, base commander Colonel Jaymack had asked a crew of volunteer engineers wearing protective clothing to enter the 707's cockpit, release the parking brake, attach a tug to the nosewheel and tow the aircraft to its new location. The passenger doors were checked closed but the hold doors were left open to help to equalise the air pressure should an explosive device be triggered by a delay timer.

A telex had informed Nigel of the plan being put together by BOAC Ops at Heathrow. A bomb disposal team with sniffer dogs was being flown in from the States and when another twenty-four hours had elapsed they would carefully remove all the bags from Foxtrot India's holds. After inspection the bags would be stored remotely, except for any which aroused suspicion, which would be destroyed in controlled explosions. BOAC had organised dispensation from the British Ministry of Aviation to allow USAF engineers at Sondrestrom to inspect the Boeing for airworthiness. If all was satisfactory the passengers would reboard—without their bags—for the onward flight to London. A Royal Air Force Britannia transport aircraft would be dispatched from RAF Brize Norton to pick up the bags and repatriate them.

It became apparent that the two main items of news that had surprised all who saw them were, in the eyes of those in the know, possibly connected. Firstly, the announcer said, a British jet airliner on route from New York to London had been forced to divert to a military air base in Greenland after an anonymous message had been received saying a bomb was on board in a passenger bag. The aircraft had safely landed in a snow storm, the announcer said, and the authorities were deciding how do deal with the baggage and when it would be safe for the jet to take the passengers home.

Meanwhile, near Mansfield, Pennsylvania, a car travelling west along Highway 6 had exploded, killing the driver and passenger. A truck travelling in the opposite direction had been blown off the road by the blast and although its trailer and load were damaged the driver escaped unharmed. The passenger in the car was a senior member of the Righteous Warriors, said the announcer, giving the names of the two victims.

At Nigel's table an earnest discussion started.

"It can't be a coincidence," said Frank. "We were told the alleged bomb was in your bag, Max. Did you know the man who was killed?"

"Vaguely. But he wasn't high ranking. If he was intending to kill me he would have been acting on the orders of someone higher up."

"Someone associated with the Hanoi plot?" asked Madeleine.

"Almost certainly. They never forgave me for going to the cops. It was revenge time. Looks like they were going to blow me up on your plane by swapping bags. Maybe they planned to do the swap in the hotel lobby but they fucked it up."

"I suppose they couldn't put obvious identifying marks on the bag with the bomb in case you noticed the difference," said Nigel. "Maybe they got confused themselves. If there were two of them maybe one did a swap and then the second one did another swap, not realising it had already been done."

"So who tipped off the authorities about the supposed bomb in the aircraft?" asked Madeleine. "Why did they do that?"

"Someone who found out there was a plot to kill me," said Max, "but didn't want it to succeed."

"Another question," said Frank. "Why did the bad guys take Max's bag away in their car? I mean, the bag they thought was Max's? Why not just dump it somewhere?"

"Well," said the American, "it would have been incriminating evidence if it was found, wouldn't it? A plane disappears over the ocean and then later on a suitcase belonging to one of the passengers turns up somewhere. Cue intensive investigation by the cops. Better to hang on to it for a while till it could have been properly disposed of—junk the contents and destroy the case. Then no one would have known what caused the plane to come down. It would be a mystery—no evidence of foul play, might have been a technical fault."

"Just had another thought," said the captain. "It looks as if the bomb was triggered by a timer. What time did the car blow up according to the news report?"

"Er . . . just before 11, wasn't it?" said Frank.

"That'll be Eastern Daylight Time," said Max.

"So, where were we at the time?" asked the captain.

"Let's work it out," said Frank. "I'll do it all on GMT then correct for local time." He pulled the typewritten menu on the table towards him and took out his pen.

"Ah . . . Captain Nixon, telex for you, sir." It was a junior USAF officer who gave him a salute and handed over the paper.

Nigel thanked him, whereupon he threw another salute, turned on his heel and briskly walked away.

"It's Ops," announced the captain. "If the bomb disposal chaps are happy they want us to get airborne tomorrow at sixteen hundred GMT, that'll be thirteen hundred, local time. I'd better start organising things."

"Another thought," said Madeleine, "if the bomb was in the car rather than our aircraft, why can't we just take off again with the bags on?"

"They're being super cautious, I suppose," said her brother. "Just in case there's another explosive device. I don't have a problem with that!"

"Interesting," said Frank, clicking his pen. "At the time the car blew up we had just started our descent."

"So if the bag swap had been successful . . ." began Madeleine.

"We might not be here talking about it," finished Nigel.

"The power of prayer," chuckled Max.

"What do you mean?" asked Madeleine.

"Once you told me we were diverting in case there was a bomb in my case I did a little serious praying. Looks like it paid off!"

CHAPTER 32

Madeleine looked at her watch. Just after nine in the evening, local time. Difficult to reconcile the hands on the watch with the view through the window of her bedroom, with the clear sun still well above the northwestern horizon, glinting on the aircraft parked on the apron. As the USAF officers had explained, at latitude sixty-seven degrees north the late spring daylight lasted twenty hours, with only four hours of twilight and darkness. Not much farther north the sun would never set until the summer turned to autumn.

After dinner in the mess Nigel's group had dispersed. Madeleine said she wanted to rest quietly in her room. Frank and Max had repaired to the officer's bar with other members of Foxtrot India's crew and passengers. Several passengers and First Officer Rick Mountford had accepted the invitation to join the USAF servicemen in the base's cinema for a showing of the Jimmy Stewart film, "Strategic Air Command". Those less enamoured of military aviation opted for relaxing in the lounge or in their rooms. The captain had taken himself off to his own room, explaining to Madeleine that he wanted to use the time constructively, going through some of his mother's paperwork. He told her that it was just as well he was going to miss the appointment with his solicitor scheduled for the next day as he was nowhere near adequately prepared. He had sent a telex to Ops asking them to arrange a new appointment for the following week. Not violating SOPs, of course, asking Ops to pass on personal messages for his passengers and his crew and himself when those messages related directly to their diversion.

Now Madeleine lay on her bed, mulling over events and the workings of fate. That bomb . . . was the whole thing a hoax, and the exploding car in Pennsylvania just a coincidence? Plausible, but the Righteous Warriors connection suggested otherwise. On the other hand, if you were planning to swap suitcases so a bomb could be planted on an aircraft, surely you'd come up with a foolproof way to distinguish them—a new scuff mark perhaps? Did they swap the label or did they make a copy? Room for confusion there unless they double-checked what they were doing. Did someone deliberately re-switch the bags to kill the car's passenger? But then there would be no need to alert the authorities about Max's plane. Was that the work of another person?

Madeleine sighed. Too complicated! Her mind tried to imagine the consequences of an explosive device detonating in a plane. The 707's fuselage was pressurised to reproduce a lower effective cabin altitude so it would have to be robustly constructed. Was it designed to tolerate a rupture? Could it have withstood the overpressure from an internal explosion? Would the damaged aircraft be controllable?

If the fuselage had disintegrated, what then? Would the occupants have survived the decompression? Would they be aware of tumbling through the freezing night sky, knowing that imminent ground impact would send them to eternity? Was that what happened to Terry when his Mosquito was shot down over Dortmund?

And what about Emma? Mother killed in a car accident, mother's partner not long after in a plane crash. Would she have been able to cope with a double bereavement?

The professor forced her mind to switch to less frightening thoughts. For example, the idea for a new post graduate project as suggested by a colleague, which might suit one or two of this year's final year students looking to start a PhD. The motivation was the trial-and-error approach that chemists followed in catalyst research. Some people were talking about examining the lifetimes of the hybrid radicals and molecules whose production the catalyst surface facilitated during chemical reactions. Lifetimes were usually measured in milliseconds, but little research had been done into actually measuring them and correlating them to product yield. The demise of the methane reformation project might be postponed or diverted by examining catalyst performance in greater detail. She'd mentioned the topic to Max who in turn said he'd pass the idea on to his Wilmington team. If they weren't interested Madeleine would have to contact John Davis at the Science Research Council to bring up the hoary old subject of finance.

John Davis. Hmm. He'd actually written her a letter apologising for the embarrassment he'd caused being caught *in flagrante* with Toni. Madeleine thanked him for his consideration and excused him on the grounds that no man would be able to resist Toni's sexual predations. When her lodger was in heat she would not be denied, Madeleine had written. And there was no reason why John and Madeleine should not continue to meet socially.

A sudden roar outside distracted Madeleine from her musings. She sat up to give herself a better view through her window. The glowing afterburners of two pairs of fighters were receding along the runway, their progress marked by a pall of sooty brown exhaust smoke. As Madeleine watched the Phantoms lifted into the blue sky, banked left in a finger four formation and disappeared towards the south, the cacophony of their engines diminishing in proportion.

The professor sank back into her supine position. John Davis. "Continue to meet socially," she had written. What if it developed into something more? If John was too reticent, should she get the ball rolling?

There was a new factor. The night in New York with Nigel. This time they felt no remorse. This was not White Waltham. They agreed that they would make love with all the enthusiasm they could muster. But it would be the last time. The taboos were too strong to ignore. They knew they loved each other and would continue to love each other passionately. But there would be no more sex. While they lay together in Nigel's bed, exhausted, they both agreed they would find other partners for carnal satisfaction but would forever give their hearts only to each other.

Would John Davis fit the bill for Madeleine? And what about Toni? Knowing her liberal attitudes to these things, Madeleine thought it unlikely she would object. As long as she didn't revive the idea of a threesome! That would be . . . would be . . . what would it be like? No . . .that was an experiment too far . . . wasn't it?

Well, at least she was still alive to think these thoughts, not a jumble of frozen body parts despoiling the Greenland icecap amongst the mangled remains of Foxtrot India and its passengers and crew.

Had Max's prayers saved them? It was many years since Madeleine had been a follower of any religion. Since her teenage years she had come to appreciate the power of scientific reasoning. You made an observation and worked out a theory to explain it. Then you tested the theory in as many scenarios as you could find. If your theory failed you did further experiments and formulated a new theory. The whole discipline of science was based on the interpretation of evidence and rigorous testing of theory.

Religion was the opposite. There were plenty of theories about God but no incontrovertible evidence of his existence. Madeleine remembered a phrase that had come up during a debate about religion during her student days. "Intellectually unsound," was the comment of a supporter of the motion, "This house considers religion is bunk." Madeleine's own input was that despite all the prayers to God asking for relief from human suffering, there seemed to be no end to wars, disease, poverty, earthquakes, floods, droughts and other worldly tribulations. She remembered that the motion was narrowly defeated.

In a way, Madeleine envied those of faith, who believed in miracles, the Virgin Birth, the Resurrection and the Ascension, all of which violated the laws of physics. And Life after Death. If only she could share their faith. But there was insufficient evidence!

And yet . . . Max had prayed that the 707 and its occupants would arrive safely and so they had. Divine intervention or

A knock on her door.

"Come in," she called, too lazy to stir.

It was Nigel, looking somewhat distracted. He walked over to the bed, holding out some sheets of paper. "I've just found these among my mother's documents. I'll give you a moment alone to read them, then I'll come back."

The captain left, closing the door on his way out.

Madeleine roused herself and sat on the edge of her bed. She started to read.

Darling Dot,

Sorry to hear you're so upset. It must have been a dreadful shock. Having given it some thought perhaps the best option is to keep the baby. I'd be happy to pay for an abortion if absolutely necessary but it would be fraught with danger, never mind the illegality. You'll have read of cases where abortions have been incorrectly performed by people with inadequate training and women have died of uncontrollable haemorrhaging or infection.

You say you don't know who the father is and it's more likely to be me than Henry because we're "together" more often than you are with him but there would be no way of knowing for sure. And you don't need to worry when you say people will wonder about your sudden increase in fertility after years of being unable to fall pregnant with Henry—after all these things are not unknown. The physical differences between me and Henry are not that marked—I'm a bit taller and my hair a bit lighter so any child I might conceive with you could be passed off as Henry's.

So, darling, tell me what you think. I say, let's let the pregnancy continue—we can't make the situation any worse and we can carry on seeing each other. You never know—the baby might actually be Henry's!

So—next Thursday? Please say yes. And carry on saying yes.
Your devoted slave,
Sebastian

23rd September 1952

Dear Mrs. Nixon,
  With regard to the recent examination of your left
breast I am happy to confirm that the swelling has been
diagnosed as an adenoma, which is a benign (non-cancerous)
growth. Although adenomas rarely develop into cancers I think
it would be a good idea to monitor your condition and for you
to see me every two to three months or at any time you think
the adenoma has changed its characteristics. At the moment no
surgical intervention is necessary.
  On a different matter, when I checked the summary record
for your family I found a discrepancy in the recorded blood
groups. The summary shows Group O for you and your ex-husband
and Group A for your son Nigel. It is impossible for husband
and wife with Group O blood to conceive a child with Group A.
  The most likely explanation is a clerical error, which
should be corrected. When convenient, please arrange an
appointment with my nurse for a blood sample to be taken.
There is no urgency about this but it is essential that your
blood group is correctly identified prior to any blood
transfusion you may require in the future. If you would be so
kind as to notify me of your son's and your ex-husband's GPs
(if you have this information) I will write to them asking
them to arrange the same.
  I hope this letter brings you reassurance.
                    Yours sincerely,
                    Arthur Black, M.D.

Open mouthed, Madeleine slowly lowered the hand that had been
holding the documents. "No . . ." she breathed.
  Another tap on the door.
  "Come in, Nigel."
  He stood at the open door, his expression troubled, uncertain.
  "You've read them, then?"
  "Yes."
  "Do you understand what they mean?"
  "Yes."
  "What do you think?"
  "Close the door and come here."

THE END

Other books by Julien Evans:

Fiction:

Chalk and Cheese
The Sommerville Case
The Damocles Plot
Flight 935 Do You Read

Non-fiction:

How Airliners Fly
Handling Light Aircraft

steemrok.com

Printed in Great Britain
by Amazon

33279016R00125